Geotechnical and Foundation Engineering

GEOTECHNICAL AND FOUNDATION ENGINEERING

Ognjen Grkinic

ARCLER
PRESS

www.arclerpress.com

Geotechnical and Foundation Engineering

Ognjen Grkinic

Arcler Press
4164 Lakeshore Road
Burlington, ON L7L 1A4
Canada
www.arclerpress.com
Email: orders@arclereducation.com

ISBN: 978-1-77956-373-6 (Hardcover)

ABOUT THE AUTHOR

Ognjen Grkinić is a physicist with a profound expertise in applied and computer physics, particularly in data science and plasma physics. He is currently pursuing his PhD at the Faculty of Physics, University of Belgrade, where his research focuses on the behavior of ionized gases and plasma within the Attosecond-Plasmas Lab. Ognjen's academic achievements include a Master's degree in applied and computer physics, where he developed advanced software for processing scientific data, earning a near-perfect GPA of 9.80/10.00. In addition to his research, Ognjen has contributed significantly to education, having taught physics and mathematics at both high school and primary school levels. His passion for teaching is matched by his meticulous approach to scientific inquiry, which is evident in the books and papers he has authored and edited in the field of physics and other natural sciences. Beyond academia, Ognjen has worked as a Multimedia Transcriber at Arcler Education Inc. and as a Data Labeling Quality Specialist at Scale AI, roles that have further refined his analytical and technical skills. Fluent in multiple languages, Ognjen is also an active volunteer, dedicating his time to educational and community initiatives. Ognjen Grkinić's diverse experiences in research, education, and professional work underscore his commitment to advancing the field of physics and contributing to the broader scientific community.

TABLE OF CONTENTS

LIST OF FIGURES

LIST OF TABLES

PREFACE

Geotechnical and Foundation Engineering are crucial for ensuring the stability and safety of structures built on the ground. This field involves understanding soil behavior, assessing ground conditions, and designing foundations to withstand various environmental and loading conditions. Techniques such as soil property investigation, soil classification, and ground improvement methods enhance the load-bearing capacity of the ground. Pile foundations, geosynthetics, and other innovative solutions provide versatile approaches to address diverse soil conditions and project requirements.

The book "Geotechnical and Foundation Engineering" comprehensively explores the intricacies of this discipline through eight detailed chapters. Beginning with an "Introduction to Geotechnical and Foundation Engineering," readers are introduced to the fundamental concepts and importance of the field. Subsequent chapters delve deeper into specific topics, such as the "Basic Characteristics and Engineering Classification of Soils," which provides a comprehensive overview of soil properties and classification systems. Chapters on "Soils Investigation" and "Geotechnical Site Investigation" detail the methods and techniques used to assess ground conditions and inform engineering decisions. "Special Ground Improvement Methods" offer innovative solutions for strengthening weak ground, while "Foundation on Soft and Filled-Up Soil" addresses the challenges of constructing foundations on less stable ground. "Pile Foundations" and "Geosynthetics and Their Applications in Geotechnical Engineering" chapters explore specialized techniques and materials utilized in foundation design, providing readers with a thorough understanding of modern practices in the field.

Through clear explanations, practical examples, and illustrative diagrams, this book serves as an invaluable resource for students, professionals, and anyone interested in the principles and applications of geotechnical and foundation engineering.

—Author

CHAPTER 1

INTRODUCTION TO GEOTECHNICAL AND FOUNDATION ENGINEERING

Contents

1.1. INTRODUCTION

Geotechnical engineering is the branch of civil engineering concerned with the engineering behavior of earth materials. In addition to its importance in civil engineering, geotechnical engineering finds applications in the petroleum, mining, military, and other engineering fields that deal with construction whether it is within the ground or on the surface. Geotechnical engineering studies the subsurface materials and conditions to determine the physical, chemical and mechanical properties of materials by using the concepts of soil and rock mechanics. It assess the stability of manmade soil deposits and natural slopes; examine the risks posed by site conditions; design structural foundations and earthworks (Holtz et al., 1981).

Foundation engineering is a subfield of geotechnical engineering that applies structural engineering, soil mechanics, and project serviceability requirements to the design and construction of foundations for off-shore structures, buildings, and other in-land structures. The main targets in the design and construction of any engineering projects are cost, time and quality. The foundation of any structure, including towers, dams, bridges, etc., is crucial therefore, it is necessary to consider the following factors to design a foundation:

i. The load that will be transferred from the superstructure to the foundation;
ii. Details about the geological and geotechnical conditions, such as the stress and behavior of the soil; and
iii. Soil layers and groundwater depth.

Geotechnical and Foundation engineering is the science of designing, constructing, operating, and maintaining engineering works by utilizing engineering techniques and principles to the chemical and physical properties of the earth materials (Becker, 1997). It covers the investigation of soil characteristics and parameters, analysis of soil properties and conditions, the prediction of behavior of soil, and the construction of engineering projects in, on, and around the ground; designing foundations along with other structures (Phoon, 2020).

It also includes the evaluation and analysis of existing structures, employing different techniques to mitigate the impact of unfavorable circumstances, and analyze their effect on the design, construction, and the operation of the engineering projects. Geotechnical engineers analyze and design structures, tunnels, retaining walls, bridges, foundations, among other engineering works using the concepts of soil and rock mechanics, engineering geology, and rock engineering (Potts et al., 2001). Geotechnical engineers are responsible for developing and carrying out strategies to deal with contaminated sites and ensuring the proper disposal of hazardous materials. Geotechnical engineering finds its applications in a wide range of projects, such as buildings design and construction, dams, tunnels, bridges, roads, and other civil engineering works (Reese et al., 2005).

1.2. GEOTECHNICAL ENGINEERING PROCESS

Geotechnical engineering processes include the assessing and analyzing of underground conditions, as well as the designing and implementation of suitable strategies to tackle these conditions. These procedures involve determining the best techniques for testing and sampling rocks, soil, and other materials as well as examining subsurface conditions (Basu et al., 2015). Furthermore, the advancement of subsurface condition models is also part of this process. Geotechnical engineers are responsible for developing and implementing techniques to handle contaminated sites and safely dispose of hazardous substances. Some of the key operations in geotechnical engineering include:

1.2.1. Site Investigation

Site investigation involves collecting and analyzing data to assess subsurface conditions and determine site suitability for a particular project (El Mountassir et al., 2018). It is a critical phase in geotechnical engineering, serving as the foundation for decision-making in construction projects. This process involves thoroughly assessing the geological and geotechnical characteristics of a site to understand its suitability for development and to identify potential risks and challenges. Geotechnical engineers employ various techniques during site investigation, including soil and rock sampling, geological mapping, and geophysical surveys. These methods provide valuable insights to soil composition, groundwater conditions, and geological features, enabling engineers to design appropriate foundations and earthworks to withstand the site's specific conditions.

1.2.2. Subsurface Modeling

Subsurface modeling involves developing numerical models of subsurface conditions enabling structural analysis and design (Tawelian & Mickovski, 2016). It involves creating detailed illustrations of the underground environment to better understand soil and rock behavior. Engineers gather data on soil properties, groundwater conditions, and geological formations through advanced techniques such as geophysical surveys, borehole logging, and laboratory testing. This information is then used to develop computerized models that simulate how soil and rock respond to various loads and environmental factors. Subsurface modeling plays a crucial role in the design and analysis of foundations, tunnels, and excavation projects, helping engineers anticipate potential challenges and optimize construction methods. By accurately predicting subsurface behavior, engineers can mitigate risks, reduce costs, and ensure the safety and longevity of infrastructure projects.

1.2.3. Foundation Design

Foundation design is the design and analysis of the foundations of different structures like buildings, bridges, and dams. It is a critical aspect of ensuring the stability and safety of structures. Foundations serve as the base upon which buildings, bridges, and other infrastructure rest, transferring their loads to the underlying soil or rock. Geotechnical engineers carefully assess the soil conditions at a site to determine the most suitable foundation type. Factors such as soil-bearing capacity, settlement potential, and groundwater conditions are evaluated to select appropriate foundation systems, including shallow foundations like spread footings or deep foundations such as piles or caissons. Through detailed analysis and design, geotechnical engineers aim to create foundations that can withstand the applied loads while minimizing settlement and ensuring long-term stability, thus providing a solid support system for structures above ground.

1.2.4. Slope Stability Analysis

Slope stability analysis helps asses the stability of dams, slopes, and other terrain (Vaníček et al., 2021). It is a critical component of geotechnical engineering, focusing on assessing the stability of natural and man-made slopes. This analysis involves evaluating the factors that influence slope stability, such as soil properties, groundwater conditions, and external forces like rainfall or seismic activity. Engineers use various methods, including analytical techniques and numerical modeling, to predict potential failures and determine appropriate mitigation measures. By understanding the mechanisms that govern slope stability, such as soil shear strength and pore water pressure, engineers can design structures and implement slope stabilization measures to minimize risks and ensure the safety of infrastructure and surrounding communities.

1.2.5. Retaining Wall Design

Retaining wall design involves the analysis and design of retaining walls to support and preserve soil, rock, or other materials (Sowers, 1993). It is important in supporting and stabilizing slopes, preventing soil erosion, and creating usable spaces in hilly terrain. Retaining walls are structures built to resist the lateral pressure of soil when there is a desired change in ground elevation. Engineers carefully consider factors such as soil type, groundwater conditions, and the magnitude of loads to determine the most suitable type of retaining wall and design parameters. Various types of retaining walls, including gravity walls, cantilever walls, and anchored walls, are designed based on the specific requirements of each project. The design process involves:

- Analyzing the stability of the wall under different loading conditions.
- Ensuring adequate drainage to prevent water buildup behind the wall.
- Incorporating reinforcement techniques when necessary to enhance structural integrity.

By employing advanced analytical methods and adhering to established design standards, geotechnical engineers create retaining wall solutions that effectively mitigate soil instability and provide long-lasting support for infrastructure projects.

1.2.6. Tunnel Design

Tunnel design involves the analysis and design of tunnels for transportation, mining and other applications (Christian, 2004). It is a complex and intricate process that requires careful consideration of various factors. Engineers must assess the geological conditions and soil properties along the tunnel route to determine the most suitable construction methods and support systems. Geological surveys and geotechnical investigations are conducted to identify potential hazards such as unstable rock formations, groundwater ingress, and soil instability. Based on these findings, engineers develop a detailed design that includes appropriate tunnel dimensions, reinforcement techniques, and drainage systems to ensure the structural integrity and safety of the tunnel. Throughout the design process, emphasis is placed on mitigating risks and addressing challenges posed by the geological environment to create a robust and resilient tunnel infrastructure.

1.2.7. Contaminated Sites Treatment

This is the process of restoring and cleaning the contaminated sites to protect the environment and human health (Bea, 2006). It focuses on addressing soil and groundwater pollution caused by hazardous substances. Geotechnical engineers play a crucial role in assessing the extent of contamination, understanding its behavior in the subsurface, and devising effective remediation strategies. Techniques such as excavation, soil washing, vapor extraction, and bioremediation are employed to mitigate the spread of contaminants and restore the site to a safe and usable condition. Geotechnical considerations, such as soil permeability, stratigraphy, and hydrogeology, are carefully analyzed to ensure the success of remediation efforts while minimizing environmental impact.

1.2.8. Construction and Operation

Geotechnical engineers work on constructing and operating engineering projects as well as developing and implementing plans for a safe disposal of potentially hazardous substances. In geotechnical engineering, the construction and operation phases are critical considerations in ensuring the stability of infrastructure projects. During the construction phase, engineers focus on managing the interaction between the engineered structures and the surrounding soil or rock mass. This involves careful planning to minimize disruptions to the natural ground conditions and implementing techniques to ensure the stability of excavations, slopes, and foundations. Once construction is complete, the operation phase begins, during which the structure's performance is monitored and maintained. This includes regular inspections to assess the impact of environmental factors and human activities on the stability of the structure. By integrating geotechnical principles into both the construction and operation phases, engineers can optimize the performance and safety of infrastructure projects throughout their lifecycle.

1.3. PRINCIPLES OF GEOTECHNICAL ENGINEERING

Geotechnical engineering principles relate to the study and design of buildings and foundations on, inside, or next to the ground by using engineering and scientific principles. These principles include study of subsurface conditions and evaluating how they affect the engineering work design, as well as selecting and designing suitable remedial measures (Cleveland & Wartman, 2006). The following are some geotechnical engineering principles:

1.3.1. Soil Mechanics

Soil mechanics is the study of the engineering and physical characteristics of soil and rocks. This field analyzes and designs bridges, foundations, and other engineering projects. It involves studying the properties of soil and rocks. It also examines how soil behaves under different conditions when designing structures like bridges and buildings. Factors such as moisture, density, and strength are considered to ensure the stability of these projects (Bishop et al., 1960).

One key aspect of soil mechanics is understanding that soil varies in composition and behavior. Properties like cohesion, friction, and permeability are examined to determine how soil will interact with structures. Assessment of how soil responds to pressure and deformation helps in creating designs that are safe and durable. By applying principles of soil mechanics, optimization of designs is possible, ensuring the long-term stability of infrastructure projects.

1.3.2. Rock Mechanics

It is a study of geometric and physical attributes of rocks. It is used in the analysis and design of ramps, tunnels, and various underground structures. It focuses on understanding the behavior of rocks and rock masses. It encompasses the study of how rocks respond to applied loads, such as those experienced during excavation, tunneling, and mining operations. By examining factors such as rock strength, deformation characteristics, and fracture mechanics, engineers gain insights into the stability and integrity of rock formations in various engineering projects. Since it guides the design of support systems and excavation methods catered to the particular characteristics of the rock mass, rock mechanics is essential in ensuring the safety and efficiency of infrastructure development. (Frank, 2007).

1.3.3. Engineering Geology

It is a study of the subsurface conditions along with application of the geological principles in the analysis, designing, and construction of engineering projects (Bogusz & Godlewski, 2019). It plays a vital role in understanding the geological conditions of a site before commencing any construction project. It involves the study of geological factors such as rock formations, soil types, and geological hazards to assess their impact on engineering works. By analyzing the geological characteristics of a site, engineers can identify potential risks, such as landslides, earthquakes, or soil erosion, that may affect the stability and safety of structures. Through techniques such as geological mapping, remote sensing, and geophysical surveys, engineering geologists provide valuable insights into the geological makeup of a site, enabling engineers to make informed decisions during the planning and design stages of a project.

1.3.4. Structural Mechanics

Structural mechanics involves studying the design and analysis of structures. It has applications in the analysis and design of bridges, buildings, dams and other engineering works (Brinkgreve, 2005). Structural mechanics is a branch of engineering that focuses on understanding the behavior of structures under various loads and conditions. It encompasses the study of how different materials, such as steel, concrete, and timber, respond to forces like compression, tension, and shear. By analyzing the internal forces and deformations within a structure, engineers can ensure its stability and safety throughout its lifespan. Structural mechanics plays a crucial role in the design, construction, and maintenance of a wide range of civil engineering projects, including buildings, bridges, dams, and towers.

1.3.5. Geodesy

Geodesy is the study of the shape and size of the Earth. Geodesy is used to analyze and design structures in areas with different terrain (Gens, 2010). It is the scientific discipline that deals with the measurement and representation of the Earth's geometric shape, orientation in space, and gravitational field. It plays a crucial role in various fields, including civil engineering, navigation, and mapping.

Geodesists use various techniques, such as satellite measurements, gravity measurements, and astronomical observations, to accurately determine the size and shape of the Earth. These measurements are essential for creating accurate maps, understanding the Earth's gravitational field, and monitoring changes in the Earth's surface over time.

1.4. HISTORY OF GEOTECHNICAL ENGINEERING

The history of geotechnical engineering can be divided into seven important time periods. Before the 18th century, people used basic geotechnical ideas for building, but they didn't study it formally. From 1700 to 1776, early theories about soil started to appear. Then, in the classical periods, from 1776 to 1910, important scientists like Charles-Augustin de Coulomb and Karl von Terzaghi developed key ideas about soil.

Modern soil mechanics started around 1910, when scientists like Terzaghi began to study soil in a systematic way. After Terzaghi's work in 1927, geotechnical engineering became its own field, with new technologies and methods improving how we understand and work with soil. This history shows how geotechnical engineering has grown and changed over time, thanks to the efforts of many important people.

1.4.1. Geotechnical Engineering Prior to the 18th Century

The earliest known usage of soil for building purposes is lost to antiquity. According to Skempton (1985), the field of geotechnical engineering as it is known today was first understood in the early 1700s. For years, the art of geotechnical engineering was based on only past experiences through a succession of experimentation without any real scientific character. Based on those experimentations, many structures were built—some of which have crumbled, while others are still standing (Turner, 1943).

Recorded history tells us that ancient civilizations flourished along the banks of rivers, such as the Nile (Egypt), the Tigris and Euphrates (Mesopotamia), the Huang Ho (Yellow River, China), and the Indus (India).. Dikes built in the Indus basin as early as 2000 B.C. protected the village of Mohenjo Dara, which was located in what became Pakistan after 1947. A large number of dykes were constructed for irrigation throughout the rule of Chan dynasty of China (1120–249 B.C.).

There's no proof that foundation stabilization or flood-induced erosion were addressed (Kerisel, 1985). For building structures, the ancient Greeks used strip-and-raft foundations and isolated pad footings. Egypt built a number of pyramids starting about 2700 B.C., the majority of which served as tombs for the Pharaohs and their consorts throughout the era of the Old and Middle Kingdoms (Santamarina & Chameau, 1987). Some of the most significant pyramids identified by the Pharaoh who gave the building order are listed in Table 1.1. There have been 138 pyramids found in Egypt as of 2008. Giza's pyramids can be seen in Figure 1.1.

When it came to building stable slopes, underground chambers, and foundations, the pyramids presented enormous challenges (Beyr, 2015). Thousands of pagodas were constructed in China after Buddhism arrived there in 68 A.D. under the Eastern Han dynasty. Silt and soft clay layers served as the foundation for many of these constructions. In certain instances, the soil's ability to support the weight of the foundation pressure exceeded, which resulted in significant structural damage (Dalby, 2015).

Table 1.1. Major Pyramids in Egypt

Pyramid/Pharaoh	Location	Reign of Pharaoh
Djoser	Saqqara	2630–2612 B.C.
Sneferu	Dashur (North)	2612–2589 B.C.
Sneferu	Dashur (South)	2612–2589 B.C.
Sneferu	Meidum	2612–2589 B.C.
Khufu	Giza	2589–2566 B.C.
Djedefre	Abu Rawash	2566–2558 B.C.
Khafre	Giza	2558–2532 B.C.
Menkaure	Giza	2532–2504 B.C.

Figure 1.1. *A view of the pyramids at Giza.*

Source: Craiyon, Creative Commons License.

One of the most famous examples of problems related to soil-bearing capacity in the construction of structures prior to the 18th century is the Leaning Tower of Pisa in Italy (Figure 1.2) (Lacasse, 2013). Under the prosperous Republic of Pisa, construction on the tower commenced in 1173 A.D. and progressed in various forms for more than 200 years. A 20-meter-diameter circular base supports the structure, which weighs roughly 15,700 metric tons overall. In the past, the tower has tipped to the east, north, west, then finally, south (Mitchell & Santamarina, 2005). The tower tilted due to ground surface compression, which was caused by a weak clay layer located approximately 11 meters below the surface, according to recent investigations. With a height of 54 meters, it became over five meters out of plumb. In 1990, the tower was shuttered due to concerns that it would topple or collapse. Excavating dirt from beneath the tower's north side has recently stabilized it (Christian, 2004).

Approximately 70 metric tons of soil was taken out in 41 distinct extractions spanning the width of the tower. The tower tilt lessened as the ground filled in the resulting space. The tower has a five-degree lean now. Although the half-degree adjustment is imperceptible, it significantly increases the stability of the structure (Skempton, 1954).

Figure 1.2. *Leaning tower of Pisa, Italy.*

Source: Freepik, Creative Commons License.

Figure 1.3 illustrates a problem that is comparable. The towers depicted in Figure 1.3 are 12th-century structures situated in Bologna, Italy. Most people call the tower on the left the Garisenda Tower. It is 48 meters tall and 4,210 metric tons in weight. It is now about 4 degrees tilted. The Asinelli Tower, standing at 97 meters high and weighing 7,300 metric tons, is the tower to the right. Its tilt is approximately 1.3° (Reynolds, 1886).

In the early 1700s, engineers and scientists started to more methodically study the characteristics and behaviors of soils due to a number of foundation-related issues they had encountered during construction centuries earlier (Rankine, 1857).

Figure 1.3. *Tilting of Garisenda Tower (left) and Asinelli Tower (right) in Bologna, Italy.*

Source: Freepik, Creative Commons License.

Four major periods can be identified in the time from 1700 to 1927 in the field of geotechnical engineering, based on both the emphasis and nature of study (Skempton, 1948):

1. Preclassical Period (1700 — 1776 A.D.)
2. Classical soil mechanics – Phase I (1776 — 1856 A.D.)
3. Classical soil mechanics – Phase II (1856 — 1910 A.D.)
4. Modern soil mechanics (1910 — 1927 A.D.) (Rankine, 1857).

These periods along with their significant developments will be discussed below:

1.4.2. Pre-Classical Period of Soil Mechanics (1700–1776)

This period concentrated on studies relating to natural slope and unit weights of various types of soils, as well as the semiempirical earth pressure theories. In 1717, Henri Gautier (1660–1737), a French royal engineer, observed the natural slopes of soils when tipped in a heap to develop retaining wall design procedures (Semmel, 2004). The natural slope is what we now refer to as the angle of repose. This study found that the natural slopes of ordinary earth and clean dry sand were 45° and 31°, respectively. Also the unit weights of ordinary earth and clean dry sand were 13.4 kN/m^3 or 85 lb/ft^3 and 18.1 kN/m^3, respectively (Algaze, 2018). There was no reporting of test results on clay. Bernard Forest (de Belidor, 1671–1761), in 1729, published a textbook for French engineers, both military and civil. He expanded on (Gautier's, 1717) original study by putting forth a theory in the book regarding lateral earth pressure upon the retaining walls. Additionally, he provided a system for classifying soil in the way that the following table illustrates (Barnard, 2002).

Classification	Unit weight (kN/m^3)
Rock	--
Firm or hard sand; compressible sand	16.7 – 18.4
Ordinary earth (dry locations)	13.4
Soft earth (primarily silt)	16.0
Clay	18.9

Francois Gadroy (1705–1759), a French engineer, reported the first laboratory model findings in 1746 on a 76-mm-high retaining wall constructed with sand backfill. At failure, Gadroy noted the presence of slip planes in the soil. J. J. Mayniel later summarized Gadroy's research in 1808. Jean Rodolphe Perronet (1708–1794), another French engineer, made a significant contribution during this time as well. He studied slope stability around 1769 and made the distinction between fills and intact ground (Bischoff, 2002).

1.4.3. Classical Soil Mechanics—Phase I (1776–1856)

The majority of the advancements in the field of geotechnical engineering during this time came from French scientists and engineers. Almost every theoretical calculation used in the pre-classical era to determine the lateral earth pressure on the retaining walls was predicated on an arbitrary failure surface in the soil (Parker, 1986). The true position of the sliding surface in the soil behind a retaining wall was discovered by Charles Augustin Coulomb (1736–1806), a French scientist, by using the concepts of calculus for maxima and minima in his well-known paper published in 1776. Coulomb applied the solid-body laws of cohesiveness and friction to this analysis. Coulomb's theory was incorporated by eminent French civil engineer Gaspard Clair Marie Riche de Prony (1755–1839) in his seminal textbook Nouvelle Architecture Hydraulique (Vol. 1). Jacques Frederic (Francais, 1820), a French engineer, and Claude Louis Marie Henri Navier (1785–1836), a French professor of applied mechanics, examined specific cases of Coulomb's work in 1820. These unique situations concerned backfills that supported surcharges and inclined backfills. An army engineer and professor of mechanics named Jean Victor Poncelet (1788–1867) expanded on Coulomb's theory in 1840 by offering a graphical method for calculating the amount of lateral earth pressure on inclined and vertical retaining walls with arbitrarily fractured polygonal ground surfaces (Baksi, 2001). Moreover, Poncelet was the first to represent the soil friction angle with the letter f. He also offered the first bearing-capacity theory that was definitive for shallow foundations. The engineering genius Alexandre (Collin, 1846) gave the instructions for deep slips in cuttings, clay slopes, and embankments in 1846. According to Collin's theory, failure occurs whenever the mobilized cohesiveness of the soil surpasses its current cohesiveness. Furthermore, he noted that arcs of cycloids could be used to roughly represent the actual failure surfaces (Lokkas et al., 2021).

Phase I of the classical soil mechanics period typically ends in 1857, the year of William John Macquorn Rankine's (1820–1872) first publication. Rankine was a civil engineering professor at the University of Glasgow. An important theory regarding earth pressure and earth mass equilibrium was presented by this study. A simplified version of Coulomb's theory is Rankine's theory (Scott, 1985).

1.4.4. Classical Soil Mechanics—Phase II (1856–1910)

In this phase, a number of experimental findings from sand-related laboratory tests were published in the literature. A publication authored by Henri Philibert Gaspard Darcy (1803–1858), a French engineer, is among the most significant and earliest. He released a study on sand filter permeability in 1856. Darcy used the results of those tests to define the term "coefficient of permeability," also known as hydraulic conductivity of soil. This parameter is still highly helpful in geotechnical engineering today (Voznesensky, 2022).

Astronomy professor Sir George Howard (Darwin, 1883) tested in a lab the moment of overturning on a hinged wall that held sand in both loose and dense compaction states. An important contribution made by Joseph Valentin (Boussinesq, 1885) in 1885 was the theory of stress distribution under loaded bearing areas in an elastic, semi-finite, homogeneous, and isotropic medium. Osborne (Reynolds, 1886) proved that sand exhibits dilatancy in 1887. John Stuart Beresford (1845–1925) and John Clibborn (1847–1938) conducted other noteworthy studies during this time that dealt with uplift pressure and water flow through sand beds. Clibborn's research was published in two publications: Technical Paper No. 97, Government of India, 1902, and Treatise on Civil Engineering, Vol. 2: Irrigation Work in India, Roorkee, 1901. Beresford did a study in 1898 on the uplift pressure on the Narora Weir on the Ganges River. It was published up in Technical Paper No. 97, Government of India, 1902.

1.4.5. Modern Soil Mechanics (1910–1927)

The basic parameters and properties of clay were established by research published during this time. The publications that are most noteworthy are discussed next (Spangler, 1948).

Albert Mauritz Atterberg, a Swedish soil scientist and chemist who lived from 1846 to 1916, defined clay-size fractions to be the percentage of particles by weight that are smaller than 2 microns. This was around 1908. He understood the significance of clay particles in soil and their plastic nature. In 1911, he defined the limits of plastic, liquid and shrinkage in order to explain the cohesive soils' consistency. The difference between the liquid and plastic limits is how he described the plasticity index (Atterberg, 1911).

The 17-meter-tall earth dam at Charmes, France, built between the years 1902 and 1906, collapsed in October 1909. Jean (Fontard, 1914), a French engineer, conducted research to identify the reason for the failure. Under those circumstances, he performed undrained double-shear tests to ascertain the shear strength parameters of clay specimen, that were 200 m thick and 0.77 m^2 in area, under constant vertical stress (Frontard, 1914). These specimens failed within a time range of 10 to 20 minutes (Bjerrum & Flodin, 1960).

Arthur Langley (Bell, 1915), an engineer from England, contributed to the design along with construction of the outer seawall at the Rosyth Dockyard. Through his work, he established connections for lateral pressure and resistance in clay, as well as the bearing capability of shallow foundations in the clay (refer to Bell, 1915). He also tested the shear strength of clay samples that had not been disturbed using shear-box tests (Schultze, 1954). Wolmar (Fellenius, 1918), a Swedish engineer, pioneered the stability analysis for saturated clay slopes under the presumption that critical surface of sliding forms an arc of a circle. These were further discussed in his published research in 1918 followed by 1926. The 1926 paper had the right numerical solution for the stability numbers of circular slip surfaces that went through the bottom of the slope (Middlebrooks, 1953).

Karl (Terzaghi, 1939–1953) an Austrian who lived from 1883 to 1963 (Figure 1.4), came up with the theory of consolidation for clay that we use today. During his tenure as a professor at the American Robert College in Istanbul, Turkey, Terzaghi formulated this theory. He conducted his research between 1919 and 1924, a period of five years and five distinct clay soils were utilized. The soils' liquid limit varied from 36 to 67, while the plasticity index fell between 18 and 38. This theory was first presented in the 1925 edition of Terzaghi's well-known book Erdbaumechanik (Massarsch & Fellenius, 2012).

Figure 1.4. *Karl Terzaghi (1883–1963).*

Source: Demanega, Creative Commons License.

1.4.6. Geotechnical Engineering After 1927

The1925 publication of Karl Terzaghi, Erdbaumechanik auf Bodenphysikalisher Grundlage, marked a significant milestone in the advancement of soil mechanics. Karl Terzaghi is widely recognized, and rightly so, as the father of the modern soil mechanics. On October 2, 1883, Terzaghi was born in Prague, the capital city of the Austrian province of Bohemia at the time. He completed his studies in 1904 at the Technische Hochschule in Graz, Austria, earning a bachelor's degree in mechanical engineering. He joined the Austrian army and served for a year after graduating. After completing a year of his service, Terzaghi decided to further his studies for another year, focusing on geological subjects. In January 1912, he graduated with a Doctor of Technical Sciences degree in Graz from his alma mater. He started working a teacher in 1916 at the Istanbul-based Imperial School of Engineers. Following the conclusion of World War I, he took on a teaching position at the American Robert College in Istanbul from 1918 to 1925. He started his research work on studying the settlement of clays and the behavior of soils, as well as the failure caused by piping in the sand under dams. His publication, Erdbaumechanik, is mainly the outcome of this research (Guler, 2017).

Terzaghi worked as a visiting lecturer at Massachusetts Institute of Technology from 1925 until 1929 after accepting the position in 1925. Throughout that period, he gained recognition as the foremost figure in an emerging field of civil engineering known as soil mechanics. In October 1929, he went back to Europe and took up the position of professor at the Technical University of Vienna. This university quickly became a hub for civil engineers who were passionate about soil mechanics. In 1939, he came back to the US and joined the faculty at Harvard University as a professor (de Prony, 1790).

The inaugural conference hosted by the International Society of Soil Mechanics and Foundation Engineering (ISSMFE) took place at Harvard University in 1936, under the leadership of Karl Terzaghi. The conference was made possible thanks to the dedication and hard work by Professor Arthur Casagrande from Harvard University in which approximately 200 people from 21 different countries were in attendance at this conference. Thanks to Terzaghi's guidance and inspiration over the past 25 years, a diverse range of topics were covered in the papers presented at the conference, including:

1. Effective stress
2. Testing with Dutch cone penetrometer
3. Preloading for settlement control
4. Machine vibration
5. Arching theory of earth pressure
6. Earthquake and soil liquefaction

7. Shear strength
8. Elastic theory and stress distribution
9. Consolidation
10. Centrifuge testing
11. Swelling clays
12. Frost action (Peck, 1985)

Terzaghi played a pivotal role in shaping the field of geotechnical engineering and soil mechanics on a global scale for the next 25 years. Ralph Peck wrote in 1985 that "during Terzaghi's lifetime, few people would have disagreed that he was not only the guiding spirit in soil mechanics, but that he was also the clearing house for research and application all over the world. In the coming years, he would work on projects across all continents except Australia and Antarctica (Navier, 1839). Thus, even in the present day, it is difficult to surpass his opinions of the state of soil mechanics, as stated in his presidential addresses and summary papers, according to Peck. At the London Institution of Civil Engineers, Terzaghi gave the 45th James Forrest Lecture in 1939. The title of his lecture was "Soil Mechanics—A New Chapter in Engineering Science." In it, he claimed that the majority of the incidents involving the foundation were no longer "acts of God" (Fellenius, 1918).

Here are some key developments in the field of geotechnical engineering and soil mechanics that have emerged since the inaugural conference of the ISSMFE in 1936:

- Karl Terzaghi's book, Theoretical Soil Mechanics, was published in 1943 by Wiley, New York.
- In 1948, the international journal of soil mechanics, Geotechnique, was first published in England.
- The 1948 release by Wiley, New York, of Karl Terzaghi and Ralph Peck's book Soil Mechanics in Engineering Practice
- In 1948, Donald W. Taylor's book Fundamentals of Soil Mechanics was published by Wiley, New York (Bishop & Henkel, 1962).

In 1948, the second annual conference of ISSMFE took place in Rotterdam, The Netherlands, following a brief pause due to World War II. Roughly 600 people took part, and the proceedings were published in seven volumes. During the event, A. W. Skempton delivered a significant paper on the concept of f = 0 for clays. Since Rotterdam, conferences of the ISSMFE have been held roughly every four years in various locations across the globe (Shan et al., 2022). Following the Rotterdam conference, there was an increase in regional conferences focusing on geotechnical engineering, including:

1. European Regional Conference on Stability of Earth Slopes, Stockholm (1954)
2. First Australia–New Zealand Conference on Shear Characteristics of Soils (1952)
3. First Pan American Conference, Mexico City (1960)
4. Research conference on Shear Strength of Cohesive Soils, Boulder, Colorado, (1960)

Between 1948 and 1960, two other significant turning points occurred: (1) the publication of a paper by W. Skempton on A and B pore pressure parameters, which improved the practicality of effective stress calculations for a variety of engineering projects; and (2) the 1957 release of a book by A. W. Bishop and J. Henkel (Arnold, London) called The Measurement of Soil Properties in the Triaxial Text (Parkhill, 2001). Computer-aided finite element and finite difference solutions were used for a variety of geotechnical engineering problems by the early 1950s. They continue to be a crucial and practical computational tool in this field. Throughout its history, this field has made significant advancements and reached a level of maturity. Geotechnical engineering has become an established field of civil engineering, with numerous civil engineers choosing it as their preferred area of expertise (Liu et al., 2011). In 1997, the ISSMFE underwent a name change to International Society of Soil Mechanics and Geotechnical Engineering (ISSMGE) in order to accurately represent its comprehensive focus. This is an important international conference for sharing information about ongoing research and new developments in geotechnical engineering (Tripathy et al., 2016). Table 1.2 provides the details of the year and the location for every conference organized by ISSMFE/ISSMGE.

Table 1.2. Details of ISSMFE (1936–1997) and ISSMGE (1997–Present) Conferences

Conference	Location	Year
I	Harvard University, Boston, U.S.A.	1936
II	Rotterdam, the Netherlands	1948
III	Zurich, Switzerland	1953
IV	London, England	1957
V	Paris, France	1961
VI	Montreal, Canada	1965
VII	Mexico City, Mexico	1969
VIII	Moscow, U.S.S.R.	1973
IX	Tokyo, Japan	1977
X	Stockholm, Sweden	1981
XI	San Francisco, U.S.A.	1985
XII	Rio de Janeiro, Brazil	1989
XIII	New Delhi, India	1994
XIV	Hamburg, Germany	1997
XV	Istanbul, Turkey	2001
XVI	Osaka, Japan	2005
XVII	Alexandria, Egypt	2009 (scheduled)

In 1960, Bishop, Alpan, Blight, and Donald conducted research and presented findings on the factors that influence the strength of cohesive soils (partially saturated.) Since then, significant progress has been made in understanding the behavior of unsaturated soils in terms of strength, compressibility, and other factors that impact the construction of earth-retaining and earth-supported structures (Qi et al., 2022).

ISSMGE has a variety of technical committees, which are responsible for organizing or co-sponsoring numerous conferences worldwide (Dewoolkar et al., 2003). A summary of those technical committees from 2010 to 2013 can be found in Table 1.3. ISSMGE also organizes International Seminars, which have become a significant event. These seminars provide a platform for practitioners, academics, and contractors to come together and benefit from each other's expertise. It doesn't matter where they come from or the resources of their respective societies, as these seminars promote a sense of unity within the ISSMGE (Fannin, 2008).

Table 1.3. List of ISSMGE Technical Committees (2010–2013)

Fundamentals
• TC101 - Laboratory Stress Strain Strength Testing of Geomaterials • TC102 - Ground Property Characterization from In-Situ Tests • TC103 - Numerical Methods • TC104 - Physical Modelling in Geotechnics • TC105 - Geo-Mechanics from Micro to Macro • TC106 - Unsaturated Soils • TC107 - Laterites and Lateritic Soils
Applications
• TC201 - Geotechnical Aspects of Dykes and Levees and Shore Protection • TC202 - Transportation Geotechnics • TC203 - Earthquake Geotechnical Engineering and Associated Problems • TC204 - Underground Construction in Soft Ground • TC205 - Safety and Serviceability in Geotechnical Design • TC206 - Interactive Geotechnical Design • TC207 - Soil-Structure Interaction and Retaining Walls • TC208 - Slope Stability in Engineering Practice • TC209 - Offshore Geotechnics • TC210 - Dams & Embankments • TC211 - Ground Improvement • TC212 - Deep Foundations • TC213 - Scour and Erosion • TC214 - Foundation Engineering for Difficult Soft Soil Conditions • TC215 - Environmental Geotechnics • TC216 - Frost Geotechnics • TC217 - Land Reclamation • TC218 - Reinforced Fill Structures
Impact on Society
• TC301 - Preservation of Historic Sites • TC302 - Forensic Geotechnical Engineering • TC303 - Coastal and River Disaster Mitigation and Rehabilitation • TC304 - Engineering Practice of Risk Assessment and Management • TC305 - Geotechnical Infrastructure for Megacities and New Capitals • TC306 - Geo-engineering Education • TC307 - Sustainability in Geotechnical Engineering • TC308 - Energy Geotechnics

1.4.7. End of an Era

A synopsis of the contributions pioneers like Arthur Casagrande, Karl Terzaghi, Laurits Bjerrum, Donald W. Taylor, and Ralph B. Peck made to modern soil mechanics was provided in the previous section. Ralph B. Peck, 95, was the last of the early titans of the field to pass away on February 18, 2008 (Zevgolis et al., 2021).

On June 23, 1912, Professor Ralph B. Peck (shown in Figure 1.5) was began his life in Winnipeg, Canada, to American parents Orwin K. and Ethel H. Peck. He went to Rensselaer Polytechnic Institute in Troy, New York, and got his B.S. in 1934 and his Ph.D. in 1937. In the years 1938 to 1939, he enrolled in courses taught by Arthur Casagrande at Harvard University, focusing on an emerging field known as "soil mechanics." During the years 1939–1943, Dr. Peck worked on the Chicago Subway Project with Karl Terzaghi, who is known as the "father" of modern soil mechanics. From 1943 until he retired in 1974, he served as a professor of foundation engineering at the University of Illinois Urbana-Champaign (Griffiths, 2014).

Figure 1.5. *Ralph B. Peck.*

Source: Ralph B. Peck, Creative Commons License.

Following retirement, he remained highly engaged in consulting work, overseeing significant geotechnical projects across 28 countries spanning five continents and 44 states in the US. He has worked on several significant consulting projects, including:

i. Rapid transit systems in Chicago, San Francisco, and Washington, D.C.
ii. James Bay Project in Quebec, Canada
iii. Alaskan pipeline system
iv. Dead sea dikes
v. Heathrow Express Rail Project (U.K.)

The Greek bridge between Rion and Antirion was his last job. The Times of the UK reported on March 13, 2008, that American civil engineer Ralph B. Peck developed a contentious construction method that would be applied to some of the world's most impressive modern engineering feats, such as the Channel Tunnel. Regarded as a pioneer in the field of soil mechanics, he played a crucial role in a series of groundbreaking earth dam and tunneling projects that defied the limits of what was believed possible (Mitchell & Santamarina, 2005).

Dr. Peck has an extensive collection of over 250 exceptional technical publications. From 1969 to 1973, he presided over the ISSMGE as president. In 1974, he was honored with the National Medal of Science by President Gerald R. Ford. For generations, geotechnical engineers worldwide looked to Professor Peck as a mentor, teacher, counselor and a friend. His attendance at the 16th ISSMGE Conference held in Osaka, Japan in 2005 would be his final major conference of that kind (Çabalar, 2007).

SUMMARY

Introduction to Geotechnical and Foundation Engineering provides a comprehensive overview of the geotechnical process, history, and applications. It delves into the evolution of geotechnical engineering, tracing its roots and highlighting key milestones that have shaped the field. The chapter emphasizes the importance of understanding geological processes and their impact on engineering projects. Through an exploration of geotechnical applications, readers gain insights into how these principles are employed in real-world scenarios, from site investigation and soil testing to the design and construction of foundations. By focusing on the geotechnical process, history, and applications, this chapter lays a solid foundation for readers to delve deeper into the complexities of geotechnical engineering throughout the book.

REFERENCES

1. Algaze, G. (2018). Entropic cities: The paradox of urbanism in ancient Mesopotamia. *Current Anthropology, 59*(1), 23–54. https://doi.org/10.1086/695769.
2. Atterberg, A. (1911). Die plastizität der Tone. *Internationale Mitteilungen für Bodenkunde, 2*(1), 4–37.
3. Baghbani, A., Choudhury, T., Costa, S., & Reiner, J. (2022). Application of artificial intelligence in geotechnical engineering: A state-of-the-art review. *Earth-Science Reviews, 228*, 103–991. https://doi.org/10.1016/j.earscirev.2021.103991.
4. Baksi, P. (2001). MEGA IV/31: Natural science notes of Marx and Engels, 1877–1883. *Nature, Society, and Thought, 14*(4), 377–390.
5. Balasubramaniam, A. S. (2021). Contributions in geotechnical engineering: Soil mechanics and foundation engineering. In *Developments in Geotechnical Engineering: From Harvard to New Delhi 1936–1994, 2*(1), 3–48.
6. Barnard, A. (2002). Hunter-gatherers: Seventeenth or eighteenth-century invention? *Archaeological Dialogues, 9*(2), 119–122. https://doi.org/10.1017/S1380203800013058.
7. Basu, D., Misra, A., & Puppala, A. J. (2015). Sustainability and geotechnical engineering: Perspectives and review. *Canadian Geotechnical Journal, 52*(1), 96–113. https://doi.org/10.1139/cgj-2014-0108.
8. Bea, R. (2006). Reliability and human factors in geotechnical engineering. *Journal of Geotechnical and Geoenvironmental Engineering, 132*(5), 631–643. https://doi.org/10.1061/(ASCE)1090-0241(2006)132:5(631).
9. Becker, D. E. (1997). Eighteenth Canadian geotechnical colloquium: Limit states design for foundations. Part I. An overview of the foundation design process. *Canadian Geotechnical Journal, 33*(6), 956–983. https://doi.org/10.1139/t96-107.
10. Bell, A. L. (1915). The lateral pressure and resistance of clay and the supporting power of clay foundations. *Minutes of the Proceedings of the Institution of Civil Engineers, 199*, 233–272.
11. Beyr, P. (2015). Military geology and geopolitics. In *International Conference on Military Technologies (ICMT) 2015, 2*(1), 1–5.
12. Bischoff, D. (2002). The invention of hunter-gatherers in seventeenth-century Europe–A comment from Upper Mesopotamia, Near East1. *Archaeological Dialogues, 9*(2), 133–136. https://doi.org/10.1017/S1380203800013071.

13. Bishop, A. W., Alpan, I., Blight, G. E., & Donald, I. B. (1960). Factors controlling the strength of partly saturated cohesive soils. *Proceedings. Research Conference on Shear Strength of Cohesive Soils, ASCE, 4*(2), 502–532.
14. Bjerrum, L., & Flodin, N. (1960). The development of soil mechanics in Sweden, 1900–1925. *Géotechnique, 10*(1), 1–18. https://doi.org/10.1680/geot.1960.10.1.1.
15. Bogusz, W., & Godlewski, T. (2019). Philosophy of geotechnical design in civil engineering–Possibilities and risks. *Bulletin of the Polish Academy of Sciences. Technical Sciences, 67*(2), 289–306. https://doi.org/10.24425/bpasts.2019.129666.
16. Boussinesq, J. V. (1885). Sur la resistance d'une sphere solide. *Comptes Rendus Hebdomadaires des Séances de l'Académie des Sciences, 100*, 935.
17. Brinkgreve, R. B. (2005). Selection of soil models and parameters for geotechnical engineering application. In *Soil constitutive models: Evaluation, selection, and calibration. Vol 1, (pp. 69–98).
18. Çabalar, A. F. (2007). Discussion of "Biological considerations in geotechnical engineering" by James K. Mitchell and J. Carlos Santamarina. *Journal of Geotechnical and Geoenvironmental Engineering, 133*(4), 485–485. https://doi.org/10.1061/(ASCE)1090-0241(2007)133:4(485).
19. Christian, J. T. (2004). Geotechnical engineering reliability: How well do we know what we are doing? *Journal of Geotechnical and Geoenvironmental Engineering, 130*(10), 985–1003. https://doi.org/10.1061/(ASCE)1090-0241(2004)130:10(985).
20. Cleveland, L. J., & Wartman, J. (2006). Principles and applications of digital photogrammetry for geotechnical engineering. *Site and Geomaterial Characterization, 3*(2), 128–135.
21. Collin, A. (1846). Recherches expérimentales sur les glissements spontanés des terrains argileux, accompagnées de considérations sur quelques principes de la mécanique terrestre par Alexandre Collin. *Carilian-Goeury, 3*(2), 12–100.
22. Dalby, S. (2015). Geoengineering: The next era of geopolitics? *Geography Compass, 9*(4), 190–201. https://doi.org/10.1111/gec3.12210.
23. Darwin, G. H. (1883). On the horizontal thrust of a mass of sand. *Minutes of the Proceedings of the Institution of Civil Engineers, 71*, 350–378.
24. de Prony, R. (1790). Nouvelle architecture hydraulique (Vol. 1796, No. 2). Paris, Firmin-Didot.

25. Dewoolkar, M. M., Goddery, T., & Znidarcic, D. (2003). Centrifuge modeling for undergraduate geotechnical engineering instruction. *Geotechnical Testing Journal, 26*(2), 201–209.
26. El Mountassir, G., Minto, J. M., van Paassen, L. A., Salifu, E., & Lunn, R. J. (2018). Applications of microbial processes in geotechnical engineering. *Advances in Applied Microbiology, 104*, 39–91. https://doi.org/10.1016/bs.aambs.2018.07.002.
27. Fannin, J. (2008). Karl Terzaghi: From theory to practice in geotechnical filter design. *Journal of Geotechnical and Geoenvironmental Engineering, 134*(3), 267–276. https://doi.org/10.1061/(ASCE)1090-0241(2008)134:3(267).
28. Fellenius, W. (1918). Kaj-och jordraset i Goteborg. *Teknisk Tidsskrift, 2*(1), 44–177.
29. Français, J. F. (1820). Recherches sur la pousseé des terres, sur la forme et les dimensions des murs derevetement et sur les talus d'excavation Research on soil pressure on the form and dimensions of revetment walls and excavation slopes. *Memoires de L'office d Genie, 4*, 157–193.
30. Frank, R. (2007). Basic principles of Eurocode 7 on geotechnical design. Proceedings of the 18th EYGEC, Ancona, Italy, 17–20.
31. Frontard, J. (1914). Notice sur l'accident de la digue de Charmes. *Annales des Ponts et Chaussées, 9e Série, 23*, 173–292.
32. Gautier, H. (1717). Dissertation sur l'epaisseur des culées des ponts. Cailleau, 34–99.
33. Gens, A. (2010). Soil–environment interactions in geotechnical engineering. *Géotechnique, 60*(1), 3–74. https://doi.org/10.1680/geot.2010.60.1.3.
34. Griffiths, J. S. (2014). Feet on the ground: Engineering geology past, present and future. *Quarterly Journal of Engineering Geology and Hydrogeology, 47*(2), 116–143. https://doi.org/10.1144/qjegh2013-047.
35. Guler, E. (2017). A material which started a new era in geotechnical engineering: Geosynthetics. *Innovative Infrastructure Solutions, 2*, 1–21. https://doi.org/10.1007/s41062-017-0083-1.
36. Holtz, R. D., Kovacs, W. D., & Sheahan, T. C. (1981). *An Introduction to Geotechnical Engineering*. Englewood Cliffs: Prentice-Hall.
37. Hussein, M. H., & Goble, G. G. (2004). A brief history of the application of stress-wave theory to piles. In *Current Practices and Future Trends in Deep Foundations* (Vol. 1, pp. 186–201).
38. Kerisel, J. (1985). The history of geotechnical engineering up until 1700.

International Conference on Soil Mechanics and Foundation Engineering, 4(3), 1–100.

39. Khalili, L. (2018). The infrastructural power of the military: The geoeconomic role of the US Army Corps of Engineers in the Arabian Peninsula. *European Journal of International Relations, 24*(4), 911–933. https://doi.org/10.1177/1354066117744850.
40. Lacasse, S. (2013). 8th Terzaghi Oration Protecting society from landslides–the role of the geotechnical engineer. In *Proceedings of the 18th International Conference on Soil Mechanics and Geotechnical Engineering, Paris, 2*(1), 15–34.
41. Liu, Z. B., Ma, X., & Dai, W. L. (2011). Experimental research on engineering property of heavy metal-contaminated kaolinite. *Applied Mechanics and Materials, 94*, 1921–1929. https://doi.org/10.4028/www.scientific.net/AMM.94.1921.
42. Lokkas, P., Chouliaras, I., Chrisanidis, T., Christodoulou, D., Papadimitriou, E., & Paschalis, E. (2021). Historical background and evolution of Soil Mechanics. *WSEAS Transactions on Advances in Engineering Education, 18*, 96–113.
43. Massarsch, K. R., & Fellenius, B. H. (2012). Early Swedish contributions to geotechnical engineering. In *Full-Scale Testing and Foundation Design: Honoring Bengt H. Fellenius, 2*(1), 239–256.
44. Middlebrooks, T. A. (1953). Earth-dam practice in the United States. *Transactions of the American Society of Civil Engineers, 118*(2), 697–722.
45. Mitchell, J. K., & Santamarina, J. C. (2005). Biological considerations in geotechnical engineering. *Journal of Geotechnical and Geoenvironmental Engineering, 131*(10), 1222–1233. https://doi.org/10.1061/(ASCE)1090-0241(2005)131:10(1222).
46. Navier, C. L. M. (1839). *Leçons sur L'Application de la Mécanique à L'Establissement des Constructions et des Machines, 2*(1), 34–116.
47. Osinubi, K. J., Eberemu, A. O., Ijimdiya, T. S., Yakubu, S. E., Gadzama, E. W., Sani, J. E., & Yohanna, P. (2020). Review of the use of microorganisms in geotechnical engineering applications. *SN Applied Sciences, 2*, 1–19. https://doi.org/10.1007/s42452-019-1967-1.
48. Oyediran, I. A., & Falae, P. O. (2018). Integrated geophysical and geotechnical methods for pre-foundation investigations. *Journal of Geology & Geophysics, 8*, 4–53. https://doi.org/10.4172/2381-8719.1000412.
49. Parker Sr, G. G. (1986). Early stage of hydrogeology in the United States, 1776 to 1912. *JAWRA Journal of the American Water Resources*

Association, 22(5), 701–716. https://doi.org/10.1111/j.1752-1688.1986.tb05872.x.

50. Parkhill, S. T. (2001). 150 Years of Geotechnical Design and Construction. In *International Engineering History and Heritage: Improving Bridges to ASCE's 150th Anniversary, 3*(1), 67–72.
51. Peck, R. B. (1985). The last sixty years. *11th International Conference on Soil Mechanics and Foundation Engineering, 3*(2), 123–133.
52. Philip, J. R. (1970). Flow in porous media. *Annual Review of Fluid Mechanics, 2*(1), 177–204. https://doi.org/10.1146/annurev.fl.02.010170.001141.
53. Phoon, K. K. (2020). The story of statistics in geotechnical engineering. *Georisk: Assessment and Management of Risk for Engineered Systems and Geohazards, 14*(1), 3–25. https://doi.org/10.1080/17499518.2020.1716256.
54. Potts, D. M., Zdravković, L., Addenbrooke, T. I., Higgins, K. G., & Kovačević, N. (2001). Finite element analysis in geotechnical engineering: Application. Vol. 2, pp. 4–27.
55. Puzrin, A. M. (2017). The Leaning Tower of St. Moritz: A structure on a creeping landslide. In *Geotechnics and Heritage,* Vol. 1, pp. 123–143.
56. Qi, Z., Xiao, J., Lu, J., Zhou, H., & Wang, H. (2022). Research on rock triaxial cyclic loading and unloading characteristics based on PFC3D. In *MATEC Web of Conferences,* Vol. 358, pp. 1–12.
57. Reese, L. C., Isenhower, W. M., & Wang, S. T. (2005). *Analysis and Design of Shallow and Deep Foundations* (Vol. 10). John Wiley & Sons.
58. Reynolds, O. (1886). Experiments showing dilatancy, a property of granular material, possibly connected with gravitation. *Proceedings of the Royal Institution of Great Britain, 11*, 12–388.
59. Santamarina, J. C., & Chameau, J. L. (1987). Expert systems for geotechnical engineers. *Journal of Computing in Civil Engineering, 1*(4), 241–252. https://doi.org/10.1061/(ASCE)0887-3801(1987)1:4(241).
60. Schultze, E. (1954). Hans-Detlef Krey and the experimental approach to soil mechanics. *Géotechnique, 4*(3), 93–96.
61. Scott, R. F. (1985). Plasticity and constitutive relations in soil mechanics. *Journal of Geotechnical Engineering, 111*(5), 559–605. https://doi.org/10.1061/(ASCE)0733-9410(1985)111:5(559).
62. Semmel, B. (2004). *The Rise of Free Trade Imperialism: Classical Political Economy the Empire of Free Trade and Imperialism.* 1750–1850. Cambridge University Press.
63. Shackelford, C. D. (2005). Environmental issues in geotechnical

engineering. In *Proceedings of the International Conference on Soil Mechanics and Geotechnical Engineering, 16*(1), 95.

64. Shahin, M. A., Jaksa, M. B., & Maier, H. R. (2001). Artificial neural network applications in geotechnical engineering. *Australian Geomechanics, 36*(1), 49–62.
65. Shan, Z., Zhang, P., & Kou, H. (2022). Mechanical and engineering behavior of MICP-treated coarse siliceous sands. *KSCE Journal of Civil Engineering, 26*(1), 79–87. https://doi.org/10.1007/s12205-021-2630-7.
66. Skempton, A. W. (1954). The pore-pressure coefficients A and B. *Géotechnique, 4*(4), 143–147. https://doi.org/10.1680/geot.1954.4.4.143.
67. Sowers, G. F. (1993). Human factors in civil and geotechnical engineering failures. *Journal of Geotechnical Engineering, 119*(2), 238–256. https://doi.org/10.1061/(ASCE)0733-9410(1993)119:2(238).
68. Spangler, M. G. (1948). Underground conduits-An appraisal of modern research. *Transactions of the American Society of Civil Engineers, 113*(1), 316–345.
69. Tawelian, L. R., & Mickovski, S. B. (2016). The implementation of geotechnical data into the BIM process. *Procedia Engineering, 143*, 734–741. https://doi.org/10.1016/j.proeng.2016.06.078.
70. Terzaghi, K. (1953). Origin and functions of soil mechanics. *Transactions of the American Society of Civil Engineers, 118*(2), 666–696.
71. Tripathy, G. R., Shirke, R. R., & Kudale, M. D. (2016). Journal of Rock Mechanics and Geotechnical Engineering. *Journal of Rock Mechanics and Geotechnical Engineering, 8*, 248–255.
72. Turner, R. (1943). Technology and geopolitics. *Military Affairs: Journal of the American Military Institute, 7*(1), 5–15.
73. Vaníček, I., Jirásko, D., & Vaníček, M. (2021). Role of Geotechnical Engineering in BIM process modelling. In *IOP Conference Series: Earth and Environmental Science, 727*(1), Article 012007. https://doi.org/10.1088/1755-1315/727/1/012007.
74. Voznesensky, E. A. (2022). Characterization and behavior of soils: Actual objectives and requirements. *Quarterly Journal of Engineering Geology and Hydrogeology, 55*(4), 125–202.
75. Zevgolis, I. E., Theocharis, A. I., & Koukouzas, N. C. (2021). Geotechnical engineering perspectives: Challenges and solutions in the transition to a post-lignite era. *Materials Proceedings, 5*(1), 17–244.
76. Zhang, W., Li, H., Li, Y., Liu, H., Chen, Y., & Ding, X. (2021). Application of deep learning algorithms in geotechnical engineering: A short critical review. *Artificial Intelligence Review, 2*(1), 1–41.

CHAPTER 2

BASIC CHARACTERISTICS AND ENGINEERING CLASSIFICATION OF SOILS

Contents

2.1. INTRODUCTION

Soil mechanics is the study of soil engineering behaviors that are relevant to the design of earth-based civil engineering structures. Nearly all of the civil engineering structures, such as buildings, bridges, highways, tunnels, earth retaining walls, embankments, basements, sub-surface waste repositories, towers, canals, and dams, are constructed in or on the surface of the earth. Every structure needs a suitable foundation in order to function properly (Kariuki et al., 2003).

The four basic types of geotechnical structures are illustrated in Figure 2.1, and most of the other cases are variations or combinations of these four types of geotechnical structures. Foundations (Figure 2.1(a)) transmit loads to the ground and the basic criterion for design is that the settlements should be relatively small. Factors that affect how a foundation is designed are the base size B, the load F, and the depth D. Relatively small loads, like wheels of a car, or a relatively large loads, like a power station, can be supported by foundations. Slopes (Figure. 2.1(b)) can be excavated or filled in, or they can form naturally through erosion. Slope angle I and depth H are the main variables, and the slope must not fail due to land sliding as per the design specification (Casagrande, 1948).

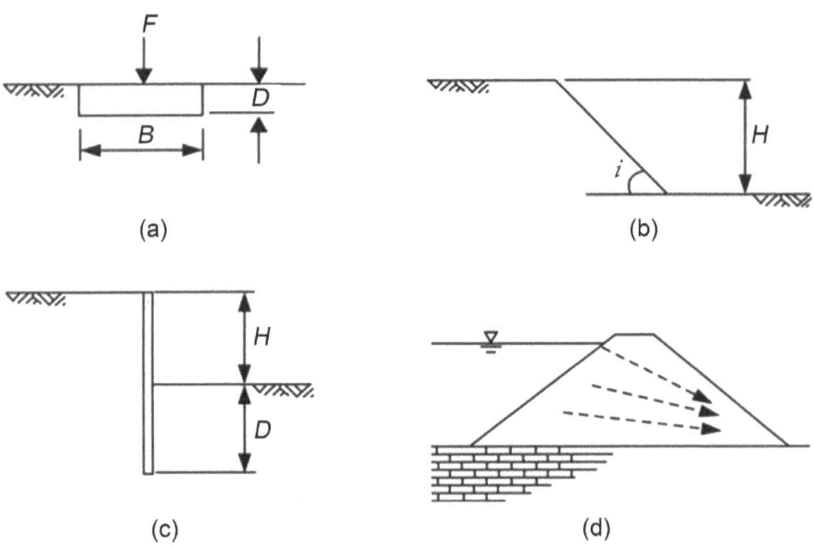

Figure 2.1. *Geotechnical structures showing (a) the Foundation, and (b) slope stability, (c) a retaining wall, and (d) the earth–fill dam.*

Source: MDPI, Creative Commons License.

Stand slopes that are deep and steep can be supported by a retaining wall (Figure 2.1(c)). The basic variables are the height of the wall H and its depth of burial D, together with the strength and stiffness of the wall and the forces in any anchors or props. The design's fundamental requirements are complex and include the wall's resistance to shearing and bending, restriction of ground movement, and overall stability. Water will seep from any structure that has different water levels, like a dam as show in (Figure. 2.1(d)) and in the areas surrounding a pumped well. Seepage controls the variation in groundwater pressure, determines a well's yield, and results in leaks through dams (Rogers et al., 2020).

2.2. SOIL COMPOSITION AND PHASE RELATIONSHIPS

2.2.1. Soil Composition

Soil is a particulate material, which means that a soil mass consists of an accumulation of individual particles that are bonded together by mechanical or attractive means, though not as strongly as for rock. To the civil engineer, soil is an uncemented or weakly cemented accumulation of mineral particles formed by the weathering of rocks, the void space between the particles containing water and/or air (Spark & Swift, 2002).

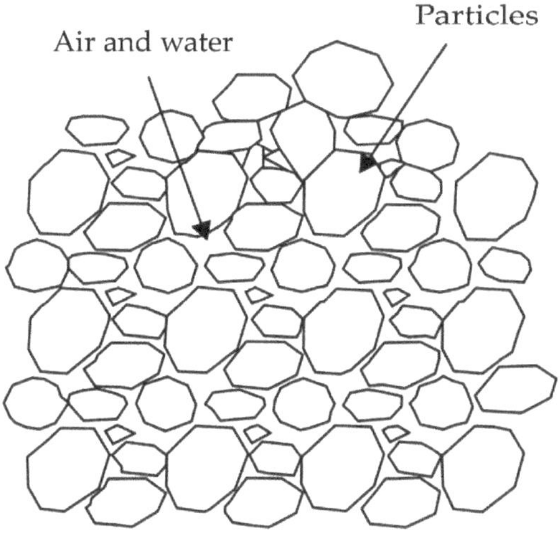

Figure 2.2. *Three-phase composition.*

Source: Engineering Libre Texts, Creative Commons License.

In nature, the void of every soil is partly or completely filled with water. Soils can be of either two phases or three phase's composition, as illustrated in Figure 2.2. A partially saturated soil has three phases, being composed of solid soil particles, pore water, and pore air. In a completely dry soil, there are two phases, namely the solid soil particles and pore air. A fully saturated soil also has two phases, being composed of solid soil particles and pore water. Below the water table, the soil is assumed to be fully saturated, although it is likely that, due to the presence of small volumes of entrapped air, the degree of saturation will be marginally below 100% (Neal et al., 1987).

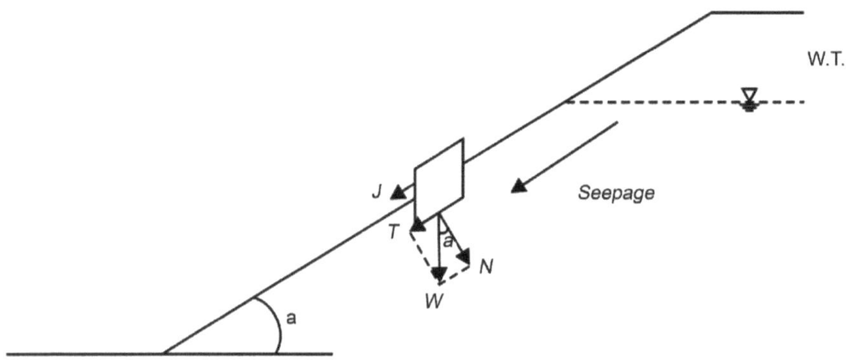

Figure 2.3. *Seepage of slope.*

Source: MDPI, Creative Commons License.

The water filling the void space may be in a state of rest or in a state of flow. Below the water table, the pore water may be static, the hydrostatic pressure depending on the depth below the water table, or may be seeping through the soil under hydraulic gradient, as illustrated in Figure 2.3. If the water is in a state of rest, the methods for solving stability and deformation problems are essentially identical with those for solving similar problems in the mechanics of solids in general. On the other hand, if the water percolates through the voids of the soil, the problems cannot be solved without previously determining the state of stress in the water contained in the voids of the soil. In this case, we are obliged to combine the mechanics of solids with applied hydraulics (Akpokodje, 1985).

Significant engineering properties of a soil deposit, such as strength and compressibility, are directly related to or at least a ected by basic factors such as volume and weight of a bulk soil consisting of solid particles or water or air. Information such as soil density (weight per unit volume), water content, void ratio, and degree of saturation are used in calculations to determine the stability of earth slopes, foundation settlement, and the bearing capacity of foundations. Therefore, we can determine the suitability of foundation types and constructions. For this reason, an understanding of the terminology and definitions related to soil composition is the key to studying soil mechanics (Gimsing et al., 2004).

2.2.2. Phase Relationships

As demonstrated in Figure. 2.4, a phase diagram can be used to represent the constituent parts of a soil. Figure 2.4(a) illustrates how bulk soil is found in nature, which is essentially an amorphous collection of air space, soil particles, and water (Nelson & Pope, 1978). For purposes of study and analysis, it is convenient to represent this soil mass by a phase or block diagram, which consists of three parts: solid particles, water or other liquid, and air or other gas, as shown in Figure 2.4(b).

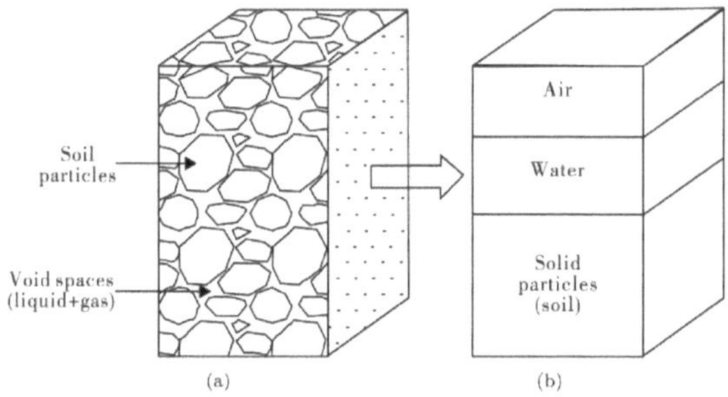

Figure 2.4. *Phase diagrams.*

Source: MDPI, Creative Commons License.

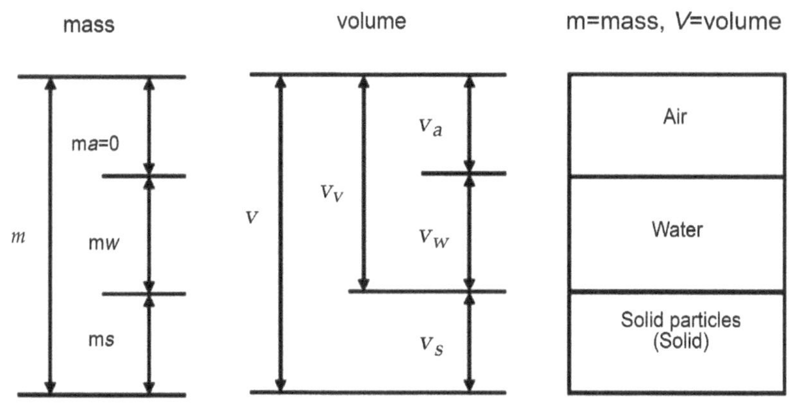

Figure 2.5. *Phase relationships between mass and volume.*

Source: Civil Engineering Bible, Creative Commons License.

The phase relationships as shown in Figure 2.5 illustrate the mass and volume of components of soil system. The relationships are summarized in Figure 2.5. The total mass m of the soil volume is taken as the sum of the mass of solids m_s and water m_w. The mass of air (in the voids) measured in air (the earth's atmosphere) is zero. The air or other gases may have a measurable weight, but it would normally be very small compared to the total mass of soil plus water and therefore can be neglected without causing serious error (Germain & Woolfson, 1968).

The total volume V of the soil bulk includes the volume occupied by solids V_s plus water V_w (or liquid) plus air V_a (or other gas). The total space occupied by water and air may collectively be indicated as the volume of voids.

$$m = m_s + m_w$$

$$V = V_s + V_w + V_a. \qquad (2.1)$$

Figure. 2.5. defines the following relationships.

i. The water content (w), or moisture content, is the ratio of the mass of water to the mass of solids in the soil. It is expressed as a percentage, i.e.,

$$w = \frac{m_w}{m_s} \text{ x } 100\%$$

The water content must be determined by test (drying method). It is determined by weighing a sample of the soil and then drying the sample in an oven at a temperature of 105–110 廉 and reweighing (Vallauri et al., 2008).

ii. The bulk density (ρ) of a soil is the ratio of the total mass to the total volume, i.e.,

$$\rho = \frac{m}{V}$$

Convenient units for density are kg/m³ or g/cm³. Water has a density of ρ_w = 1000 kg/m³ (1.00 g/cm³). For a completely dry soil ($S_r = 0$),

$$\rho_d = \frac{ms}{V}$$

For a fully saturated soil ($S_r = 1$),

$$\rho_{sat} = \frac{ms + Vv\rho w}{V}$$

The buoyant density (ρ') can be calculated by

$$\rho' = \frac{ms - Vs\rho w}{V}$$

The unit weight (γ) of a soil is the ratio of the total weight (a force) to the total volume, i.e.,

$$\gamma \aleph \frac{\pm}{V} \quad \frac{\quad}{v} \quad \rho g$$

The soil unit weight can be expressed in kN/m^3

iii. The specific gravity of the soil particles (d_s) can be calculated as;

$$ds = \frac{ms \quad Vv \quad w}{\pm} = \frac{\rho s}{\rho w}$$

where ρ_s is the particle density. If the units of ρ_s are g/cm^3, then ρ_s and d_s are numerically equal (Brussaard et al., 2007).

iv. The void ratio (e) is the ratio of the volume of voids to the volume of solids, i.e.,

$$e = \frac{Vv}{Vs}$$

The porosity (n) is the ratio of the volume of voids to the total volume of the soil, i.e.,

$$n = \frac{Vv}{V} \times 100\%$$

'n' and 'e' are inter-related as follows:

$$n = \frac{n}{1-n}$$

$$n = \frac{e}{1+e} \times 100\%$$

v. The degree of saturation (S_r) is the ratio of the volume of water to the total volume of void space. It is also expressed as a percentage, i.e.,

$$S_r = \frac{Vw}{Vv} \times 100\%$$

The degree of saturation can range between the limits of zero for a completely dry soil and for a fully saturated soil it is 1 (or 100%) (Ansell & Boorman, 1971).

From the definition of void ratio, if the volume of solids is 1 unit, then the volume of voids is e units. The mass of solids is then d_s and, from the definition of water content, the mass of water is $d_s w \rho_w$. The volume of water is thus $d_s w$. These volumes and masses are represented in Figure 2.6. The following relationships can be obtained:

$$e = \frac{d_s(1+w)\rho_w}{\rho} - 1$$

$$S_r = \frac{d_s w}{e}$$

$$\rho = \frac{d_s(1+w)\rho_w}{1+e}$$

$$\rho_{\text{sat}} = \frac{d_s + e}{1+e}\rho_w$$

$$\rho_d = \frac{d_s}{1+e}\rho_w.$$

The formulae and symbols for physical indexes are shown in Table 2.1.

Table 2.1. Practical Calculation Formulas of Physical Indexes

Index and the symbol	Expression or conversion formula	Unit
Specific gravity, d_s	$d_s = \frac{m_s}{m_w} = \frac{V_s\rho_s}{V_s\rho_w} = \frac{\rho_s}{\rho_w}$	Dimensionless quantity
Water content, w	$w = \frac{m_w}{m_s}$ x 100%	%
Density, ρ	$\rho = \frac{d_s(1+w)\rho_w}{1+e} = \frac{d_s + S_r e}{1+e}\rho_w$	g/cm³
Unit weight, γ	$\gamma = \frac{d_s(1+w)\gamma_w}{1+e} = \frac{d_s + S_r e}{1+e}\gamma_w$	kN/m³
Void ration, e	$e = \frac{d_s(1+w)\rho_w}{\rho} - 1,\ e = \frac{d_s(1+w)\gamma_w}{\gamma} - 1$ $e = \frac{d_s\rho_w}{\rho_d} - 1,\ e = \frac{d_s\gamma_w}{\gamma_d} - 1$ $e = d_s w\ (S_r = 100\%)$	Dimensionless quantity
Porosity, n	$n = \frac{e}{1+e}$ x 100%	%
Degree of saturation, S_r	$S_r = \frac{d_s w}{e}$	%
Dry density, ρ_d	$\rho_d = \frac{d_s}{1+e}\rho_w,\ \rho_d = \frac{\rho}{1+w}$	g/cm³
Dry unit weight, γ_d	$\gamma_d = \frac{d_s}{1+e}\gamma_w,\ \gamma_d = \frac{\gamma}{1+w}$	kN/m³
Saturated density, ρ_{aat}	$\rho_{\text{aat}} = \frac{d_s + e}{1+e}\rho_w$	g/cm³
Saturated unit weight, γ_{aat}	$\gamma_{\text{aat}} = \frac{d_s + e}{1+e}\gamma_w$	kN/m³
Buoyant density, ρ'	$\rho' = \frac{d_s - 1}{1+e}\rho_w,\ \rho' = \rho_{\text{aat}} - \rho_w$ $\rho' = (d_s - 1)(1 - n)\rho_w$	g/cm³
Buoyant unit weight, γ'	$\gamma' = \frac{d_s - 1}{1+e}\gamma_w,\ \gamma' = \gamma_{\text{aat}} - \gamma_w$ $\gamma' = (d_s - 1)(1 - n)\gamma_w$	kN/m³

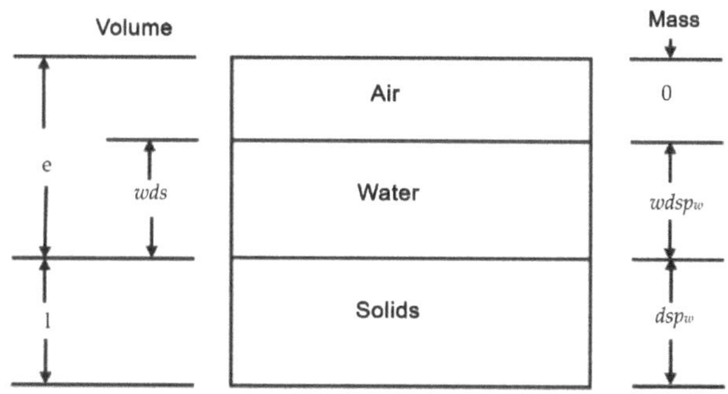

Figure 2.6. *Relationship of volumes and masses.*

Source: Civil Engineering Bible, Creative Commons License.

2.3. SOIL FABRIC

The specific arrangement of soil particles forms the soil fabric at the micro scale. Particle size, shape, arrangement, and, most importantly, the mutual connecting condition—a synthetic combination of soil grain characteristics—are all considered components of the soil fabric. The three main categories of soil fabric types are depicted in Figure 2.7 (Merkus & Merkus, 2009).

Crushed stone, gravel, sand, and some other coarse grain soil all belong to single grain fabric, which is formed in sedimentation. As illustrated in Figure. 2.7(a), it can be classified into two types based on how it is arranged: tight and loose. A tighter structure can be achieved with a wider range of particle sizes present. In relation to its gravity, a grain of gravel or sand has a large grain size and a relatively small relative surface energy and surface area. There is virtually no inter-particle force between particles and only strongly bound water on the surface of the particles. Thus, this kind of soil is also called cohesionless soil (Yimsiri & Soga, 2010).

Silt is a component of honeycomb fabric, as shown in Figure. 2.7(b). When fine-grained soils come into contact with already-sunk soils during the sedimentation process, they will remain at the points of contact rather than sink because the interparticle force is greater than the gravitational force. As a result, the honeycomb fabric with large voids forms, as shown in Figure. 2.7(b). This type of soil is very unstable and will distort significantly when submerged in water or when subjected to outside forces. The clay particles are part of the flocculent fabric, as Figure. 2.7(c) illustrates (Fonseca et al., 2013).

Soil with clay mineral particles (with $d < 0.002$ mm), clay (with $d < 0.005$ mm), and colloidal particles is very fine and has thin beds. This gives it a lot of surface area and keeps it suspended in water. In the event that the interparticle reaction is perceived as an attractive force, the particles will readily combine and gradually form a small aggregate of soil particles, thereby increasing the mass and causing it to sink (Bender & Barenberg, 1978).

When a link touches the other, they attract each other, and gradually a big link can be formed. The soil grain of this type of fabric exhibits high compressibility and low strength due to its random arrangement and large pore space, while also making it sensitive to disturbance. However, for cementation and compaction—which are the primary sources of cohesive force in clay—the linkage strength between soil grains is growing (Brewer & Sleeman, 1960).

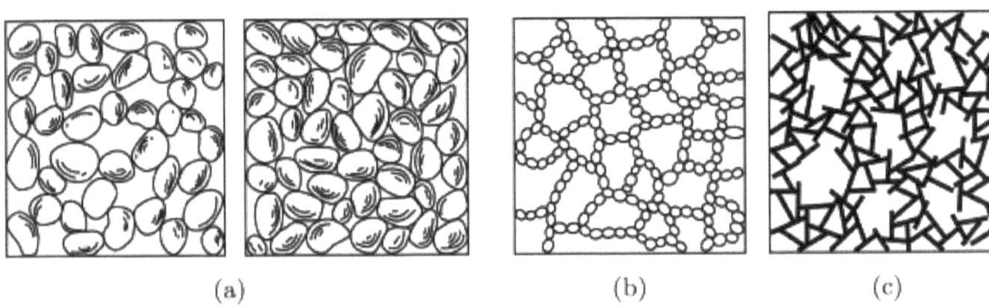

Figure 2.7. *Soil fabric showing (a) A single grain fabric; (b) The honey comb fabric, and finally; (c) The flocculent fabric.*

Source: Brainkart, Creative Commons License.

2.4. PARTICLE SIZE ANALYSIS

The particle size analysis of a soil sample involves determining the percentage by mass of particles within the di erent size ranges. Soils contain particles ranging in size from under 0.001 mm to more than 100 mm. Size ranges specified by the British, Chinese, and Japanese Standards are shown in Figure. 2.8 (Chu & Gan, 2004).

The same terms are also used to describe particular types of soil. The sieving technique can be used to ascertain the coarse-grained soil's particle size distribution (PSD). The process of sedimentation can be used to ascertain the PSD in either fine-grained portion of a coarse-grained soil or a fine-grained soil (Barth & Sun, 1989).

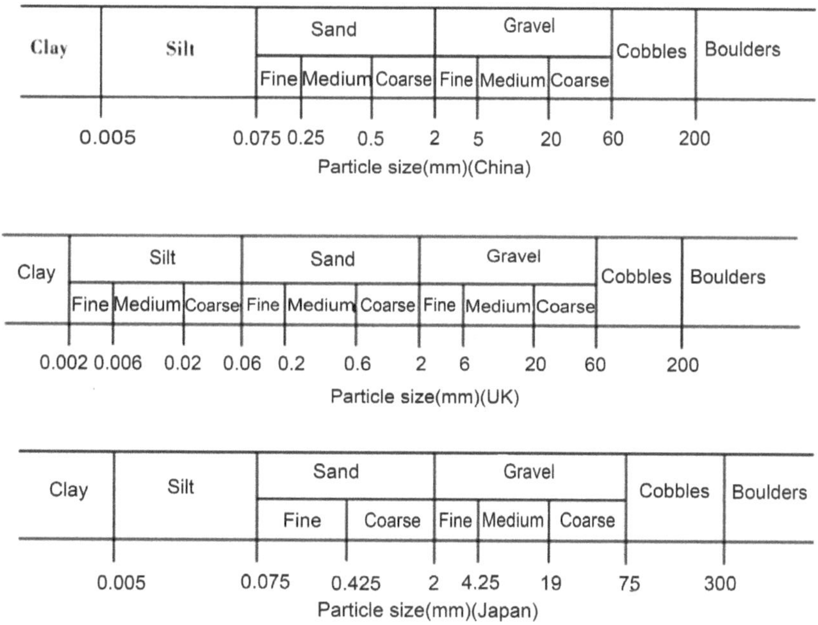

Figure 2.8. *The particle size ranges in different standards.*

Source: MDPI, Creative Commons License.

2.4.1. Method of Sieving

The soil sample is passed through a series of standard test sieves having successively smaller mesh sizes. The mass of soil retained in each sieve is determined and the cumulative percentage by mass passing each sieve is calculated. The particle size distribution of a soil is presented as a curve on a semilogarithmic plot. The ordinate is the percentage by mass of particles smaller than the size given by the abscissa. Examples of particle size distribution curves appear in Figure 2.9. The particle size corresponding to any specified value on the "percentage smaller" scale can be read from the particle size distribution curve (Chen, 2007).

The size in which 10% of the particles are smaller than that size is denoted by D_{10}. Other sizes such as D_{30} and D_{60} can be defined in a similar way. The size D_{10} is defined as the e ective size. The general slope and shape of the distribution curve can be described by means of the coe cient of uniformity (C_u) and the coe cient of curvature (C_c), defined as follows:

$$C_u = \frac{D_{60}}{D_{10}}$$

$$C_c = \frac{D_{30}^2}{D_{60}D_{10}},$$

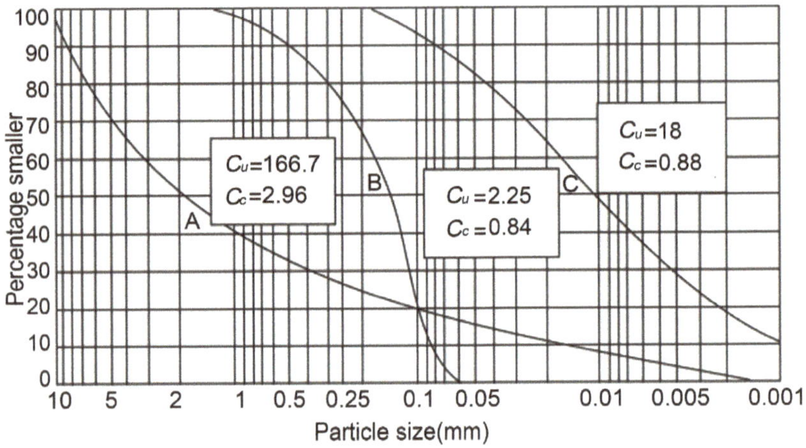

Figure 2.9. Particle size distribution curves.

Source: Slide Share, Creative Commons License.

Where; C_u indicates the slope of the particle size distribution curve, the higher the C_u value, the larger the range of particle sizes. C_c indicates continuous characteristics of the slope, a well-graded soil has $1 < C_c < 3$ and $C_u > 5$.

2.4.2. Method of Sedimentation

This method is based on Stokes' law which governs the velocity at which spherical particles settle in a suspension: the larger the particle, the greater is the settling velocity and vice versa (Zhao et al., 2013).

2.5. DENSITY PROPERTIES FOR GRANULAR SOILS

Gravel, crushed stone, stone, and sand soil are examples of cohesionless soils; they all have a single grain structure without cohesive force or cementation effect. The primary factor influencing this type of soil's engineering properties is its compactness (Wasti, 1987). Therefore, a more compacted soil will have a higher bearing capacity, a smaller compressibility, and ultimately, much better stability. Water has minimal impact on the engineering properties of this type of soil due to the coarse soil particles and small relative surface area. There is no weakly bound liquid on the surface—only strongly bound water exists (Fragaszy & Siddiqi, 1990).

2.5.1. Void Ratio

Figure 2.10 shows two of the many possible ways that a system of equal-sized spheres can be packed. The dense packings represent the densest possible state for such a system. Looser systems than the simple cubic packing can be obtained by carefully constructing arches within the packing, but the simple cubic packing is the loosest of the stable arrangements. The void ratio and porosity of these simple packings can be computed from the geometry of the packings, and the results are given in Table 2.2.

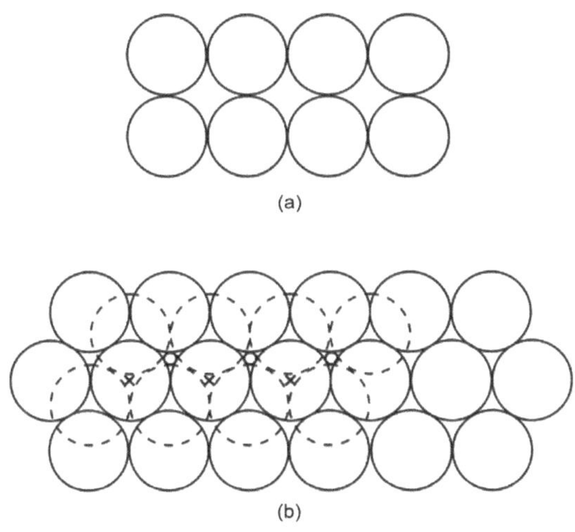

Figure 2.10. *A uniform sphere arrangement, (a) Elevation and plan views: simple cubic packing; (b) Dense packing in the plan view. First layer: solid circles; second layer: dashed circles; third layer: face-centered cubic array; close-packed hexagonal array, location of sphere centers.*

Source: NIMbios, Creative Commons License.

The possibility of forming a loose particles arrangement increases with decreasing minimum density, which is related to the presence of a more angular and narrower range of particle sizes in the soil, making it more nearly uniform. The maximum density (i.e., the spaces between the larger particles that can be filled by smaller particles) increases with the particle size range present (Åberg, 1992).

Table 2.3 describes granular soil density based on void ratio, based on engineering experience.

Table 2.2. Maximum and Minimum Densities for Granular Soils

	Void ratio		Porosity		Dry unit weight (pcf)	
Description	e_{max}	e_{min}	n_{max}	n_{min}	$\gamma_{d\,max}$	$\gamma_{d\,min}$
Uniform spheres	0.92	0.35	47.6	26.0	—	—
Standard Ottawa sand	0.80	0.50	44	33	92	110
Clean uniform sand	1.0	0.40	50	29	83	118
Uniform inorganic silt	1.1	0.40	52	29	80	118
Silty sand	0.90	0.30	47	23	87	127
Fine to coarse sand	0.95	0.20	49	17	85	138
Micaceous sand	1.2	0.40	55	29	76	120
Silty sand and gravel	0.85	0.14	46	12	89	146

Table 2.3. Division of Compactness of Sandy Soils in Terms of Void Ratio

	Compactness			
Soil type	Dense	Medium dense	Less dense	Loose
Gravel, coarse or medium sand	$e < 0.6$	$0.60 \leq e \leq 0.75$	$0.75 < e \leq 0.85$	$e > 0.85$
Fine sand, silt	$e < 0.7$	$0.70 \leq e \leq 0.85$	$0.85 < e \leq 0.95$	$e > 0.95$

Values of water content for natural granular soils vary from less than 0.1% for air-dry sands to more than 40% for saturated, loose sand (Lade et al., 1998).

2.5.2. Relative Density

Relative density (D_r) is a useful metric for characterizing a natural granular soil's density:

$$D_r = \frac{e_{max} - e}{e_{max} - e_{min}} \text{ x } 100\% = \frac{\gamma_{d\,max}}{\gamma_d} \text{ x } \frac{\gamma_d - \gamma_{d\,min}}{\gamma_{d\,max} - \gamma_{d\,min}} \text{ x } 100\%$$

Where; γ_{dmax} is the dry unit weight of soil in the most dense condition, γ_{dmin} is the dry unit weight of soil in the maximum loose condition, γ_d is the in-place dry unit weight, e_{max} is the void ratio of soil in the loosest condition, e_{min} is the void ratio of soil in the densest condition, and e is the in-place void ratio (Vaid & Negussey, 1984).

The engineering experience indicates that relative density D_r can be used to categorize the compactness of the sandy soils:

$0 < D_r \leq 0.33$ loose

$0.33 < D_r \leq 0.67$ medium dense

$0.67 < D_r \leq 1.0$ dense.

A variety of tests have been proposed to measure the maximum and minimum void ratios (Kolbuszewski, 1948). In order to find the maximum density, a vibration test is typically used. The test to determine minimum density usually involves pouring oven-dried soil into a container. Unfortunately, the details of these tests have not been entirely standardized, and values of the maximum density and minimum density for a given granular soil depend on the procedure used to determine them. By using special measures, one can obtain densities greater than the so-called maximum density. Densities considerably less than the so-called minimum density can be obtained, especially with very fine sands and silts, by slowly sedimenting the soil into water or by fluffing the soil with just a little moisture present (Liu et al., 2013).

Table 2.4. Standard Penetration Test

Relative density	Dense	Medium dense	Less dense	Loose
Penetration resistance, N	$30 < N$	$15 < N \leq 30$	$10 < N \leq 15$	≤ 10

2.5.3. Standard Penetration Test

Using void ratio e and relative density Dr to express the compactness of soils has many defects. Besides a lot of experimental influencing factors, it is very hard to get intact specimens of the granular soils. Thus, many field experiments have been proposed (i.e., standard penetration test and the cone penetration test) to get the compactness of soils.

According to *Code for Design of Building Foundation* (GB 50007- 2011), China, Table 2,4 presents a correlation of standard penetration resistance with relative density for sand.

The standard penetration test is a very valuable method of soil investigation. It should, however, be used only as a guide, because there are many reasons why the results are only approximate (Aggour & Radding, 2001).

2.6. PLASTICITY PROPERTIES OF SOILS

2.6.1. Plastic Behavior and Consistency Limits

Plasticity is a very important characteristic of fine-grained soil. In general, depending on its water content, a soil may exist in one of the liquid, plastic, semi-solid, and solid states. If a soil is in the liquid state initially, it will eventually transition to the plastic and semi-solid states as its water content decreases, and its volume will gradually decrease until it reaches the solid state and vice versa. The shrinkage limit, or w_s, is the water content at the stage of transition between the semi-solid and solid states. The plastic limit w_p is the boundary between the plastic and semi-solid states (Moreno & Alonso, 2017). The liquid limit (w_L) is the percentage of water that indicates the transition from a liquid to plastic state. The liquid limit, plastic limit, and shrinkage limit are also called consistency limits (Figure. 2.11).

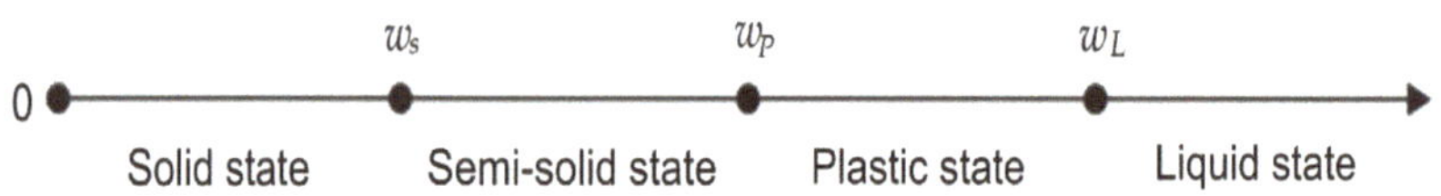

Figure 2.11. *Consistency limits.*

Source: Geoengineer.org, Creative Commons License.

The plasticity index I_p, or plastic and liquid limits of water's content range, is defined as follows:

$I_p = w_L - w_P$

It has been discovered that using both the plasticity index and liquid limit values is preferable for accurately assessing a soil's plasticity characteristics (Yukselen et al., 2008). The liquidity index (I_L) can be used to represent a soil's natural water content (w) in relation to the plastic and liquid limits, i.e.,

$$I_L = \frac{w - w_p}{w_L - w_p}.$$

When the water content is close to the plastic limit (a low or near-zero IL value), the cohesive strength will be relatively high and the sensitivity will remain low. The sensitivity rises as the natural water content gets closer to the liquid limit. The arbitrary test procedures are used to determine the plastic and liquid limits (Bonatti & Mohr, 2022).

2.6.2. Tests for Liquid and Plastic Limits

2.6.2.1. Determination of the Plastic Limit

There are two methods to determine the plastic limit: Rubbing method and liquid-plastic limit combined device method.

Not only is the rubbing method easy to use, but the apparatus is also very simple. First, we need to make several specimens. Then, take a specimen and roll it into a ball and then we put it on a ground glass sheet to roll it evenly into a thread of soil with our palms until the diameter of the thread reaches 3 mm (Figure 2.12). This procedure of rolling continues until the thread starts to crumble or there is no shadow mark on the glass just as the diameter of 3 mm is reached; at this point, the water content of the specimen is determined. The same process is carried out in other specimens and the average water content is stated as the plastic limit of the soil. In spite of the seeming arbitrary nature of this test procedure, an experienced technician can obtain a reasonable plastic limit result for the engineering construction (Rashid et al., 2008).

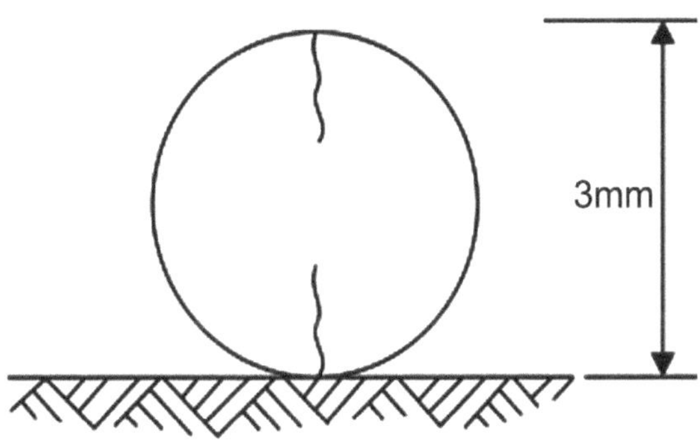

Figure 2.12. Rubbing method.

Source: Geoengineer.org, Creative Commons License.

Figure. 2.13 shows a liquid-plastic combine tester. Initially, three types of specimens with varying water contents are created. The three carefully prepared specimens are then combined and placed inside the soil cup. Next, using the electromagnetic fall-cone test, which was first described in the Chinese Standard for Soil Test Method (GB/T 50123–1999), the cone is released to precisely penetrate the soil paste for 5 seconds and record the depth (Ishaque et al., 2013). The specimens undergo three iterations of the penetration procedure. The water content and penetration depth are plotted as a log-log plot, with three points in a line. Thus, the value of the plasticity limit is the water content that corresponds to a 2 mm penetration (Bobrowski & Griekspoor, 1992).

2.6.2.2. Determination of the Liquid Limit

There are two methods for the liquid limit test: liquid-plastic limit combined device method and Casagrande method, namely dish-type liquid limit device method. Although the liquid-plastic limit combined device method is preferred, the Casagrande method of determining liquid limit is still widely used. The apparatus used is shown in Figure 2.14 and consists basically of a metal dish which may be raised by rotating a cam and then allowed to fall through at a height of 10 mm on to a hard rubber block (Sherwood & Ryley, 1970).

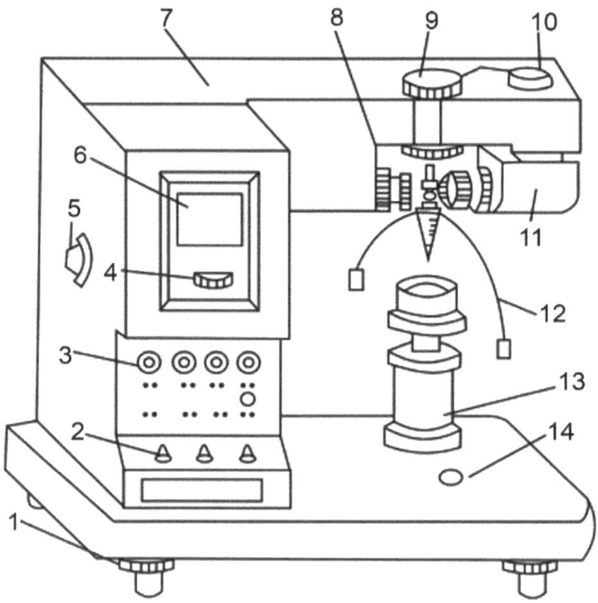

Figure 2.13. *Liquid-plastic limit combined device.*

Note: The following components are included: (1) a horizontal adjustment screw; (2) a pilot switch; (3) an indicate lamp; (4) a nil-line adjustment screw; (5) a retro-reflector adjustment screw; (6) a screen; (7) a chassis; (8) an objective lens adjustment screw; (9) an electromagnetic apparatus; (10) a luminous source adjustment screw; (11) a luminous source; (12) a cone apparatus; (13) a elevator platform, and (14) a spirit bubble.

The soil is dried and mixed in the same way as in the previously described method. Some of the soil paste is placed in the dish and leveled-o parallelly with the rubber base until the depth of the specimen is 10 mm.

The standard grooving tool is then drawn through the soil paste to form the groove. By turning the handle (at 2 rvs/s), the cup is raised and dropped on to the rubber base until the lower part of the groove has closed up over a length of 13 mm. The number of blows (number of revolutions) required for this is recorded. The dish is then refilled with the same paste mixture and the grove-closing procedure is repeated several times and an average number of blows are obtained for that mixture. After this, a small portion of the paste is taken and its water content is found. The whole procedure is then repeated with paste mixture having di erent water contents for five or six times in all.

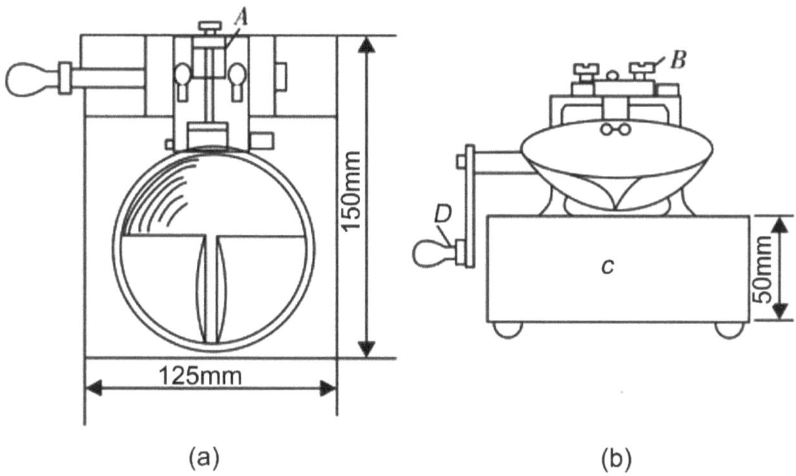

Figure 2.14. *Dish-type liquid limit device. (a) Vertical view; (b) Front view.*

Notes: A is the adjustment plate; B is the screw; C is the base; and D is the handle.

2.6.3. Two Important Indexes and the Engineering Application

2.6.3.1. Plasticity Index, I_p

It can be seen from Figure 2.11 that if the value of plasticity limit w_L and liquid limit w_P has big difference, the range of plasticity of the soil will be wider, the weakly bound water on the surfaces of soil particles will be thicker, the specific surface area of the soil particles will be bigger and the surface adsorption capability of soil particles will be greater. Thus, plasticity index, to a large extent, indicates the amount of clay particles, colloidal particles, and clay mineral contents of the soil (Sridharan et al., 1999).

According to Code for Design of Building Foundation (GB 50007–2011) and Code for Rock and Soil Classification of Railway Engineering (TB 10077–2001), China, divide the cohesive soils on the basis of the range of plasticity index I_p.

$10 < I_p \leq 17$ Silty clay

$17 < I_p$ Clay

Table 2.5. Division of Hardness of Cohesive Soil into Five State on the Basis of I_L

Degree of hardness	Solid	Stiff-plastic	Malleable	Soft-plastic	Flowing
Liquidity index	$I_L \leq 0$	$0 < I_L \leq 0.25$	$0.25 < I_L \leq 0.75$	$0.75 < I_L \leq 1.0$	$1.0 < I_L$

Table 2.6. Division of Hardness of Cohesive Soil into Four States on the Basis of I_L.

Degree of hardness	Solid	Stiff-plastic	Soft-plastic	Flowing
Liquidity index	$I_L \leq 0$	$0 < I_L \leq 0.5$	$0.5 < I_L \leq 1.0$	$1.0 < I_L$

2.6.3.2. Liquidity index, I_L

Figure 2.11 illustrates that if the natural water content falls below the plastic limit ($w \leq w_p$), the soil will remain in a solid state and the liquidity index will become negative ($I_L \leq 0$); conversely, if the natural water content exceeds the liquid limit ($w \geq w_L$), the soil will be in a flow condition and the liquidity index will become equal to or greater than 1.0 ($I_L \geq 1$). Therefore, the cohesive soil's degree of hardness is indicated by the liquidity index (Vardanega & Haigh, 2014). According to Code for Investigation of Geotechnical Engineering (GB 50021–2001) and Code for Design of Building Foundation (GB 50007–2011), China, divide the cohesive soils into five states on the basis of liquidity index, IL, shown in Table 2.5. According to Code for Rock and Soil Classification of Railway Engineering (TB 10077–2001), China, divide the cohesive soils into four states on the basis of liquidity index, IL, shown in Table 2.6 (Guillaume, 2015).

Tables 2.5 and 2.6 demonstrate that the closer the soil mass is to its liquid state, the higher the value of the liquidity index, or I_L.

2.7. SOIL COMPACTION

Compaction is the process of increasing the density of a soil by packing the particles closer together with the reduction in the volume of air: there is no significant change in the volume of water of the soil. In the construction of embankments, loose soil is placed in layers ranging from 75 mm to 450 mm in thickness, each layer being compacted to a specified standard by means of rollers, vibrators, or rammers. In general, the higher the degree of compaction, the higher will the strength be and the lower the compressibility of the soil (Nawaz et al., 2013).

The degree of compaction of a soil is measured in terms of dry density, i.e., the mass of the solid only per unit volume of soil. If the bulk density of the soil is ρ and the water content is w, then it is apparent that the dry density is given by (Rossiter, 2007).

$$\rho d = \frac{\rho}{1+w}$$

The compaction characteristics of a soil can be assessed by means of the standard compaction test. The soil is compacted in a cylindrical mold using a standard compactive effort. In the test, the volume of the mold is 947 mm^3 and the soil (with all particles larger than 5 mm removed) is compacted by a rammer consisting of a 2.5 kg mass falling freely through 305 mm: the soil is compacted in three equal layers, each layer receiving 25 blows with the rammer. In the modified A.A.S.H.O test, the mold is the same as used in the above test, but the rammer consists of a 4.5 kg mass falling though 457 mm: the soil (with all particles larger than 40 mm removed) is compacted in five layers, each layer receiving 56 blows with the rammer (Batey, 2009).

The effectiveness of this process depends on these factors:

i. *The* maximum dry density can't be obtained if there is just a little or a lot of water in the soil.

ii. The energy supplied by the compaction equipment (also called the "compactive effort").

iii. The nature and type of soil (e.g., clay or sand; plastic or non-plastic; uniform or well-graded or uniform).

iv. The large soil particles (Hamza & Anderson, 2005).

After compaction using one of the two standard methods, the bulk density and water content of the soil are determined and the dry density is calculated. The procedure is carried out no fewer than five times for a soil, increasing the sample's water content each time. Plotting dry density against water content yields a curve similar to the one in Figure. 2.15. This curve demonstrates that a maximum value of dry density can be obtained for a given compaction method (i.e., compactive effort) at a specific water content value, called the optimum water content (w_{op}). Most soils have a tendency to be stiff and difficult to compact at low water contents. An increase in water content makes the soil more workable, which promotes compaction and raises dry densities—a highly advantageous property for engineering construction (Taylor & Brar, 1991). On the other hand, at high water contents, the dry density falls as the water content rises and the proportion of water occupying the soil volume increases. Therefore, dry density is a crucial quality inspection metric for compaction (Kozlowski, 1999).

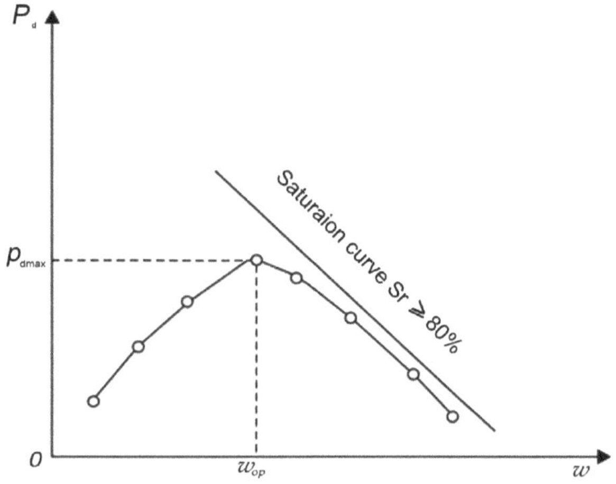

Figure 2.15. *Soil compaction curve.*

2.8. ENGINEERING CLASSIFICATION OF SOILS

The direct approach to the solution of a soil engineering problem consists of first measuring the soil property needed and then employing this measured value in some rational expression to determine the answer to the problem (Kariuki et al., 2003).

Measuring fundamental soil properties like permeability, compressibility, and strength can be di cult, time consuming, and expensive. In many soil engineering problems, such as pavement design, there are no rational expressions available for the analysis for the solution. For these reasons, sorting soils into groups showing similar behavior may be very helpful. Such sorting is soil classification (Polidori, 2015).

There are various classification standards that are used in China: Specification of Soil Test (SL 237–1999); Test Methods of Soils for Highway Engineering (JTG E40–2007); Code for Design of Building Foundation (GB 50007–2011); Standard for Engineering Classification of Soil (GB/T 50145–2007).

2.8.1. Crushed Stones, Gravels

It describes soils where more than 50% of the total mass weight is made up of particles larger than 2 mm in size. This type of soil's gradation and particle size have a significant influence on engineering construction. Therefore, Table 2.7 illustrates the classification of this type of soil based on particle size and gradation (Khanal & Tamrakar, 2009).

Table 2.7. Classification of Crushed Stones and Gravels

Soil type	Soil particle shape	Gradation
Boulder	Round forms dominant	Particles (size > 200 mm) take up more than 50% of the whole mass weight.
Block stone	Angular forms dominant	
Cobble	Round forms dominant	Particles (size > 20 mm) take up more than 50% of the whole mass weight.
Crushed stone	Angular forms dominant	
Round gravel	Round forms dominant	Particles (size > 2mm) take up more than 50% of the whole mass weight.
Angular gravel	Angular forms dominant	

2.8.2. Sand

It refers to the soils of which the particles from size > 2 mm take up less than 50% of the whole mass weight to size > 0.075 mm which take up more than 50% of the whole mass weight. For this kind of soil, the classification code is similar to the crushed stones and gravels, as shown in Table 2.8 (Mehta & Barker, 1994). This type of soil can also be categorized using the standard penetration test, relative density and void ratio.

Table 2.8. Classification of Sand in the Basis of Gradation

Soil type	Gradation
Gravelly sand	Particles (size > 2 mm) take up 25-50% of the whole mass weight.
Coarse sand	Particles (size > 0.5 mm) take up 50% of the whole mass weight.
Medium sand	Particles (size > 0.25 mm) take up 50% of the whole mass weight.
Fine sand	Particles (size > 0.075 mm) take up 85% of the whole mass weight.
Silty sand	Particles (size > 0.075 mm) take up 50% of the whole mass weight.

2.8.3. Silty Soil

It refers soils where the plasticity index (Ip) is less than 10 and the particles of size greater than 0.075 mm account for less than 50% of the total mass weight (Andrews & Martin, 2000).

SUMMARY

In this chapter the fundamental soil properties and their significance in engineering applications are explored. The chapter outlines key characteristics such as grain size distribution, mineral composition, and moisture content, which profoundly influence soil behavior. Additionally, it delves into soil classification systems essential for categorizing soils based on their engineering properties. Understanding these basic soil characteristics and classifications is crucial for engineers and geotechnical professionals in assessing soil stability, foundation design, and construction feasibility. This chapter serves as a foundational guide for comprehending soil mechanics and its implications in engineering projects.

REFERENCES

1. Åberg, B. (1992). Void ratio of noncohesive soils and similar materials. *Journal of Geotechnical Engineering, 118*(9), 1315–1334.
2. Aggour, M. S., & Radding, W. R. (2001). Standard penetration test (SPT) correction. Report No. MD02-007B48, Maryland State Highway Administration, Baltimore, 87, 23–114.
3. Akpokodje, E. (1985). The engineering classification of some Australian arid zone soils. *Bulletin of Engineering Geology & the Environment, 31*(1), 22–118.
4. Andrews, D. C., & Martin, G. R. (2000). Criteria for liquefaction of silty soils. In *Proc., 12th World Conf. on Earthquake Engineering, Vol. 1*, (pp. 1–8).
5. Ansell, H. G., & Boorman, R. S. (1971). Phase Relationships in the In S System. *Journal of the Electrochemical Society, 118*(1), 133.
6. Balawi, S., & Abot, J. L. (2008). The effect of honeycomb relative density on its effective in-plane elastic moduli: An experimental study. *Composite Structures, 84*(4), 293–299.
7. Barth, H. G., & Sun, S. T. (1989). Particle size analysis. *Analytical Chemistry, 61*(12), 143–152.
8. Batey, T. (2009). Soil compaction and soil management–A review. *Soil Use and Management, 25*(4), 335–345.
9. Bender, D. A., & Barenberg, E. J. (1978). Design and behavior of soil-fabric-aggregate systems. *Transportation Research Record, 6–71*(2), 1–12.
10. Beyer, L., Blume, H. P., Elsner, D. C., & Willnow, A. (1995). Soil organic matter composition and microbial activity in urban soils. *Science of the Total Environment, 168*(3), 267–278.
11. Bobrowski Jr, L. J., & Griekspoor, D. M. (1992). Determination of the plastic limit of a soil by means of a rolling device. *Geotechnical Testing Journal, 15*(3), 284–287.
12. Bonatti, C., & Mohr, D. (2022). On the importance of self-consistency in recurrent neural network models representing elasto-plastic solids. *Journal of the Mechanics and Physics of Solids, 158*, Article 104697.
13. Brewer, R., & Sleeman, J. R. (1960). Soil structure and fabric: Their definition and description. *Journal of Soil Science, 11*(1), 172–185.
14. Brussaard, L., Pulleman, M. M., Ouédraogo, É., Mando, A., & Six, J. (2007). Soil fauna and soil function in the fabric of the food web. *Pedobiologia, 50*(6), 447–462.

15. Campbell, D. J. (1976). Plastic limit determination using a drop-cone penetrometer. *Journal of Soil Science, 27*(3), 295–300.
16. Casagrande, A. (1948). Classification and identification of soils. *Transactions of the American Society of Civil Engineers, 113*(1), 901–930.
17. Chang, C. S., Wang, J. Y., & Ge, L. (2015). Modeling of minimum void ratio for sand–Silt mixtures. *Engineering Geology, 196*, 293–304.
18. Chen, X. (2007). Large sample sieve estimation of semi-nonparametric models. *Handbook of Econometrics, 6*, 5549–5632.
19. Chepil, W. S., & Bisal, F. (1943). A rotary sieve method for determining the size distribution of soil clods. *Soil Science, 56*(2), 95–100.
20. Chu, J., & Gan, C. L. (2004). Effect of void ratio on K 0 of loose sand. *Geotechnique, 54*(4), 285–288.
21. Cubrinovski, M., & Ishihara, K. (2002). Maximum and minimum void ratio characteristics of sands. *Soils and Foundations, 42*(6), 65–78.
22. Feng, T. W. (2001). A linear log d log w model for the determination of consistency limits of soils. *Canadian Geotechnical Journal, 38*(6), 1335–1342.
23. Fletcher, G. F. (1965). Standard penetration test: Its uses and abuses. *Journal of the Soil Mechanics and Foundations Division, 91*(4), 67–75.
24. Fonseca, J., O'Sullivan, C., COOP, M. R., & Lee, P. D. (2013). Quantifying the evolution of soil fabric during shearing using directional parameters. *Géotechnique, 63*(6), 487–499.
25. Fragaszy, R. J., Su, W., & Siddiqi, F. H. (1990). Effects of oversize particles on the density of clean granular soils. *Geotechnical Testing Journal, 13*(2), 106–114.
26. Gee, G. W., & Or, D. (2002). 2.4 Particle size analysis. *Methods of Soil Analysis: Part 4 Physical Methods, 5*, 255–293.
27. Germain, G. T., & Woolfson, M. M. (1968). On the application of phase relationships to complex structures. *Acta Crystallographic Section B: Structural Crystallography and Crystal Chemistry, 24*(1), 91–96.
28. Gimsing, A. L., Borggaard, O. K., & Bang, M. (2004). Influence of soil composition on adsorption of glyphosate and phosphate by contrasting Danish surface soils. *European Journal of Soil Science, 55*(1), 183–191.
29. Grainger, P. (1984). The classification of mudrocks for engineering purposes. *Quarterly Journal of Engineering Geology and Hydrogeology, 17*(4), 381–387.
30. Guillaume, F. (2015). The LIX: A model-independent liquidity index. *Journal of Banking & Finance, 58*, 214–231.

31. Hamza, M. A., & Anderson, W. K. (2005). Soil compaction in cropping systems: A review of the nature, causes and possible solutions. *Soil and Tillage Research, 82*(2), 121–145.
32. Harison, J. A. (1988). Using the BS cone penetrometer for the determination of the plastic limit of soils. *Geotechnique, 38*(3), 433–438.
33. Ishaque, F., Hoque, M. N., & Rashid, M. A. (2013). Determination of plastic limit of some selected soils using rolling device. *Progressive Agriculture, 21*(1–2), 187–194.
34. Kariuki, P. C., Van Der Meer, F., & Siderius, W. (2003). Classification of soils based on engineering indices and spectral data. *International Journal of Remote Sensing, 24*(12), 2567–2574.
35. Kozlowski, T. T. (1999). Soil compaction and growth of woody plants. *Scandinavian Journal of Forest Research, 14*(6), 596–619.
36. Lade, P. V., Liggio, C. D., & Yamamuro, J. A. (1998). Effects of non-plastic fines on minimum and maximum void ratios of sand. *Geotechnical Testing Journal, 21*, 336–347.
37. Liu, S., Yamada, M., Collier, N., & Sugiyama, M. (2013). Change-point detection in time-series data by relative density-ratio estimation. *Neural Networks, 43*, 72–83.
38. Mehta, A., & Barker, G. C. (1994). The dynamics of sand. *Reports on Progress in Physics, 57*(4), 383.
39. Merkus, H. G., & Merkus, H. G. (2009). Sedimentation techniques. *Particle Size Measurements: Fundamentals, Practice, Quality*, 319–348.
40. Moreno-Maroto, J. M., & Alonso-Azcárate, J. (2017). Plastic limit and other consistency parameters by a bending method and interpretation of plasticity classification in soils. *Geotechnical Testing Journal, 40*(3), 467–482.
41. Nakase, A., Kamei, T., & Kusakabe, O. (1988). Constitutive parameters estimated by plasticity index. *Journal of Geotechnical Engineering, 114*(7), 844–858.
42. Nakhaei, A., Marandi, S. M., Kermani, S. S., & Bagheripour, M. H. (2012). Dynamic properties of granular soils mixed with granulated rubber. *Soil Dynamics and Earthquake Engineering, 43*, 124–132.
43. Nawaz, M. F., Bourrie, G., & Trolard, F. (2013). Soil compaction impact and modelling: A review. *Agronomy for Sustainable Development, 33*, 291–309.
44. Neal, R. H., Sposito, G., Holtzclaw, K. M., & Traina, S. J. (1987). Selenite

adsorption on alluvial soils: I. Soil composition and pH effects. *Soil Science Society of America Journal, 51*(5), 1161–1165.

45. Nelson, R. C., & Pope, G. A. (1978). Phase relationships in chemical flooding. *Society of Petroleum Engineers Journal, 18*(05), 325–338.
46. Pecharsky, V. K., & Gschneidner Jr, K. A. (1997). Phase relationships and crystallography in the pseudobinary system Gd5Si4–Gd5Ge4. *Journal of Alloys and Compounds, 260*(1–2), 98–106.
47. Polidori, E. (2015). Proposal for a new classification of common inorganic soils for engineering purposes. *Geotechnical and Geological Engineering, 33*, 1569–1579.
48. Rashid, A. S. A., Katimon, A., & Noor, N. M. (2008). Determination of plastic limit of soil using modified methods. *Malaysian Journal of Civil Engineering, 20*(2), 33–-199.
49. Reeve, M. J., Hall, D. G. M., & Bullock, P. (1980). The effect of soil composition and environmental factors on the shrinkage of some clayey British soils. *Journal of Soil Science, 31*(3), 429–442.
50. Rogers, C. D. F., Dijkstra, T. A., & Smalley, I. J. (2020). Keynote lecture: Classification of arid soils for engineering purposes: An engineering approach. In *Engineering Characteristics of Arid Soils, Vol. 2*, (pp. 99–133).
51. Rossiter, D. G. (2007). Classification of urban and industrial soils in the world reference base for soil resources. *Journal of Soils and Sediments, 7*, 96-100.
52. Salot, C., Gotteland, P., & Villard, P. (2009). Influence of relative density on granular materials behavior: DEM simulations of triaxial tests. *Granular Matter, 11*(4), 221–236.
53. Scott, H. G. (1975). Phase relationships in the zirconia-yttria system. *Journal of Materials Science, 10*, 1527–1535.
54. Sherwood, P. T., & Ryley, M. D. (1970). An investigation of a cone-penetrometer method for the determination of the liquid limit. *Géotechnique, 20*(2), 203–208.
55. Spark, K. M., & Swift, R. S. (2002). Effect of soil composition and dissolved organic matter on pesticide sorption. *Science of the Total Environment, 298*(1–3), 147–161.
56. Sridharan, A., Nagaraj, H. B., & Prakash, K. (1999). Determination of the plasticity index from flow index. *Geotechnical Testing Journal, 22*(2), 175–181.

57. Tamrakar, N. K. (2009). Evaluation of quality of crushed limestone and siltstone for road aggregates. *Bulletin of Department of Geology, 12*, 20–42.
58. Taylor, H. I. M., & Brar, G. S. (1991). Effect of soil compaction on root development. *Soil and Tillage Research, 19*(2–3), 111–119.
59. Unger, P. W., & Kaspar, T. C. (1994). Soil compaction and root growth: A review. *Agronomy Journal, 86*(5), 759–766.
60. Vaid, Y. P., & Negussey, D. (1984). Relative density of pulvinated sand samples. *Soils and Foundations, 24*(2), 101–105.
61. Vallauri, D., Adrián, I. A., & Chrysanthou, A. (2008). TiC–TiB_2 composites: A review of phase relationships, processing and properties. *Journal of the European Ceramic Society, 28*(8), 1697–1713.
62. Vardanega, P. J., & Haigh, S. K. (2014). The undrained strength–liquidity index relationship. *Canadian Geotechnical Journal, 51*(9), 1073–1086.
63. Wasti, Y. (1987). Liquid and plastic limits as determined from the fall cone and the Casagrande methods. *Geotechnical Testing Journal, 10*(1), 26–30.
64. Yimsiri, S., & Soga, K. (2010). DEM analysis of soil fabric effects on behavior of sand. *Géotechnique, 60*(6), 483–495.
65. Yukselen-Aksoy, Y., Kaya, A., & Ören, A. H. (2008). Seawater effect on consistency limits and compressibility characteristics of clays. *Engineering Geology, 102*(1–2), 54–61.
66. Zhao, H., Brautigam, C. A., Ghirlando, R., & Schuck, P. (2013). Overview of current methods in sedimentation velocity and sedimentation equilibrium analytical ultracentrifugation. *Current Protocols in Protein Science, 71*(1), 20–12.

CHAPTER 3

SOILS INVESTIGATION

Contents

3.1. INTRODUCTION

Geological forces and processes often result in inhomogeneous and discontinuous formations that significantly influence the stability and costs of civil engineering works. Important geological factors that must be taken into account in geotechnical engineering include the type and methods of construction, the investigation required to characterize a site economically, and natural geological hazards like groundwater conditions, volcanic activity, and earthquakes. Many structural failures that claimed lives and property have been caused by unrealized geological conditions (Bailey et al., 2002). Examine the geology of a county's possible building site, as depicted in Figure 3.1. It takes the use of geophysical techniques and a number of closely spaced boreholes to map all the geological features. It's challenging to determine the exact dimensions of every geological feature. For example, to design a safe and cost-effective foundation for a skyscraper, you need to be aware of the geological conditions in the vicinity of, and under the building (Xin & Gustin, 2007).

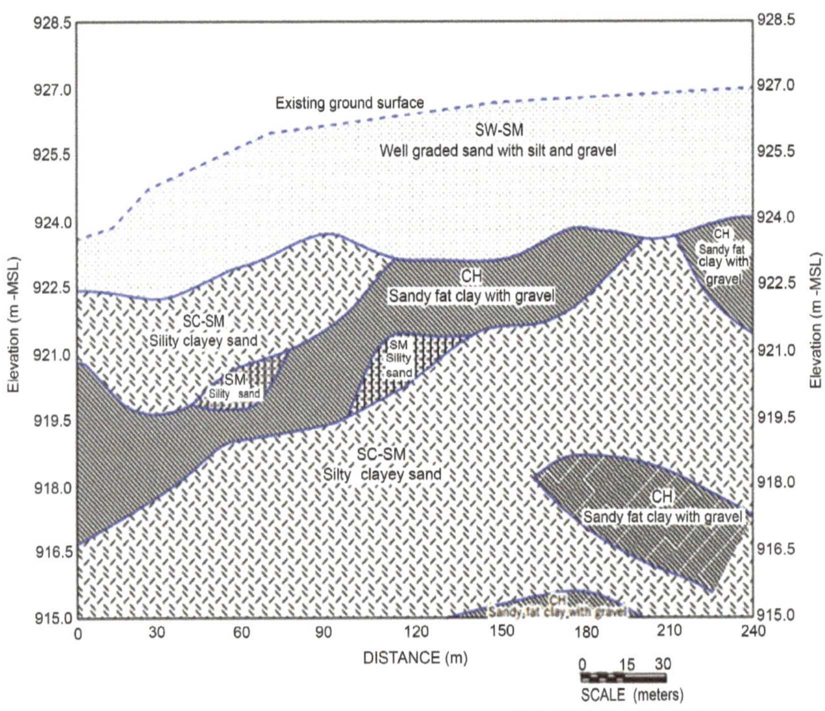

Figure 3.1. *The soil profile of a proposed site of construction.*

Source: Bartelby.com, Creative Commons License.

Soil investigation is an important part of designing and building a proposed structural system, such as a building, highway, road or dam, etc. During a proposed site's soil investigation, the soils are identified, examined, and recovered (Wan & Fell, 2004). Since it would be unfeasible to undertake a thorough investigation of the entire site, soil investigations is typically limited to a portion of the proposed site. The costs and performance of structures built at a site will be significantly impacted by judgments and estimates based on data from a small number of observations as well as lab and field test results (Alqadad et al., 2017).

3.2. PURPOSES OF A SOILS INVESTIGATION

Soil investigation is needed in gathering data for designing and building, environmental impact assessments, and for due diligence in the project. Project due diligence involves looking into a potential project so that the owner can make better business decisions (Hilgard, 1894). A soils investigation is conducted for the following reasons:

i. To assess the site's overall suitability for the proposed project.
ii. To enable an adequate and economical design to be made.
iii. To identify and consider potential challenges arising from site and different local conditions during construction (Lark, 1999).

3.3. PHASES IN THE SOILS INVESTIGATION PROCESS

The scope of a soils investigation depends on the type, size, and importance of the structure; the client; the engineer's familiarity with the soils at the site; and local building codes. Extensive soil investigation is typically needed for structures which may be sensitive to settlement, such as high-use buildings and machine foundations, than for residential foundations (Rogasik et al., 1999). In order to save money, a client might want to take a bigger risk than usual and place restrictions on the kind and scope of the investigation of site. Any attempt to shrink the scope of any soils investigation outside what is desirable for taking reasonable risks for comparable projects within or on comparable ground conditions should be approached with caution. If the geotechnical engineer knows the site well, he or she may do just a basic soils investigation to confirm what they already know. Provisions defining the scope of site investigations can be found in certain local building codes. It is imperative that the proposed site be visited (Lin et al., 2012).

An investigation into soils consists of three parts. Before designing, the first part is completed. The second part is completed in the design process. The third element is completed while building is underway. For contingencies, the second and third elements are required. The first part is usually carried out in stages and is more comprehensive (Gong et al., 2014). These phases are outlined below:

- **Phase I:** This phase is sometimes called as "desk study." It entails gathering the relevant data, including a plan of the site; loading conditions; the scale, significance and kind of the structure; previous geotechnical reports; any maps, including aerial photos, topographical maps, geologic maps, satellite imagery, and still photos; and newspaper clippings. Online resources offer a wide variety of maps that contain info on climate, contours, geology, elevation, hydrology and regional seismicity (Lindsay & Moreno, 1960). Geographic information system or GIS for short, is a technological instrument utilized to observe, exchange, comprehend, analyze, and visually represent data in a manner that emphasizes patterns, trends, and interconnections. It uses complex hardware and software along digital data to gather, sort, analyze, and show spatial information. GIS data consists of discrete objects like roads and continuous fields like elevation. These can be kept as vector or raster objects in storage. Maps, terrain, and three-dimensional buildings can all be viewed with Google Earth (http://earth.google.com). With Google Earth, project maps can also be made (Bui et al., 1990).
- **Phase II:** An overview of the site's geology and topography through an initial reconnaissance or site visit. The data collected in the first phase must be brought to the site during the visit in order to be compared with the current state of the site. The visit notes ought to contain the following information:

i. Photographs of the site and its neighborhood.

ii. Access to the site for workers and equipment.

iii. Sketches of all fences, walk-ways, utility posts, drainage systems, driveways, and other structures (Joseph et al., 2010).

iv. All available utility services including electricity and water.

v. Topographic images with all of the current structures, ponds, fills, cuts, and depressions in the ground.

vi. Every building's condition that is currently there or even close by. Include any exterior or interior cracks, any noticeable tilt, the construction type (such as framed stucco or brick building), any signs of frost damage, mold, and any unusual features in your notes.

vii. Geological characteristics found in any exposed region, like a road cut.

viii. In order to better understand the site, a few trial pits, trenches, and boreholes may occasionally be dug (Kaiser & Zech, 1997).

- **Phase III:** This step entails detailed soils exploration. Objectives of this stage include:

i. Establish the geological structure that should encompass the extent, sequence and thickness of the soil strata.

ii. To ascertain the state of the groundwater.

iii. To collect undisturbed disturbed samples for laboratory testing.

iv. To carry out in-situ tests in the field (Alani et al., 2013).

- **Phase IV:** Testing in the lab. The following are the goals of laboratory testing:

i. Soil classification.

ii. To ascertain the parameters related to settlement, permeability, stress-strain response, compactibility, failure stresses/strains, and soil strength. Not every one of these might be needed for a particular project (Konoplev et al., 1996).

Step V: Prepare a report. A detailed account of the soils in the site, the exploration techniques used, all in-situ and lab tests processes and findings, soil stratigraphy, and location of groundwater must all be included in the report. Any unusual soil and groundwater condition, such as areas that are susceptible to frost or are wet, soil, and water-bearing strata that could cause problems during construction should all be mentioned and/or explained.

3.4. SOILS EXPLORATION PROGRAM

A soils exploration program usually involves test pits and/or soil borings (boreholes). During the site visit (Phase II), you should work out most of the soils exploration program. A detailed soils exploration consists of:

1. Assessing the requirements and extent of geophysical exploration.
2. The approximate location of every test pit and/or borehole.
3. The test pit or borehole numbers (Wu & Wong, 1981)
4. The intended depth of every test pit or borehole.
5. Techniques and protocols to advance the boreholes.
6. Guidelines for sampling the first borehole, at the very least. The quantity of samples and potential locations must be specified in the

sampling instructions. Following the initial borehole, modifications to the sampling guidelines are common (Hirst et al., 2000).

7. Determining the kinds of in-situ tests that are necessary.
8. Groundwater observation requirements (Lynch, 2011).

3.4.1. Soils Exploration Methods

Several techniques can be employed to investigate the soils at a location:

3.4.1.1. Geophysical methods

These are the nondestructive methods for obtaining geographical data on soils, rocks, environmental, hydrological, and other conditions. Common techniques are:

- **Ground-Penetrating Radar (GPR):** GPR or Georadar, is a high-resolution and frequency electromagnetic wave technique (10 emd 1000 MHz) that is also used for imaging ground and soil structures. Radar pulses produced by the pulse generator are transmitted and recovered using an antenna (Figure 3.2). The returned pulse is then processed to produce images of the soil profile. Finding buried objects and imaging soil profiles are the two main geotechnical applications. With minimal soil disturbance, GPR generates continuous-resolution images of a soil's profile. GPR is not appropriate for wet clays and silts that are very conductive (>15 milliohms/m). Depth causes the GPR resolution to decrease (Bristow & Jol, 2003).
- **Seismic Surveys**: Seismic investigations make use of the variation in velocity at which surface waves propagate through distinct materials. By measuring the magnitude and duration of seismic waves—basically, compression waves (P waves)—at a location some distance from the wave's source, subsurface interfaces can be identified. When using seismic methods, the most crucial parameter is the propagation velocity. The propagation velocity is determined by the elastic characteristics and densities of the geological materials. A seismic wave's energy is transferred by diffraction, refraction, or reflection when it crosses a boundary between two elastic media (Gordon et al., 2003). Geotechnical site characterization makes use of seismic refraction and reflection.

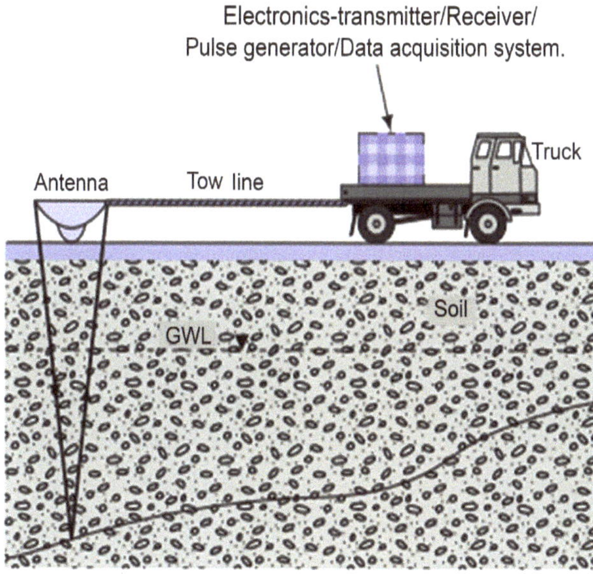

Figure 3.2. *A ground-penetrating radar (GPR) equipment.*

Source: Slide Share, Creative Commons License.

In seismic reflection tests, the travel times of waves reflected from subsurface interfaces are measured by geophones (Figure 3.3a). Geophones are motion-sensitive transducers that convert ground motion to electric signals. The travel time is related to the interfaces' depth, size, and shape (Gisiner, 2016). The angle of reflection of the waves is a function of the material density contrast. Seismic reflection is used when high resolution of soil profile is required, especially at large depths (>50 m).

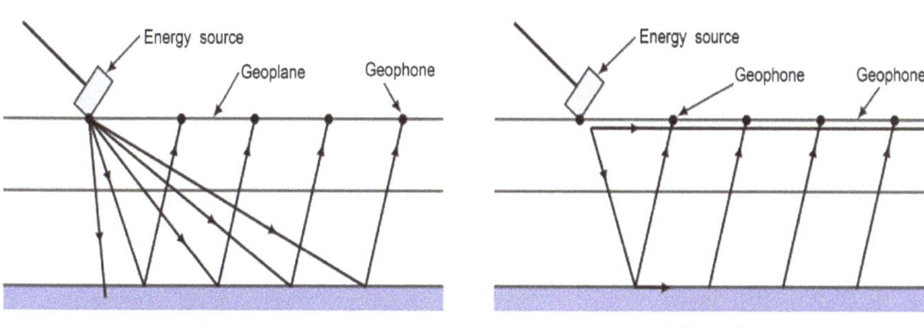

Figure 3.3. Transmission of wave in (a) seismic reflection; and (b) refraction tests.

Source: Slide Share, Creative Commons License.

Seismic refraction surveys and seismic reflection surveys are nearly identical, with the exception of the measuring location of refraction waves and placing the source geophone farther away (Figure 3.3b). The latter allows seismic waves, which are mainly horizontal instead of vertical, to be recorded. Usually, only the first P waves are observed in refraction surveys. The thickness and depth of soil profile, and presence of buried structures can all be ascertained using the seismic refraction method (Slabbekoorn et al., 2019). To produce seismic waves at shallow investigation depths, a sledgehammer is used to pound the ground surface; and a small explosive charge is employed for deeper depths. Seismic methods can be affected by vibrations and noise. Different filtering methods are applied to lessen vibration and background noise. Spatial changes in low-velocity materials are mapped using multichannel analysis of surface waves (MASW). Figure 3.4 displays an interpreted soil profile derived from MASW (Gausland, 2000).

Crosshole seismic testing are used to obtain data on the soil layers stiffness. One borehole contains the seismic source, and another nearby borehole contains the geophone (Figure 3.5). S (shear) and P wave velocities can be determined using arrival times and geophone distances. The soil stiffness's are then computed using these (Scheidegger & Willmore, 1957).

To identify layering and assess the layers' strength, downhole seismic testing is utilized. The seismic source lies on the ground, and the geophones are in a hole in the ground (Figure 3.6).

- **Electrical Resistivity:** Measurements of electrical resistivity can be used to identify and quantify the depth of groundwater, identify clays, and determine the conductivity of groundwater. Changes in temperature and moisture content affect soil resistivity, which is expressed in ohm-centimeters (ohm-cm). Soil resistivity generally decreases as soil moisture content rises. In clays, both the surface-charged particles and the pore fluid provide electrical pathways, whereas in sands, the pore fluid acts as the sole electrical pathway. Wet fine-grained soils typically have substantially lower resistivity's than wet coarse-grained soils. When a soil is dry, its resistivity can differ by numerous orders of magnitude from when it is wet (Samouëlian et al., 2005).

The subsurface resistivity can be measured by putting four electrodes at equal distances in the ground in line, sending a measured AC current to the two electrodes on the outside, and checking the AC voltage between the two electrodes on the inside. As the measured voltage is divided by the measured current, one can determine the measured resistance. The apparent resistivity is then calculated by multiplying this resistance by a geometric factor which takes into account the distance between each electrode (McLachlan et al., 1990).

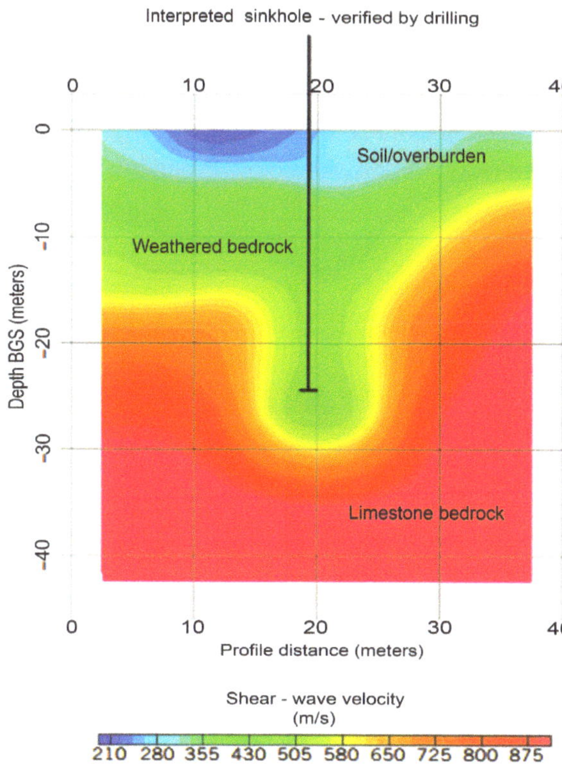

Figure 3.4. Surface wave data from seismic tests using a multichannel analysis of the soil profile.

Source: Parsan Biz, Creative Commons License.

At shallow depths (<10 m), electrode spacings of 0.75, 1.5, 3.0, 6.0, and 12.0 m are commonly employed. For deeper investigations, larger electrode spacings of 1.5, 3.0, 6.0, 15.0, 30.0, 100.0, and 150.0 m are usually utilized. Usually, the investigation depth is less than the electrode spacing maximum (Matula, 1979). To enhance electrical contact, water is injected down the electrode holes when the electrodes are pushed into the ground. Usually, multiple electrode spacings are measured successively at one location to create a sub-surface resistivity profile. Figure 3.7 displays a soil profile derived from resistivity measurements (Layssi et al., 2015).

Other geophysical techniques relevant to geotechnical engineering include:

i. Other geophysical techniques relevant to geotechnical engineering include:

ii. Gamma density, also known as gamma-gamma, is a measure of electron density that can be used to calculate the porosity or overall density of soil.

iii. The hydrogen density is measured by neutron porosity. It is applied to the estimation of porosity below groundwater level.

iv. Seismic velocity is measured by Sonic-VDL. Measuring soil stiffness and determining bedrock elevation are useful applications of this technique.

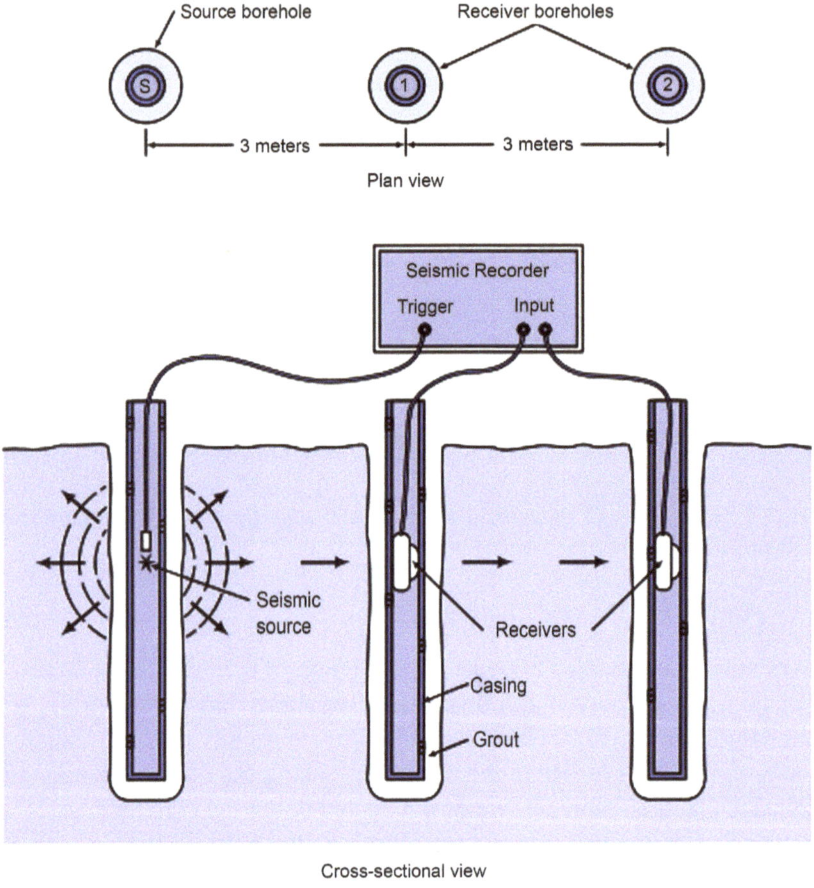

Figure 3.5. Typical setup for a cross-hole seismic survey.

Source: Cone Tec, Creative Commons License.

v. Microgravity is a useful tool for identifying cavities because it can identify variations in subsurface densities. At specific locations on the surface of the earth, a gravimeter is utilized to measure minute variations in gravity. These variations, which are associated with changes in density, are known as gravity anomalies (Montgomery, 1971).

vi. The main application of frequency domain electromagnetics (FDE), a continuous-wave field technique, is the mapping of lateral variations of subsurface structures. One frequency is utilized by a fixed transmitter. The measurement of the electromagnetic current's phase and magnitude gives the soil's electrical conductivity.

vii. Another continuous-wave field technique that measures increases in electrical conductivity is called very low frequency electromagnetics (VLFE), which measures the distortion of very low frequency waves. VLFE is a very helpful tool for fracture detection (Basinski et al., 1963).

viii. The rate at which pulsating currents decay over time is measured by time domain electromagnetics (TDE). TDE is helpful in figuring out how a soil's conductivity changes with depth. For instance, you can use TDE to get variations in soil saturation with depth.

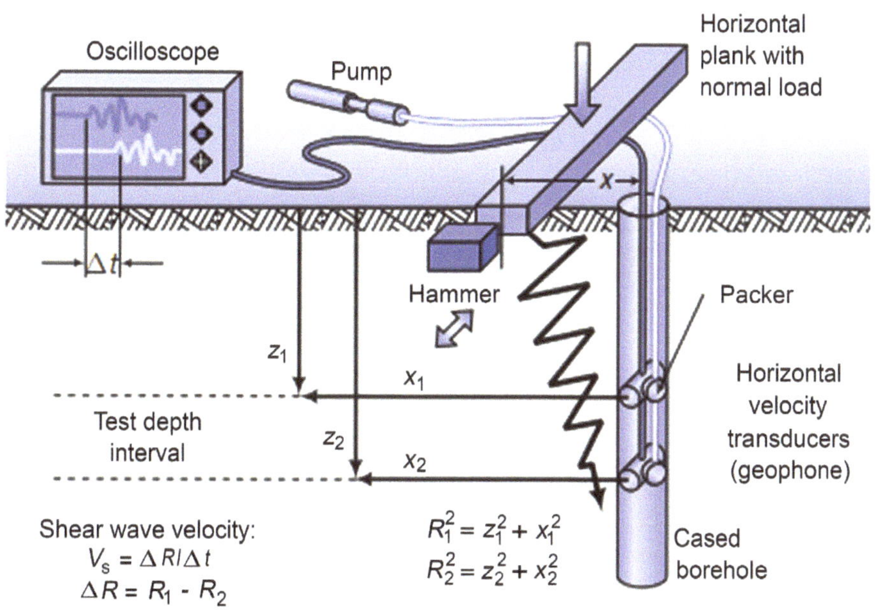

Figure 3.6. Downhole seismic survey.

Source: Cone Tec, Creative Commons License.

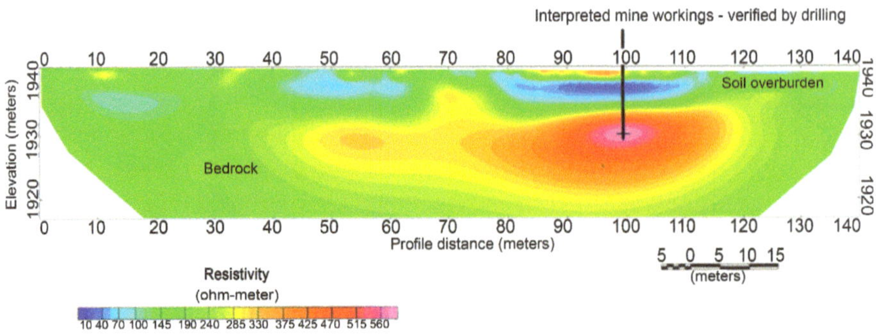

Figure 3.7. *Electrical resistivity test results for the soil profile.*

Source: Parsan Biz, Creative Commons License.

3.4.1.2. Destructive Methods

Destructive techniques require disturbing the soil, frequently to considerable depths. These techniques enable in situ testing (destructive tests), soil sample recovery, and soil mass inspection (Chen & Sun, 1991). The typical techniques are:

i. **Test or Trial Pits:** A pit is excavated manually with shovels or with the aid of a backhoe or other machinery (Sengul, 2014). This technique can produce very good soil stratigraphy at shallow depths (Figure 3.8).

Figure 3.8. *A test pit.*

Source: Seylone, Creative Commons License.

ii. **Hand or Power Augers:** These are instruments (Figure 3.9) that are used to rapidly drill a hole in the ground that is between 100 and 250 mm in diameter. Examining the soil and taking undisturbed samples is possible (certain disturbance is inevitable) specimens for laboratory examinations (Mai et al., 1998).

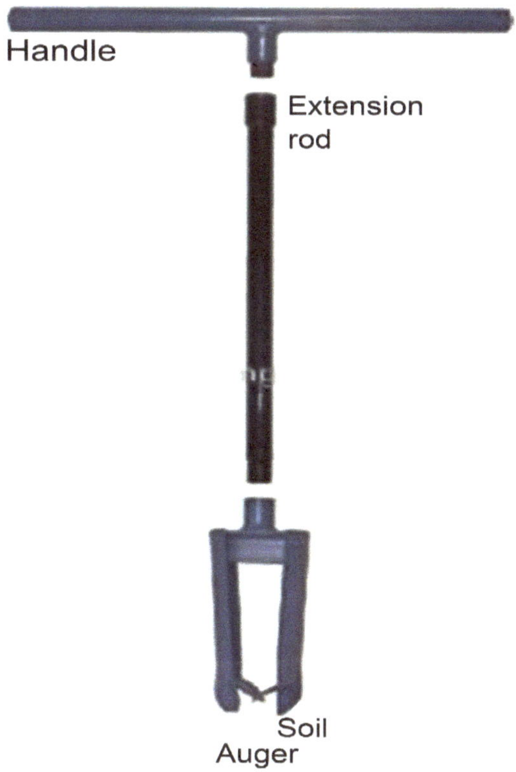

Figure 3.9. *A hand-powered auger.*

Source: Seylone, Creative Commons License.

iii. **Wash Boring:** To remove dirt from a borehole, water is pumped through a hollow rod which either is or is not fitted with a drill bit. The types of soil can be estimated from the washings (Williams & Drexler, 1981).

iv. **Rotary Rigs:** This technique comprises mechanical tools that are used to excavate ground, gather samples of soil, and enable in situ examinations (Figure 3.10).

Figure 3.10. *A rotary drill.*

Source: Zong-Xian Zhang, Creative Commons License.

v. **Direct-Push:** A hydraulic jack moves a split-spoon-shaped tube (mandrel), with a plastic insert, that is 30–50 mm in diameter and 1.5–5 m long. This lets the soil be sampled continuously (Ivanchev, 2022).

vi. **Sonic Drilling:** Soil or rock cores measuring 100–200 mm in diameter are cut using mechanical vibrations operating at frequencies ranging from 50 Hz to 150 Hz.

Table 3.1. Enlists the various advantages and disadvantages of these techniques.

3.4.2. Soil Identification in the Field

Inspection is used to determine the predominant types of soil in the field based on texture. Sands and gravels have visible individual particles and are gritty. When pressure is applied, water rises to the surface and silts crumble easily. Due to their extremely slow water migration, clays fail this test. When wet, clays are greasy, smooth, and sticky to the touch; however, when dry, they become extremely strong and hard (Batey & McKenzie, 2006).

Table 3.1. Advantages and Disadvantages of Soil Exploration Methods

Method	Advantages	Disadvantages
Geophysical methods Ground penetration radar Seismic surveys Electrical resistivity Microgravity Electromagnetic survey (e.g., FDE, VLFE, TDE)	▪ Nondestructive ▪ Quick ▪ Provide stratigraphy, groundwater, and relative wetness ▪ Relatively inexpensive ▪ Provide subsurface geologic information with which to plan detailed soils investigations	▪ No soil samples ▪ Limited design parameters ▪ Site may not have enough real estate to conduct tests adequately ▪ Much of the information is qualitative
Test pits A pit is dug either by hand or by a backhoe.	▪ Cost-effective ▪ Provide detailed information on stratigraphy ▪ Large quantities of disturbed soils are available for testing ▪ Large blocks of undisturbed samples can be carved out from the pits ▪ Field tests can be conducted at the bottom ofthe pit	▪ Depth limited to about 6m ▪ Deep pits uneconomical ▪ Excavation below groundwater and into rock difficult and costly ▪ Too many pits may scar site and require backfill soils
Hand augers The auger is rotated by turning and pushing down on the handlebar.	▪ Cost-effective ▪ Not dependent on terrain ▪ Portable ▪ Low headroom required ▪ Used in uncased holes ▪ Groundwater location can easily be identified and measured	▪ Depth limited to about 6m ▪ Labor-intensive ▪ Undisturbed samples can be taken only for soft clay deposits ▪ Cannot be used in rock, stiff clays, dry sand, or caliche soils

3.4.3. Number and Depths of Boreholes

It is both economically and practically unfeasible to explore the entire project site. Making decisions regarding the quantity, placement, and depth of borings is necessary in order to supply enough data for construction and design. The zone of soil where the structural loads would have an impact should be covered by the number and depth of borings. There isn't a set rule to abide by. The majority of the time, experience based on the ground's geological makeup, the expected structural loads, the structure's importance, and the equipment's accessibility determines the quantity and depth of borings. The minimum quantity and depth of borings are specified by regulatory organizations and building codes (Keawsawasvong & Shiau, 2021).

There should be enough boreholes in the site to identify any variations in the soils. In the event that the exact positions of all the loads on the structure's footprint are known (which is frequently not the case), one ought to think about boring a minimum of one borehole where the largest load is located (Ruud, 1982). For a building area of approximately 250 m^2, it is recommended to drill at least of three boreholes, and for approx. 1000 m^2 building, approximately five boreholes are enough. Table 3.2 provides guidelines on the least number of boreholes required for due diligence and for buildings in subdivisions (Magri, 1987). The following provides general guidelines regarding the depth of boreholes:

1. For compressible soils, like clays, the borings should go down to a depth of embedment that is at least one to three times the width of the intended foundation, or till the stress increment caused by the heaviest foundation load is below 10%, whichever is higher.
2. Borings should reach a depth of 5–7 meters in dense and coarse-grained soils and extremely stiff clays to demonstrate that the stratum is sufficiently thick.
3. Borings ought to go at least three meters into the rock.
4. Borings must go through very soft layers beneath the intended structure or any fills.
5. Boreholes should be at least 6 meters deep, unless they are in bedrock or extremely dense material (Aydın & Sisman, 2015).

Table 3.3 lists general rules for the bare minimum of boreholes, how often they should be drilled, and how deep they should be for common geotechnical structures.

Table 3.2. Minimum Number of Boreholes Required for Subdivisions and Buildings According to Area

Buildings		Subdivisions	
Area (m^2)	No. of boreholes (min.)	Area (m^2)	No. of boreholes (min.)
100	2	4000	2
250	3	8000	3
500	4	20,000	4
1000	5	40,000	5
2000	6	80,000	7
5000	7	400,000	15
6000	8		
8000	9		
10,000	10		

3.4.4. Soil Sampling

The goal of soil sampling is to get soils that are the right size and have the least disturbance for lab tests and observations. Typically, to collect soil samples, drill rods are used to secure a thin-walled and open-ended tube known as a Shelby tube or a sampling tube, which is then forced down into the soil (Cline, 1944).

Table 3.3. Minimum Borehole Quantity, Frequency, and Depth Guidelines for Common Geo-structures

Geostructure	Minimum number of boreholes	Minimum depth
Shallow foundation for buildings	1, but generally boreholes are placed at node points along grids of sizes varying from 15 m × 15 m to 40 m × 40 m	5 m or 1*B* to 3*B*, where *B* is the foundation width
Deep (pile) foundation for buildings	Same as shallow foundations	25–30 m; if bedrock is encountered, drill 3 m into it
Bridge	Abutments: 2 Piers: 2	25–30 m; if bedrock is encountered, drill 3 m into it
Retaining walls	length < 30 m: 1 length > 30 m: 1 every 30 m, or 1 to 2 times the height of the wall	1 to 2 times the wall height Walls located on bedrock: 3 m into bedrock
Cut slopes	Along length of slope: 1 every 60 m; if the soil does not vary significantly, 1 every 120 m On slope: 3	6 m below the bottom of the cut slope
Embankments including roadway (highway, motorway)	1 every 60 m; if the soil does not vary significantly, 1 every 120 m	The greater of 2× height or 6 m

Hopefully the soil is still inside the tube when it is carefully pulled out. During the sampling process, soil disturbances can arise from various sources, including friction between the sampling tube and the soil (Meyer & Brill Jr, 1988) the thickness of the tube's wall, the cutting edge's sharpness, and improper handling and care of the tube during transit. It is best to push the sampling tube into the ground rather than drive it in order to reduce friction (Forster, 1995).

Commonly used sampling tubes are made with minimal sampling disturbances in mind. The recovery ratio, which is expressed as L/z, where L is the sample length and z is the distance the sampler was pushed, is one way to assess the impact of sampler wall thickness. Higher wall thickness causes more sampling disturbance and a greater recovery ratio.

The piston sampler is a typical type of soil sampler. It is a seamless steel tube with thin walls that ranges in diameter from 50 to 100 mm, with the latter being the most popular. The tube has a wall thickness of approximately 1.75 mm, and it is long (600–1000 mm), with the latter being the most popular (Figure 3.11a). The tube is attached to a hydraulic cylinder. For fine-grained soils, especially soft clays, this sampler is very helpful. There is also the "standard" sampler, which is sometimes called the split-spoon sampler or split-barrel sampler. It comes with about 35 mm inside diameter and about 50 mm outside diameter (Figure 3.11b). The sampler features a split barrel that is secured by a cap at the top and a screw-on driving shoe on the bottom. In certain countries, the sampler is filled with a steel liner; however, in the US, this liner is typically not used (Boone et al., 1999).

As a result, the diameter of the soil sample is larger. For lined and unlined samplers, the outcomes of SPT (see in situ tests later in this chapter) differ. Higher driving stresses are possible with the standard sampler's thicker wall than with the Shelby tube, but higher soil disturbance levels result from this advantage. Samples of split spoons are disturbed. They are employed in tests of classification and visual inspection (Kerry & Oliver, 2004).

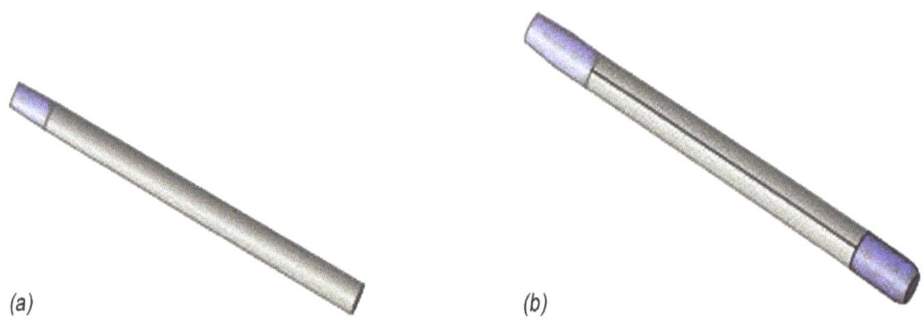

Figure 3.11. *(a). a thin-walled tube; and (b). a split-tube sampler.*

Source: Zong-Xian Zhang, Creative Commons License.

3.4.5. Groundwater Conditions

Digging a hole into a mass of soil in which every void is completely saturated with water will reveal water in the hole up to some point. This water level is referred as the groundwater table or groundwater level (Foster et al., 2017).

There is atmospheric pressure above the groundwater level, which is sometimes called the "free surface." Either ▾ or Δ can be used to represent groundwater level. The failure to identify or disclose the presence of groundwater at a site is a major cause of construction-related mishaps, legal disputes, and overspending. The water table always varies in response to geological factors, human activity (such as drawdown during construction, and pumping groundwater from wells), and environmental factors (such as droughts, monsoons, winter rains, and rainfall patterns) (Christensen et al., 2000).

An aquifer is a body of water-bearing soil that is located below the groundwater table. It is possible for aquifers to be semiconfined, confined, or unconfined (Figure 3.12). The groundwater level in an unconfined aquifer is allowed to fluctuate up and down based on the availability. The groundwater level typically rises in the winter and early spring. The level of groundwater decreases during drier months (Rivett et al., 2008).

Geological formations that divide aquifers can occasionally impede the flow of groundwater. Aquicludes are the formations that are impermeable, including nonporous rocks and/or fine-grained soils like clay. The formations are referred to as aquitards if they are semi-impermeable.

An confined aquifer is the one that contains water and is bounded above and below by aquicludes, which are impermeable geological formations. An unconfined aquifer's water is under pressure due to its confinement. Water can rise above the surface if either of these impermeable formations—typically the top—is broken. During well drilling, water rose to more than 50 meters above the surface of some unconfined aquifers. Seasonal climate variations do not directly impact confined aquifers, also known as artesian aquifers. While some infiltration from the surrounding rock or soil does occur, there isn't actually a true confined aquifer.

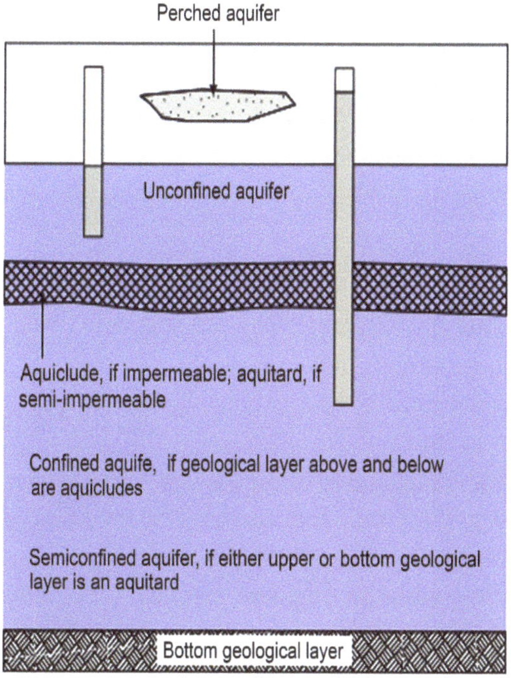

Figure 3.12. *Representation of perched, unconfined, and confined aquifers; aquitard and aquiclude.*

Source: Zekâi Şen, Creative Commons License.

Rarely do geological formations exhibit continuity, especially with alluvial aquifers. The aquifer frequently consists of zones, fingerings, or lenses made of materials that are semi-impermeable and impermeable. Occasionally, zone(s) of water may gather inside an unsaturated geological formation. The groundwater levels inside these are referred to as perched water tables, and they are referred to as perched aquifers. These perched aquifers could lead to instability, construction issues, and legal issues if they are not found and mentioned in the soils report (Hanson, 2008). In addition to determining the groundwater level plus any unique circumstances (such as an artesian condition), one should also determine the potential range of variations in the groundwater level (Cherry et al., 1979).

3.4.6. Types of In-Situ or Field Tests

Numerous in situ testing tools have been developed over time to assess the deformation and strength properties of the soil and to characterize it. The most widely used devices are:

- Vane shear test (VST)
- Standard penetration test (SPT)
- Cone penetrometer test (CPT)
- Pressuremeter test (PMT)
- Flat plate dilatometer (DMT)

In situ testing has the following benefits, particularly when utilizing SPT, VST, and CPT:

i. Soil profiling that is continuous or semicontinuous (Wroth, 1984).
ii. Depending on the kind of in situ test employed, the soil response given in its natural state.

In-situ tests are generally less expensive than lab tests. The drawbacks consist of:

i. The loading rate and stress/strain states applied during the specific in situ test might not be indicative of the loading rate and stress/strain states imposed by the structure.
ii. It is impossible to regulate boundary conditions and drainage (Mayne, 2015).
iii. With the pressure meter serving as a notable exception, design soil parameters cannot be directly obtained. It is necessary to employ or create empirical correlations through meticulously regulated laboratory experiments. Frequently, there are weak correlations (low R-squared values or regression coefficient).

3.4.6.1. Vane Shear Test (VST)

The shear vane device consists of four thin metal blades welded orthogonally (90°) to a rod (Figure 3.13) . Typically, the vane is pushed to the required depth from the bottom of a borehole. A torque head device attached onto the rod and situated above the soil surface applies a torque at a rate of roughly 6° per minute (Olmedo et al., 2020).

Once the max torque (Tmax) is found, the shear vane is turned eight to ten times more to find the residual torque (Tres). The soil sensitivity, St, is obtained by dividing the maximum torque by the residual torque, where;

$$S_t = \frac{T_{max}}{T_{res}}$$

The sensitivity is a way to measure how much soil disturbance lowers its undrained shear strength. Vane shear or undrained strength versus depth is how the results of a vane shear test are presented (Xu & Bian, 2018). A VST result example is displayed in Figure 3.14. The VST can be completed quickly, easily, and affordably, and the necessary equipment is widely accessible. Soil remolds when the vane is inserted. When the blade thickness is higher, the soil weakens and remolds more. The thickness of the blade must not surpass 5% of the diameter of the vane.

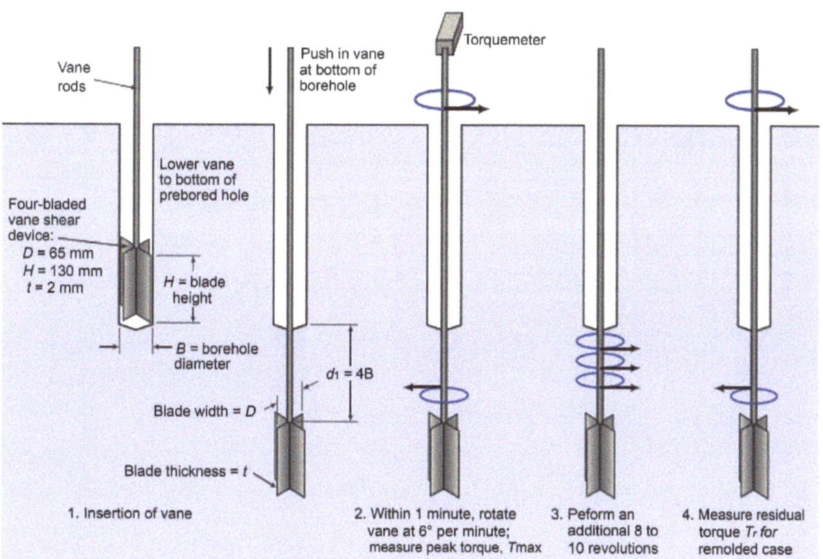

Figure 3.13. *Vane shear tester.*

Source: Zekâi Şen, Creative Commons License.

The rate of rotation typically affects the soil resistance. Calibration errors, uneven rotation, and excessive friction are some of the errors that can occur in torque measurements. The VST can't be used on very stiff clays and coarse-grained soils (Gupta et al., 2016).

FLELD VANE SHEAR TEST RECORD

FV-No:	FV-16
Phase	
Date Started:	
Date Finished:	
Time Started:	
Time Finished:	
Contractor:	

Vane Information

Type of Machine:	GEONOR-H10
Type:	Rectangular
Vane Dia., mm:	65
Vane Height., mm:	130
Loading, deg/sec.:	0.2

Field Vane Test Information

No. of Test, if:	11
Test Interval, #/m:	0.5
Final Test Depth, m:	5.5
Refusal Depth, m:	5.7

FV Test Coord. & Elev.

East, m	North, m	Elev., m
600018.5	11179098.7	-1.25

Notes:
1. Casing,vane and vane shoe equipment coupled with torque rods were used to perform vane shear test.
2. Peak undrained shear strength (s_u) herein refers to 'raw field' shear strength.

Test Depth (m)	Intact			Remolded			Sensitivity (-)
	Max. Torque (N.m)	Peak s_u (kPa)	Peak Reading (sec.)	Max. Torque (N.m)	Peak s_u (kPa)	Peak Reading (sec.)	
0.5	29.5	29.3	90	16.7	16.6	60	1.8
1.0	29.5	29.3	90	16.1	16.0	60	1.8
1.5	22.3	22.1	75	12.8	12.7	45	1.7
2.0	26.2	26.0	90	13.4	13.3	45	2.0
2.5	24.0	23.8	90	12.8	12.7	60	1.9
3.0	21.2	21.0	75	11.2	11.1	45	1.9
3.5	18.9	18.8	75	9.5	9.5	45	2.0
4.0	16.7	16.6	75	9.0	8.9	45	1.9
4.5	16.7	16.6	75	9.5	9.5	45	1.7
5.0	19.5	19.4	90	10.6	10.5	60	1.8
5.5	17.8	17.7	90	9.5	9.5	45	1.9

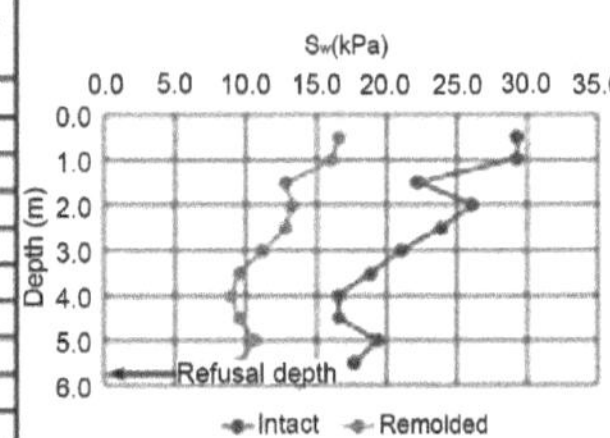

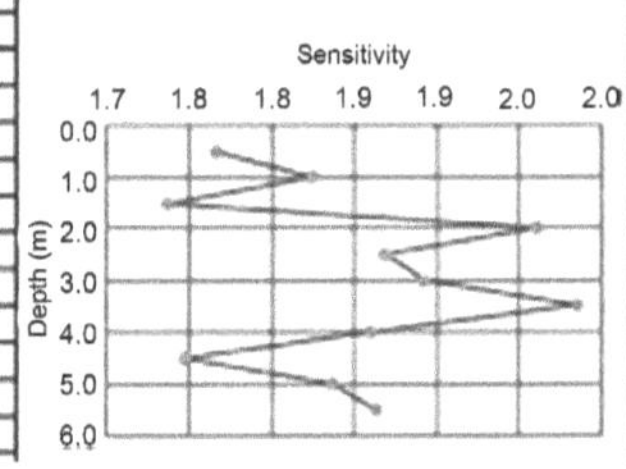

Figure 3.14. *An example of a VST result.*

Source: Zekâi Şen, Creative Commons License.

3.4.6.2. Standard penetration test (SPT)

Developed circa 1927, SPT is arguably the most widely used field test. A standard split-spoon sampler is driven into the ground for the SPT by blows from a drop hammer with a mass of 63.5 kg that falls 760 mm (Figure 3.15). At the bottom of a borehole, the sampler sits at 152 mm in the soil (Boulanger & Idriss, 2012).

The number of blows (N) needed to drive it a further 304 mm is then counted. The number of 'N' is called the standard penetration number. When a barrier or extremely dense and coarse-grained soil is encountered, 50 more blows per 25 mm of penetration may be made. The SPT records would show "refusal" in this case. SPTs are typically performed in practice at intervals of 0.6 meters. In Figure 3.16, an SPT record example is displayed (Mahmoud, 2013).

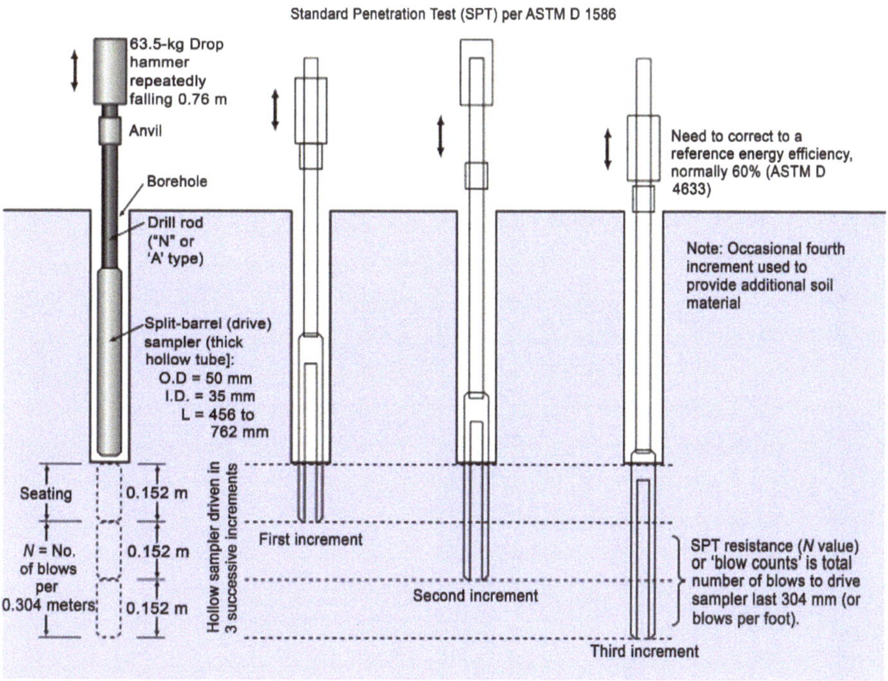

Figure 3.15. *The increments and the general driving sequence of SPT.*

Source: Khaled Chowdhury et al., Creative Commons License.

The N values are adjusted in a number of ways to take rod length, overburden pressure, and energy losses along other factors into consideration. Most of the time, the N values are adjusted so that the rod energy ratio is 60%. The energy supplied to the split spoon sampler divided by the hammer's free-falling energy is known as the rod energy ratio (Ameratunga et al., 2016). The updated N values are provided as N60 and are indicated as:

$$N_{60} = N\left(\frac{ER_r}{60}\right) = NC_E$$

where CE is the correction factor for rod energy ratio of 60% and ERr is the energy ratio. Table 3.4 provides the correction factors for the apparatus (60%

rod energy ratio correction), borehole diameter, sampler type, and rod lengths (Fletcher, 1965).

The correction factors listed in Table 3.4 can be written as a composite correction factor, CRSBEN as follows:

$$CRSBEN = CRCSCBCECN$$

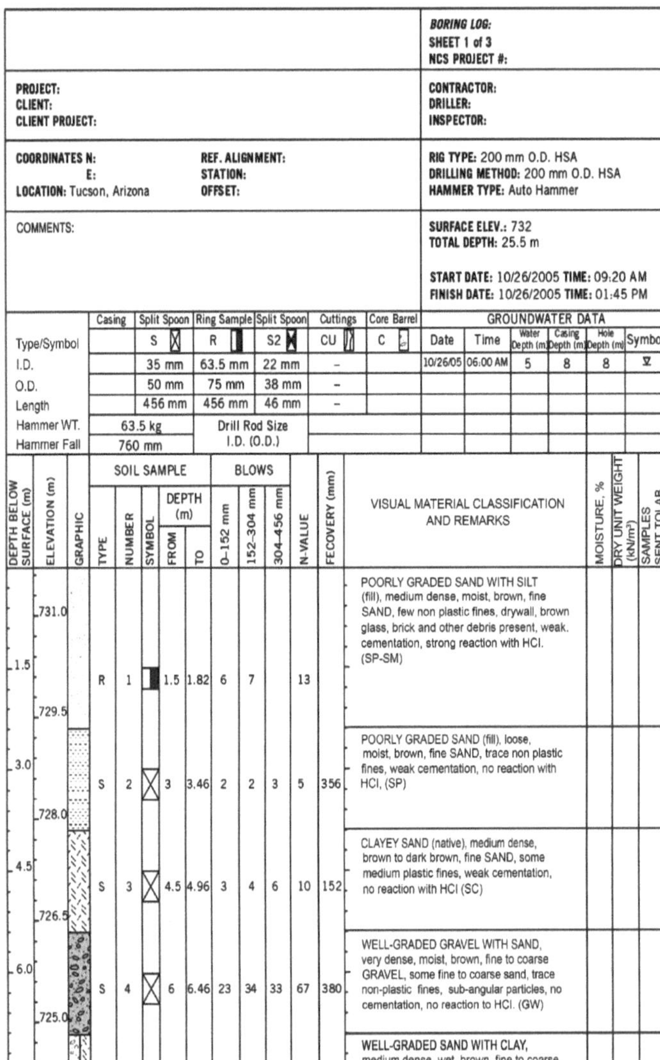

Figure 3.16. *Example of an SPT record.*

Source: Khaled Chowdhury et al., Creative Commons License.

In this case, the correlation factors for rod length are denoted by CR, sampler type by CS, borehole diameter by CB, overburden pressure by CN, or the effective soil pressure, which is calculated by multiplying the unit weight of soil by the depth (minus the groundwater pressure at that depth), and rod energy by CE. The correlation factor CN is:

$$C_N = \sqrt{\frac{98.5}{\sigma'_{zo}}};\quad C_N \leq 2; \text{ the unit for } \sigma'_{zo} \text{ is kPa}$$

$N_{1,60}$, or corrected N value, is;

$$N_{1,60} = C_{RSBEN}N$$

Any corrected N value need to be provided in whole numbers preferrably.

The SPT is an extremely valuable instrument for locating bedrock and identifying stratigraphic changes. In addition, the split spoon sampler can be used to describe the soil profile by examining the soil and extracting disturbed samples for lab analysis (Rogers, 2006).

The SPT can be completed quickly and easily. The equipment is easy to find and can penetrate dense substances. The bearing capacity, settlement of foundations, and engineering attributes of soils have all been linked to SPT results. However, the majority of those correlations are weak (r-squared value < 60% for the regression coefficient). They only apply to soils in the test conditions that were used to create the correlations. Errors can come from a variety of sources, such as using non-standard equipment or performing tests improperly. Testing errors include incorrect ways of lifting and lowering the hammer, not cleaning the borehole's bottom properly before the test starts, and not maintaining the groundwater level if one is found. N values that result from these mistakes are not indicative of the soil (Ataee et al., 2019). It is not a good idea to use SPT tests on boulders, coarse gravel, silts, soft clays, and mixed soils that have boulders, clays, cobbles, and silts. In an SPT, the surrounding soil is subjected to dynamic stresses. Exceedingly high water pressure in the void spaces is produced in non-free-draining soils like clays and mixed soils and clays. When the SPT sampler is driven into non-free-draining soils, water acts like a rigid substance (like concrete) because it is incompressible. As a result, neither soil strength nor any other soil parameter can be linked to the reported N value in such soils. SPTs are not performed continuously, so significant soil layers—especially porous and weak soils—may be missed by the test. SPT needs to be carried out in tandem with additional testing techniques, like the cone penetration test (Yusof & Zabidi, 2018).

Table 3.4. Correction Factors for Rod Length, Type of Sampler, and Size of Borehole

Correction factor	Item	Correction factor
C_R	Rod length (below anvil)	$C_R = 0.8$; $L \leq 4$m $C_R = 0.05L + 0.61$; $4\text{m} < L \leq 6$m $C_R = -0.0004L^2 + 0.017L + 0.83$; $6\text{m} < L < 20$m, $C_R = 1$; $L \geq 20$m L = rod length
C_S	Standard sampler	$C_S = 1.0$
	US sampler without liners	$C_S = 1.2$
C_B	Borehole diameter:	
	65 mm to 115 mm	$C_B = 1.0$
	152 mm	$C_B = 1.05$
	200 mm	$C_B = 1.15$
C_E	Equipment:	
	Safety hammer (rope, without Japanese "throw" release)	$C_E = 0.7$ to 1.2
	Donut hammer (rope, without Japanese "throw" release)	$C_E = 0.5$ to 1.0
	Donut hammer (rope, with Japanese "throw" release)	$C_E = 1.1$ to 1.4
	Automatic trip hammer (donut or safety type)	$C_E = 0.8$ to 1.4

Table 3.5. Compactness of Coarse-Grained Soils Based on N Values.

N	γ (kN/m^3)	D_r (%)	Compactness
0–4	11–13	0–20	Very loose
4–10	13–16	20–40	Loose
10–30	16–19	40–70	Medium
30–50	19–21	70–85	Dense
>50	>21	>85	Very dense

Table 3.6. Consistency of Saturated Fine- Grained Soils Based on SPT

Description	N
Very soft	0–2
Soft	2–4
Medium stiff	4–8
Stiff	8–15
Very stiff	15–30
Hard	>30

Table 3.5 shows how compact the coarse-grained soils are according to their N values. Table 3.6 provides the SPT-based fine-grained soil consistency. These tables' values are approximations meant to give a general idea of the ground's characteristics (Al-Jabban, 2013).

3.4.6.3. Cone Penetrometer Test (CPT)

The cone penetrometer (Figure 3.17a) is a rod-attached cone with a diameter of approximately 36 mm, a base area of approximately 10 cm^2, and a 60° cone angle. A roughly 150 cm^2 surface area outer sleeve encloses the rod slightly above the base of the cone. The thrusts needed to push the cone along with the sleeve inside the ground at a speed of 2 cm/s are measured separately to ensure the cone tip or end resistance (qc) and sleeve resistance or side friction (fs) can be estimated properly. The sleeve resistance is frequently presented using the friction ratio, or Rf = 100 fs /qc. The cone penetrometer was created initially for pile design, but it has also been used to calculate the settlement of a variety of foundations and to approximate the bearing capacity. There isn't a cone geometry that is recognized globally. Cones with various geometries are therefore used. These various cones will produce different outcomes. Cone penetrometer testing procedures and the cone geometry depicted in this textbook are commonly utilized in the United States (Rogers, 2006).

As shown in Figure 3.17b, the piezocone (uCPT or CPTu) is a type of electric cone penetrometer that has porous parts incorporated into the cone or the sleeve to measure porewater pressure. It is the porous elements' location that determines the measured porewater pressure. One common tool for determining the force of penetration is a load cell. For soil profiling, the piezocone is an incredibly helpful tool. According to research, the piezocone offers helpful information for estimating the soil's consolidation characteristics, bearing capacity, and shear strength. Figure 3.17d displays the CPT field configuration, while Figure 3.18 displays typical piezocone results.

The vision cone (VisCPT or VisCPTu) and the seismic cone (SCPT) are two more CPT variations. Geophones are positioned within the cone of the SCPT. Surface disturbances are caused by hammers, and the geophones (typically three) record the seismic waves that result. After that, the data is analyzed to provide parameters for soil strength and damping characteristics (Stegmann et al., 2006).

The CPT probe of VisCPTu and VisCPT is equipped with miniature cameras that continuously capture images of the soil surrounding the cone. It is possible to infer the texture of the soil through image processing. Liquefiable soils can also be found using the VisCPTu.

Most of the time, CPT and CPTu are used to figure out the soil stratigraphy from the cone resistance values measured. qc is typically greater than 4500 kPa in sands and less than 2000 kPa in fine-grained soils. Figure 3.19 displays a widely used classification chart that has been suggested. This chart utilizes CPT data pertaining to depths of 30 meters maximum. For depths greater than

30 meters, extra caution must be taken when applying soil classification based on Figure 3.19 since it is usually seen that f_s and q_c increase as depth increases (Hashemi & Nikudel, 2016).

The average values of the soil resistance are given over a length of roughly ten cone diameters, which is about 5 diameters above and below the tip. It doesn't matter which CPT probe is used. Cone measurements of soil resistance may not show individual layers in layered soils, especially in thin layers (less than 5 cone diameters) (Tanaka et al., 2003).

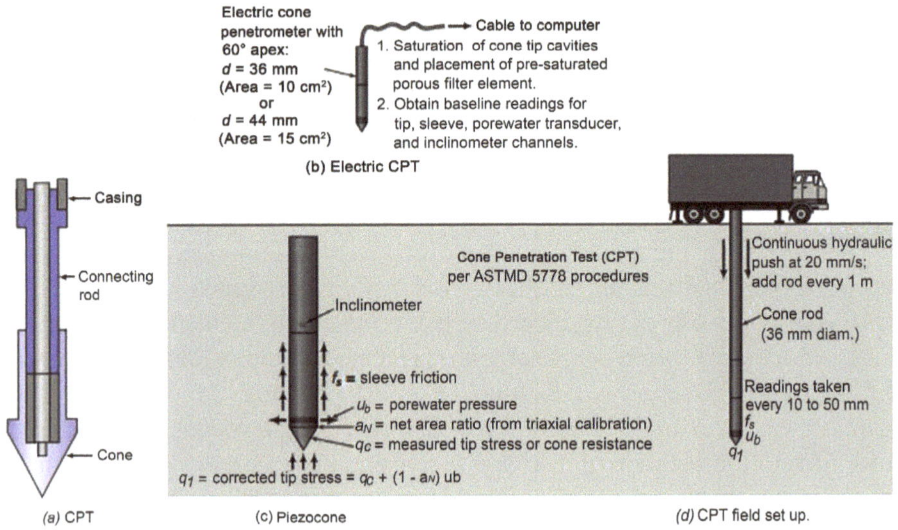

Figure 3.17. (a) Apparatus of CPT; (b) electric CPT; (c) piezocone; and (d) CPT field set up.

Source: Tine Missiaen et al., Creative Commons License.

Numerous soil factors, including soil fabric and type, soil minerology and density, stratigraphy and stress levels, affect the cone resistance. Empirical relationships have been established for deformation and strength parameters by correlating CPT results with laboratory tests. Researchers have also connected the results of CPT to those of other field exams, specifically SPT. In actuality, the cone data are examined, and big values are disregarded in accordance with experience. Figure 3.20 shows the drawing of a combined cone resistance (tip resistance) distribution. In this instance, three soil layers or zones are distinguished, each of which is thought to have a consistent cone resistance (Mohammad et al., 1999). For instance, zone 2's cone resistance is roughly 1900 kPa.

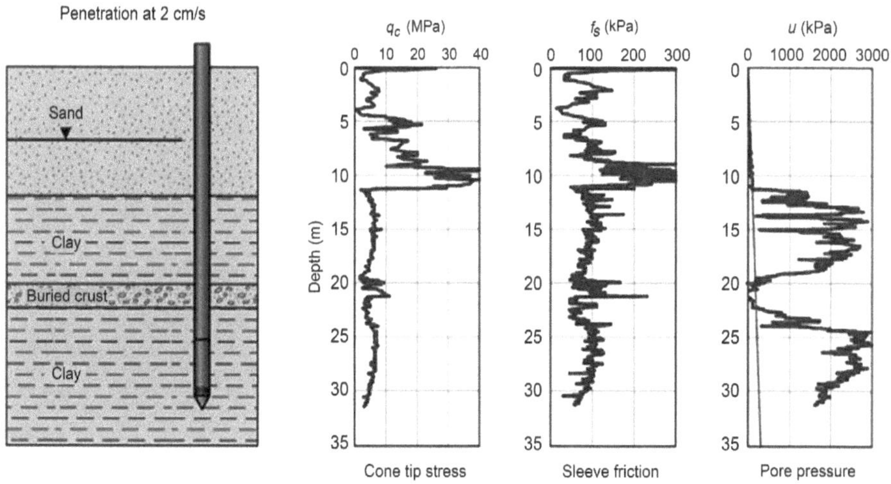

Figure 3.18. *Penetration results of piezocone at 2 cm/s.*

Source: Acta Geotechnica et al., Creative Commons License.

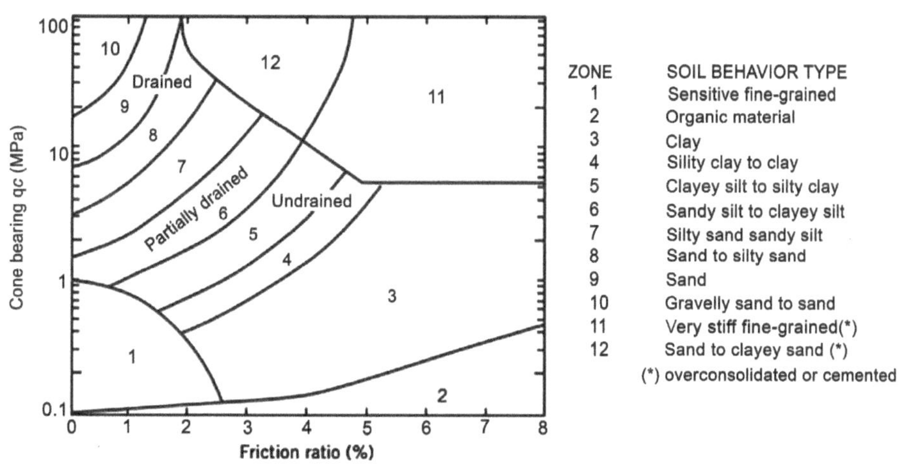

Figure 3.19. *Classification of soil based on friction ratio and cone tip resistance.*

Source: Erick Baziw et al., Creative Commons License.

When compared to SPT, CPT is faster and has fewer performance errors. It has the ability to continuously record soil conditions. CPT is ineffective in mixed soils that contain silts, clays, cobbles and boulders, as well as dense and coarse-grained soils (such as boulders and coarse gravel) (Robertson, 2016).

When the cone tip comes into contact with something dense, it can get damaged. The more complex VisCPT, SCPT, and uCPT tests typically call for experts to administer and interpret the findings.

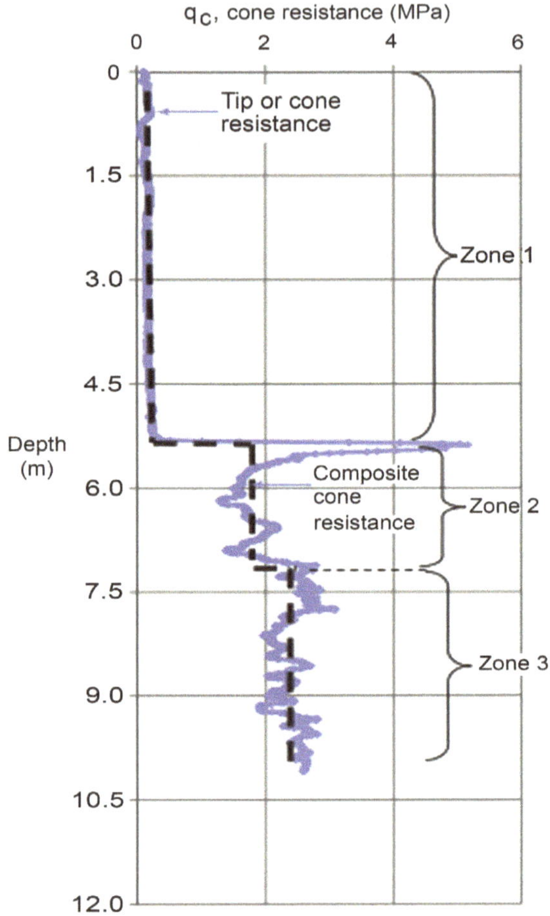

Figure 3.20. Graph to estimate the composite cone tip resistance.

Source: Muthusamy Karthikeyan et al., Creative Commons License.

3.4.6.4. Pressuremeter

A probe called the Menard pressuremeter (as seen in Figure 3.21) is put into an unlined borehole to the right depth, and pressure is then applied to a measuring cell on the probe. The pressure that is being applied is comparable to a cylindrical cavity expanding. At regular intervals, the pressure is increased gradually, and at each stage, the volume is recorded. Next, a pressure-volume change curve is created, from which it is possible to estimate the horizontal stress, elastic and shear modulus, friction angle, and undrained shear strength. This test is less accessible and more expensive than the CPT and flat plate dilatometer. The interpretation of the test results to determine the shear strength and modulus is uncertain because the drainage condition is unknown. The dimensions displayed in Figure 3.21 are not universal (Houlsby & Withers, 1988).

3.4.6.5. Flat Plate Dilatometer (DMT)

The flat plate dilatometer comprises of a tapered blade shown in Figure 3.22. It's dimensions are not consistent across the globe. The dilatometer is a flat face with a flexible steel membrane which pushes soil laterally when inflated. Drill rods connect to the blade, and a drill rig moves it into the soil at about 2 cm/s. Typically, tests are carried out every 200 mm. Pneumatic pressures are measured for three purposes: (1) for bringing the membrane flush with the surface of soil; (2) to laterally push the soil 1.1 mm; and (3) to determine the point at which the membrane returns to its initial position (Konrad, 1988).

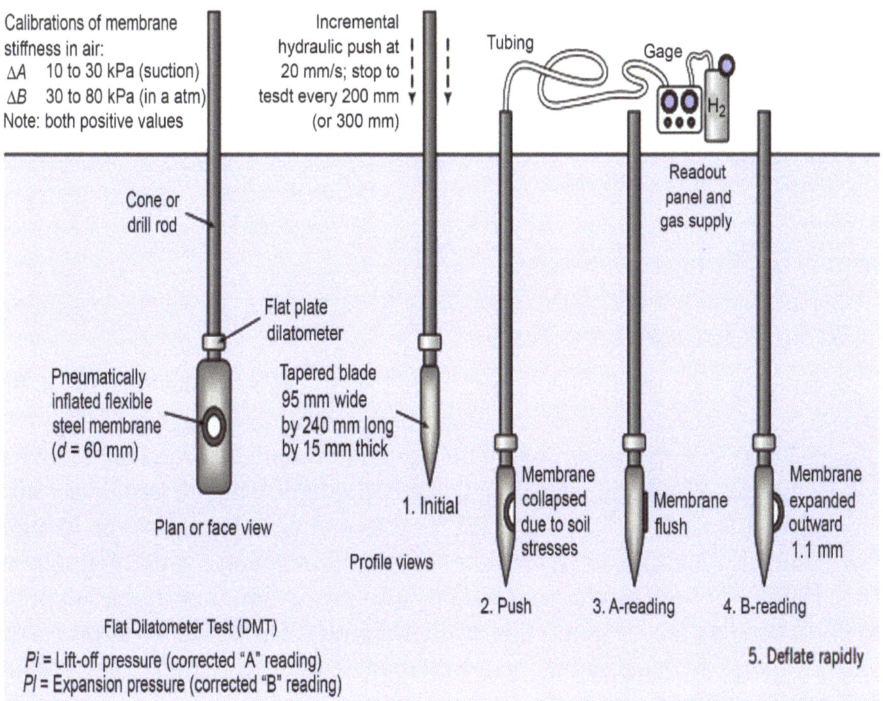

Figure 3.21. *Pressuremeter test CPT.*

Source: geocipta.co, Creative Commons License.

Dilatometer results have been connected to elastic modulus, over consolidation ratio, lateral earth pressures, and undrained shear strength. DMT is an easy and quick procedure. It is less expensive than the pressuremeter test and offers accurate estimates of horizontal stress. Before the test is conducted, dilatometers significantly remold the soil; therefore, care should be taken when interpreting the results. The dilatometer test works best on sands and clays (Monaco et al., 2005).

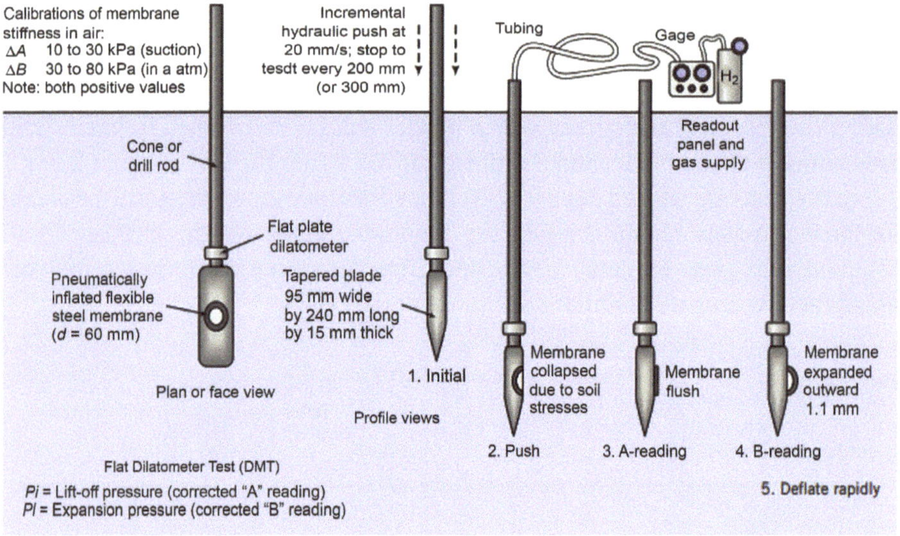

Figure 3.22. *Flat plate dilatometer.*

3.4.7. Soils Laboratory Tests

Field samples are usually taken and tested in the lab to find out about the material's physical and mechanical properties, such as its strength and ability to deform. These specifications govern the design of foundations and the selection of soils for use in building materials. Disturbed samples, like those taken with a standard sampler, are typically utilized for tests and visual examination in order to ascertain physical characteristics like shape, grain size and plasticity (Rampino et al., 1999). For both mechanical and physical properties, undisturbed samples are used, such as those from a thin-walled sampler. Disturbances in processes from sampling to handling to transportation leading to sample preparation have a significant impact on test results, particularly those that deal with the mechanical properties. Thus, caution needs to be taken to preserve the soil samples' undamaged state. To stop moisture loss from occurring, soil samples are frequently coated with wax (Spagnoli & Shimobe, 2020).

3.4.8. Types of Laboratory Tests

To categorize soils and establish stiffness, settlement, and strength characteristics for design and construction, laboratory testing is required. Compared to in situ tests, they provide greater control over the test conditions that are applied to the soil. Test samples in laboratories are always disturbed, and the level of disturbance can have a major effect on the test outcomes. To minimize testing disruptions, appropriate precautions must be taken. There are two classes of laboratory tests: Class I tests are designed to ascertain the physical attributes, while Class II tests serve to ascertain the mechanical attributes (Namikawa & Koseki, 2007).

Table 3.7. An Overview of Lab Tests Conducted to Evaluate Physical Properties

Physical Properties	Test objective	Parameters determined	Purpose
Specific gravity	To determine the specific gravity of soils	G_s	To calculate soil density (unit weight)
Grain size determination	To determine the grain size distribution	D_{10}, D_{50} $Cu = \frac{D_{60}}{D_{10}}$ $CC = \frac{D_{30^2}}{D_{10}D_{60}}$	Soil classification
Water content	To determine the water content of a soil	w	Qualitative information on strength and deformation
Index test	To determine the water content at which soil changes phases	PL, LL, PI, SL, LI	Soil classification; qualitative information on strength and settlement
Compaction	To determine the maximum dry density and optimum water content	$(\gamma_d)_{max}$, w_{opt}	Specification of compaction in the field
Permeability	To determine the hydraulic conductivity	k	Estimate of flow of water and seepage forces; stability analysis
Maximum and minimum dry Density	To determine the maximum and minimum dry density of coarse-grained soil	e_{max} e_{min}	Soil classification

3.5. SOILS REPORT

The site investigation report needs to be written in a clear, succinct, and accurate manner. The following should be included in the report, at the very least:

1. A document approving the investigation, usually a letter.
2. A one-page synopsis of the work completed and recommendations.

3. Work scope.
4. An overview of the location (Whitman, 1970).
5. Information about the kinds of investigations carried out, data on soil and groundwater, including laboratory and field test results, the investigation's assumptions and limitations, and potential construction-related challenges. Typically, soil boring logs have been utilized to compile soil data. Generally speaking, the boring log ought to include the following:

   i. The project name, address, and street name.
   ii. Boring location: offset and station.
   iii. The date boring was done (Loch, 1982).
   iv. Elevated surface.
   v. Each stratum's thickness and depth, along with the fill pattern, to enable rapid identification of various soil types. The soils report needs to have a legend for the fill pattern.
   vi. Sample or test numbers, along with the depths where the samples or the in situ tests took place (Frankenberger & Tabatabai, 1980).
   vii. Each stratum's classification of soil.
   viii. The distance to the water (if any).
   ix. Examination and interpretation concerning the gathered information.
   x. Design and construction recommendations, along with a discussion of any unique requirements.

SUMMARY

A soils investigation is required to identify and characterize the soils at a project site. Phases of this type of investigation may include borehole surveys, field and lab testing, and geophysical investigations. Following this investigation, the client typically requests a meticulously crafted report.

REFERENCES

1. Alani, A. M., Aboutalebi, M., & Kilic, G. (2013). Applications of ground penetrating radar (GPR) in bridge deck monitoring and assessment. *Journal of Applied Geophysics, 97*, 45–54.
2. Al-Jabban, W. J. M. (2013). Estimation of standard penetration test (SPT) of Hilla city-Iraq by using GPS coordination. *Jordan Journal of Civil Engineering, 7*(2), 133–145.
3. Alqadad, A., Shahrour, I., & Sukik, A. (2017). Smart system for safe and optimal soil investigation in urban areas. *Underground Space, 2*(4), 220–226.
4. Ameratunga, J., Sivakugan, N., Das, B. M., Ameratunga, J., Sivakugan, N., & Das, B. M. (2016). Standard penetration test. Correlations of Soil and Rock Properties in Geotechnical Engineering. *2*(1), 87–113.
5. Ataee, O., Moghaddas, N. H., & Lashkaripour, G. R. (2019). Estimating shear wave velocity of soil using standard penetration test (SPT) blow counts in Mashhad city. *Journal of Earth System Science, 128*(3), 66–199.
6. Aydın, M., & Sisman, A. (2015). Experimental and computational investigation of multi-U-tube boreholes. *Applied Energy, 145*, 163–171.
7. Bailey, V. L., Smith, J. L., & Bolton Jr, H. (2002). Fungal-to-bacterial ratios in soils investigated for enhanced C sequestration. *Soil Biology and Biochemistry, 34*(7), 997–1007.
8. Basinski, Z. S., Dugdale, J. S., & Howie, A. (1963). The electrical resistivity of dislocations. *Philosophical Magazine, 8*(96), 1989–1997.
9. Batey, T., & McKenzie, D. C. (2006). Soil compaction: Identification directly in the field. *Soil Use and Management, 22*(2), 123–131.
10. Boone, R. D., Grigal, D. F., Sollins, P., Ahrens, R. J., & Armstrong, D. E. (1999). Soil sampling, preparation, archiving, and quality control. Standard soil methods for long-term ecological research. *3*(2), 3–28.
11. Boulanger, R. W., & Idriss, I. M. (2012). Probabilistic standard penetration test–based liquefaction–Triggering procedure. *Journal of Geotechnical and Geoenvironmental Engineering, 138*(10), 1185–1195.
12. Bristow, C. S., & Jol, H. M. (2003). An introduction to ground penetrating radar (GPR) in sediments. *Geological Society, London, Special Publications, 211*(1), 1–7.
13. Bui, E. N., Loeppert, R. H., & Wilding, L. P. (1990). Carbonate phases in calcareous soils of the western United States. *Soil Science Society of America Journal, 54*(1), 39–45.

14. Chen, P., & Sun, Z. (1991). A review of non-destructive methods for quality evaluation and sorting of agricultural products. *Journal of Agricultural Engineering Research, 49*, 85–98.

15. Cherry, J. A., Shaikh, A. U., Tallman, D. E., & Nicholson, R. V. (1979). Arsenic species as an indicator of redox conditions in groundwater. In J. P. Riley & R. Chester (Eds.), *Developments in Water Science, 12*, (pp. 373–392).

16. Christensen, T. H., Bjerg, P. L., Banwart, S. A., Jakobsen, R., Heron, G., & Albrechtsen, H. J. (2000). Characterization of redox conditions in groundwater contaminant plumes. *Journal of Contaminant Hydrology, 45*(3–4), 165–241.

17. Cline, M. G. (1944). Principles of soil sampling. *Soil Science, 58*(4), 275–288.

18. Emmett, B. A., Reynolds, B., Chamberlain, P. M., Rowe, E., Spurgeon, D., Brittain, S. A., & Woods, C. (2010). Countryside survey: Soils report from 2007. *Vol 1*, 23–33.

19. Fletcher, G. F. (1965). Standard penetration test: Its uses and abuses. *Journal of the Soil Mechanics and Foundations Division, 91*(4), 67–75.

20. Forster, J. C. (1995). Soil sampling, handling, storage and analysis. In M. A. Oliver & J. M. C. Oliver (Eds.), *Methods in Applied Soil Microbiology and Biochemistry, 4*(2), (pp. 49–121).

21. Foster, T., Brozović, N., & Butler, A. P. (2017). Effects of initial aquifer conditions on economic benefits from groundwater conservation. *Water Resources Research, 53*(1), 744–762.

22. Frankenberger Jr, W. T., & Tabatabai, M. A. (1980). Amidase activity in soils: I. Method of assay. *Soil Science Society of America Journal, 44*(2), 282–287.

23. Gausland, I. (2000). Impact of seismic surveys on marine life. In *SPE International Conference and Exhibition on Health, Safety, Environment, and Sustainability, 3*(2), (pp. SPE–61127).

24. Gisiner, R. C. (2016). Sound and marine seismic surveys. *Acoust. Today, 12*(4), 10–18.

25. Gong, W., Luo, Z., Juang, C. H., Huang, H., Zhang, J., & Wang, L. (2014). Optimization of site exploration program for improved prediction of tunneling-induced ground settlement in clays. *Computers and Geotechnics, 56*, 69–79.

26. Gordon, J., Gillespie, D., Potter, J., Frantzis, A., Simmonds, M. P., Swift, R., & Thompson, D. (2003). A review of the effects of seismic surveys on

marine mammals. *Marine Technology Society Journal, 37*(4), 16–34.

27. Gunnarsson, O., Calandra, M., & Han, J. E. (2003). Colloquium: Saturation of electrical resistivity. *Reviews of Modern Physics, 75*(4), 10–85.
28. Gupta, T., Chakraborty, T., Abdel-Rahman, K., & Achmus, M. (2016). Large deformation finite element analysis of vane shear tests. *Geotechnical and Geological Engineering, 34*, 1669–1676.
29. Hanson, B., Hopmans, J. W., & Šimůnek, J. (2008). Leaching with subsurface drip irrigation under saline, shallow groundwater conditions. *Vadose Zone Journal, 7*(2), 810–818.
30. Hashemi, M., & Nikudel, M. R. (2016). Application of dynamic cone penetrometer test for assessment of liquefaction potential. *Engineering Geology, 208*, 51–62.
31. Hilgard, E. W. (1894). The chemical and physical investigation of soils. *Journal of the American Chemical Society, 16*(1), 34–47.
32. Hirst*, A. G., & Rodhouse, P. G. (2000). Impacts of geophysical seismic surveying on fishing success. *Reviews in Fish Biology and Fisheries, 10*, 113–118.
33. Houlsby, G. T., & Withers, N. J. (1988). Analysis of the cone pressure meter test in clay. *Geotechnique, 38*(4), 575–587.
34. Ivanchev, I. (2022). Investigation with non-destructive and destructive methods for assessment of concrete compressive strength. *Applied Sciences, 12*(23), 12172.
35. Joseph, S. D., Camps-Arbestain, M., Lin, Y., Munroe, P., Chia, C. H., Hook, J., & Amonette, J. E. (2010). An investigation into the reactions of biochar in soil. *Soil Research, 48*(7), 501–515.
36. Kaiser, K., & Zech, W. (1997). Competitive sorption of dissolved organic matter fractions to soils and related mineral phases. *Soil Science Society of America Journal, 61*(1), 64–69.
37. Keawsawasvong, S., & Shiau, J. (2021). Instability of boreholes with slurry. *International Journal of Geosynthetics and Ground Engineering, 7*, 1–11.
38. Kerry, R., & Oliver, M. A. (2004). Average variograms to guide soil sampling. *International Journal of Applied Earth Observation and Geoinformation, 5*(4), 307–325.
39. Konoplev, A. V., Bulgakov, A. A., Popov, V. E., Hilton, J., & Comans, R. N. J. (1996). Long-term investigation of 137Cs fixation by soils. *Radiation Protection Dosimetry, 64*(1–2), 15–18.
40. Konrad, J. M. (1988). Interpretation of flat plate dilatometer tests in sands

in terms of the state parameter. *Geotechnique, 38*(2), 263–277.

41. Lark, R. M. (1999). Soil–landform relationships at within-field scales: An investigation using continuous classification. *Geoderma, 92*(3-4), 141–165.
42. Layssi, H., Ghods, P., Alizadeh, A. R., & Salehi, M. (2015). Electrical resistivity of concrete. *Concrete International, 37*(5), 41–46.
43. Lin, Y., Munroe, P., Joseph, S., Kimber, S., & Van Zwieten, L. (2012). Nanoscale organo-mineral reactions of biochars in ferrosol: An investigation using microscopy. *Plant and Soil, 357*, 369–380.
44. Lindsay, W. L., & Moreno, E. C. (1960). Phosphate phase equilibria in soils. *Soil Science Society of America Journal, 24*(3), 177–182.
45. Loch, J. P. (1982). State-of-the-art report–frost action in soils. In J. DiMaggio (Ed.), *Developments in Geotechnical Engineering* (Vol. 28, pp. 213–224). Elsevier.
46. Lynch, J. P. (2011). Root phenes for enhanced soil exploration and phosphorus acquisition: Tools for future crops. *Plant Physiology, 156*(3), 1041–1049.
47. Magri, E. J. (1987). Economic optimization of the number of boreholes and deflections in deep gold exploration. *Journal of the Southern African Institute of Mining and Metallurgy, 87*(10), 307–321.
48. Mahmoud, M. A. A. N. (2013). Reliability of using standard penetration test (SPT) in predicting properties of silty clay with sand soil. *International Journal of Civil & Structural Engineering, 3*(3), 545–556.
49. Mai, K., Mäder, E., & Mühle, M. (1998). Interphase characterization in composites with new non-destructive methods. *Composites Part A: Applied Science and Manufacturing, 29*(9–10), 1111–1119.
50. Matula, R. A. (1979). Electrical resistivity of copper, gold, palladium, and silver. *Journal of Physical and Chemical Reference Data, 8*(4), 1147–1298.
51. Mayne, P. W. (2015). In-situ geocharacterization of soils in the year 2016 and beyond. *Advances in Soil Mechanics, 5*, 139–161.
52. McLachlan, D. S., Blaszkiewicz, M., & Newnham, R. E. (1990). Electrical resistivity of composites. *Journal of the American Ceramic Society, 73*(8), 2187–2203.
53. Meyer, P. D., & Brill Jr, E. D. (1988). A method for locating wells in a groundwater monitoring network under conditions of uncertainty. *Water Resources Research, 24*(8), 1277–1282.
54. Mohammad, L. N., Titi, H. H., & Herath, A. (1999). Evaluation of

resilient modulus of subgrade soil by cone penetration test. *Transportation Research Record, 1652*(1), 236–245.

55. Monaco, P., Marchetti, S., Totani, G., & Calabrese, M. (2005). Sand liquefiable assessment by flat dilatometer test (DMT). In *Proceedings of the 16th International Conference on Soil Mechanics and Geotechnical Engineering* (Vol. 1, pp. 2693–2698).
56. Montgomery, H. C. (1971). Method for measuring electrical resistivity of anisotropic materials. *Journal of Applied Physics, 42*(7), 2971–2975.
57. Namikawa, T., & Koseki, J. (2007). Evaluation of tensile strength of cement-treated sand based on several types of laboratory tests. *Soils and Foundations, 47*(4), 657–674.
58. Olmedo, N. A., Fisseha, B., Wilson, W., Barczyk, M., Zhang, H., & Lipsett, M. G. (2020). An automated vane shear test tool for environmental monitoring with unmanned ground vehicles. *Journal of Terramechanics, 91*, 53–63.
59. Rampino, C., Mancuso, C., & Vinale, F. (1999). Laboratory testing on an unsaturated soil: Equipment, procedures, and first experimental results. *Canadian Geotechnical Journal, 36*(1), 1–12.
60. Rivett, M. O., Buss, S. R., Morgan, P., Smith, J. W., & Bemment, C. D. (2008). Nitrate attenuation in groundwater: A review of biogeochemical controlling processes. *Water Research, 42*(16), 4215–4232.
61. Robertson, P. K. (2016). Cone penetration test (CPT)-based soil behavior type (SBT) classification system—An update. *Canadian Geotechnical Journal, 53*(12), 1910–1927.
62. Rogasik, H., Crawford, J. W., Wendroth, O., Young, I. M., Joschko, M., & Ritz, K. (1999). Discrimination of soil phases by dual energy X-ray tomography. *Soil Science Society of America Journal, 63*(4), 741–751.
63. Rogers, J. D. (2006). Subsurface exploration using the standard penetration test and the cone penetrometer test. *Environmental & Engineering Geoscience, 12*(2), 161–179.
64. Ruud, C. O. (1982). A review of selected non-destructive methods for residual stress measurement. *NDT International, 15*(1), 15–23.
65. Samouëlian, A., Cousin, I., Tabbagh, A., Bruand, A., & Richard, G. (2005). Electrical resistivity survey in soil science: A review. *Soil and Tillage Research, 83*(2), 173–193.
66. Scheidegger, A. E., & Willmore, P. L. (1957). The use of a least squares method for the interpretation of data from seismic surveys. *Geophysics, 22*(1), 9–21.

67. Sengul, O. (2014). Use of electrical resistivity as an indicator for durability. *Construction and Building Materials, 73*, 434–441.
68. Slabbekoorn, H., Dalen, J., de Haan, D., Winter, H. V., Radford, C., Ainslie, M. A., & Harwood, J. (2019). Population level consequences of seismic surveys on fishes: An interdisciplinary challenge. *Fish and Fisheries, 20*(4), 653–685.
69. Spagnoli, G., & Shimobe, S. (2020). An overview on the compaction characteristics of soils by laboratory tests. *Engineering Geology, 278*, 105–830.
70. Stegmann, S., Mörz, T., & Kopf, A. (2006). Initial results of a new Free Fall-Cone Penetrometer (FF-CPT) for geotechnical in situ characterization of soft marine sediments. *Norwegian Journal of Geology/Norsk Geologisk Forening, 86*(3), 123–234.
71. Tanaka, H., Tanaka, M., Suzuki, S., & Sakagami, T. (2003). Development of a new cone penetrometer and its application to great depths of Pleistocene clays. *Soils and Foundations, 43*(6), 51–61.
72. Wan, C. F., & Fell, R. (2004). Investigation of rate of erosion of soils in embankment dams. *Journal of Geotechnical and Geoenvironmental Engineering, 130*(4), 373–380.
73. Whitman, R. V. (1970). The response of soils to dynamic loadings; report 26, final report. Contract Report, *3*(2), 3–26.
74. Williams, E. J., & Drexler, J. S. (1981). A non-destructive method for determining peanut pod maturity. *Peanut Science, 8*(2), 134–141.
75. Wroth, C. P. (1984). The interpretation of in situ soil tests. *Geotechnique, 34*(4), 449–489.
76. Wu, T. H., & Wong, K. (1981). Probabilistic soil exploration: Case history. *Journal of the Geotechnical Engineering Division, 107*(12), 1693–1711.
77. Xin, M., & Gustin, M. S. (2007). Gaseous elemental mercury exchange with low mercury containing soils: Investigation of controlling factors. *Applied Geochemistry, 22*(7), 1451–1466.
78. Xu, J., & Bian, X. (2018). DEM Modeling of Vane Shear Test in Cohesive Solids. *Environmental Vibrations and Transportation Geodynamics, 7*, 227–232.
79. Yusof, N. Q. A. M., & Zabidi, H. (2018). Reliability of using standard penetration test (SPT) in predicting properties of soil. *Journal of Physics: Conference Series, 1082*(1), 012094.

CHAPTER 4

GEOTECHNICAL SITE INVESTIGATION

Contents

4.1. INTRODUCTION

Geotechnical engineering problems are different for each type of project and geological condition. Therefore, it is important to understand the goals and features of construction well enough to create a comprehensive plan for site investigation that will enable the best engineering solution to be implemented and help resolve any potential geotechnical issues that may arise prior to, during, or after the building phase (Jaksa et al., 2003). In order to stabilize structures, foundations either transfer loads from the superstructure to the soil or rock or help stabilize temporary excavations. It is well known that there is a greater risk of foundation construction for subsurface structures than for surface structures throughout the building phase. It is also known that careful site investigation conducted as part of the integral site investigation process can significantly lower foundation design costs (Simons et al., 2002).

In general, construction, subterranean excavation, slope engineering, roads, railroads, dams etc., are all included in the field of foundation engineering. The main parts of a building's foundation are the temporary stabilization structures and the permanent substructures that are used during excavations. For excavations, a variety of temporary stabilization techniques are available, including retaining walls, ground anchors, and internal bracings. One option for a permanent sub-structure is to build a mat, pile, or caisson foundation once the temporary excavation face has stabilized (Parsons & Frost, 2002). The subsequent superstructure is built on top of the stable subsurface structure. Nonetheless, because the majority of foundation projects are underground, they are not readily apparent. Planning and designing are crucial to a geotechnical engineer because of subterranean groundwater or geological conditions, as well as their variable distribution characteristics. Each project should begin with a geological survey to examine such conditions of various sites. This survey should also be used to carefully consider any potential issues or damage modes that may arise during construction and even after it is finished, so that proactive measures can be taken to address them (Moh, 2004).

The primary responsibility of geotechnical engineers is to provide the optimal solution for engineering design. To find the best solution, considerations like construction feasibility, safety, and economy are made. For an engineering project, the cost-optimal plan must be determined, but safety must always come first. It is necessary to analyze the safety of various scenarios, such as the during construction and the long-term safety under conditions of earthquake and heavy rain following the construction. In this chapter, several instances of proper site investigation failing to be carried out as part of geotechnical design and resulting in damages are provided to remind readers of the significance of site investigation (Wang & Jiang, 2023).

The issue with excavating a basement in soft clay is depicted in Figure 4.1. This type of clay has a very soft layer that is categorized as either CH (high-plasticity clay) or CL (low-plasticity clay) according to the unified soil classification system, where the SPT-N value is 4 at max. They decided to use a braced retaining system. Nevertheless, during excavation, there was a major tilt in nearby buildings due to an excavation failure. The failure was caused by an inadequate designing of the braced retaining system due to a poor grasp of the properties of the soil beneath the surface. Improved site investigation is necessary to identify the key design parameters, including soil strength, compressibility, and groundwater condition, in order to prevent such tragedies. After that, a reasonable stability analysis and the design of a secure braced retaining system need to be carried out (Yang et al., 2019).

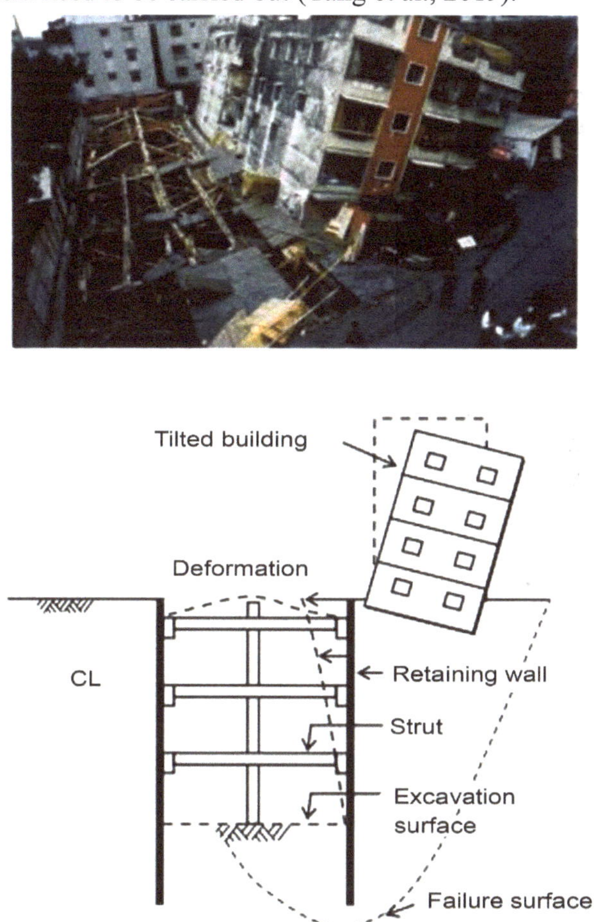

Figure 4.1. *An inadequate support system while in excavation caused the adjacent building to tilt. Source: Sayed Mohamed Ahmed, Creative Commons License.*

Several three-story houses constructed on loose sandy ground are depicted in Figure 4.2. The SPT-N value of the soil profile is less than 10, and it is categorized as SP (poorly graded sand) as well as SM (silty sand). The houses weren't safely founded for a few years after they were built (Park, 2013). During an earthquake, the shallow-founded houses' foundations suffered severe soil liquefaction. This, coupled with the lateral spread caused by the liquefaction-induced bearing capacity loss within the nearby retaining wall, resulted in the houses' permanent inclination. When there is seismic motion, loose sandy soils are more susceptible to soil liquefaction because of increased pore water pressure, which significantly lowers the soil's effective stress. We learned from it that, in order to consider building on soil liquefiable sites, laboratory testing of the soils' general physical properties should be done in addition to a more thorough site investigation to determine the groundwater depth and the SPT-N values of the soil. Building in seismic risk areas has proven to be essential for ensuring a reasonable level of safety against soil liquefaction (Crisp et al., 2020).

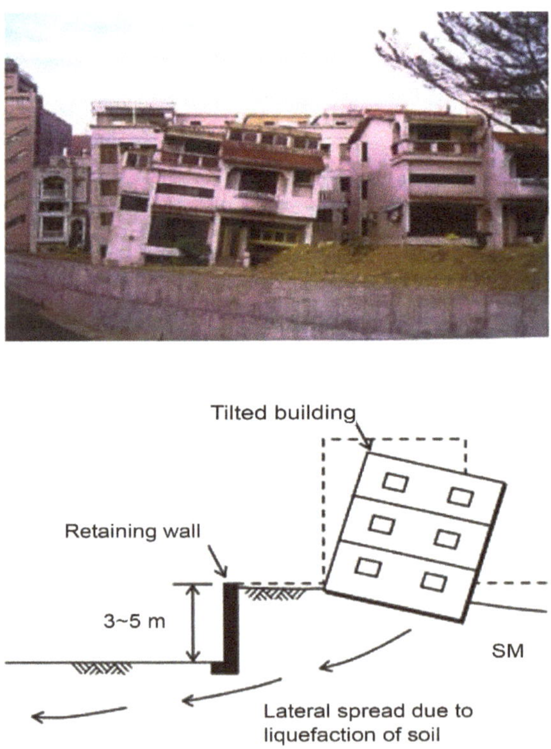

Figure 4.2. *Building tilt resulting from loose silty sand liquefaction (N <10).*

Source: Sayed Mohamed Ahmed, Creative Commons License.

A northern Taiwanese hillside community is the subject of Figure 4.3. Typhoon Winnie in August 1997 in Taiwan caused the anchored retaining wall to fail. The site is situated on a dip slope where shale and sandstone are interbedded. It has been noted that slope sliding happened as the downslope force increases due to rising groundwater exceeding the resistance force, which causes the anchored retaining wall to fail instantly (Purdy et al., 2022). In addition, the rapidly collapsing the dip slope rock mass to crash into the houses, causing damage, which killed 28 people, and tilting them. The fault with the anchored retaining wall was that it was not sufficiently understood how groundwater and geology work (Guan & Wang, 2020).

Slope stability can be greatly impacted by the type of discontinuity and the groundwater conditions. As a result, this failure demonstrated the necessity of thoroughly investigating the site to obtain precise data regarding the type of rock, discontinuity, groundwater level and strength before designing a retaining wall for artificial slopes. After that, a reliable stability study of the retaining wall and slope can be carried out to guarantee slope safety in the event of an earthquake or heavy rain, thereby averting significant landslides (Arnous, 2013).

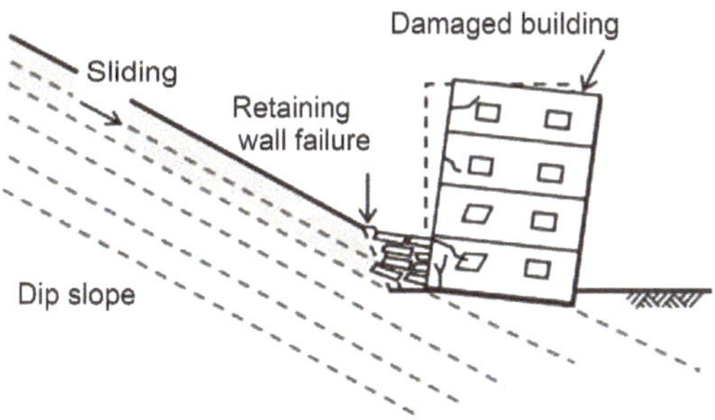

Figure 4.3. *Residential community on a hillside, Lincoln Community's dip slope slides.*

Source: Sayed Mohamed Ahmed, Creative Commons License.

4.2. PLAN FOR SUBSURFACE EXPLORATION

For each site, specific background data is needed for the building foundation design along with basement excavation. This information includes the groundwater and geological conditions, strength of the geotechnical material, and more (Furgale et al., 2010). Because every site has its own geological conditions and engineering project is different in terms of things like excavation depth, building height, and the condition of the house next door, every case presents a unique engineering problem that needs to be solved in its own subsurface exploration plan. For more information on this topic, readers can refer to the relevant reference (Singh & Chauhan, 2002).

4.2.1. Investigation Methods

A detailed summary of the methods used for field investigation is provided in Table 4.1. These methods include data collection, surface geology surveys, groundwater investigation, subsurface exploration (geophysical prospecting and soil boring), lab and in-situ tests, and finally monitoring. In subsurface exploration, soil boring, surface geological surveys, and data collection are typically carried out as part of a civil engineering project (Sklet, 2004).

Information about topography, geology, and aerial photos are all part of the data collection process. Subsurface geological surveys entail examining the topography, geology, and landforms of the surrounding areas. Both are employed in the preliminary development of a study area geological model for the purpose of identifying possible issues and organizing more thorough research (Mahmood, 1987). A list of often used investigation techniques in engineering practice is also provided in Table 4.1. Timelines for using the method depend on the cost, complexity, and scale of the building project. A thorough field study is conducted in the construction site, allowing accurate geological conditions to be recorded (Roy & Apparao, 1971).

4.2.1.1. Site Investigation Layout

Site investigation plan preparation is the first step towards assessing the investigation area. For instance, one of the main frequently used techniques when carrying out a construction project is soil boring (Tam et al., 2002). A number of things need to be taken into account:

4.2.1.2. Investigation Area

The whole construction site is often included in the investigation area. On the other hand, a larger area of the buildings may need to be inspected for possible construction damage if the area is surrounded by nearby structures. The investigation area in this instance will be expanded to encompass the whole significant construction area. The investigation should also take into account the kind of foundation used by the nearby buildings, the geological situation, and the material along with the purpose of the pipeline (Diez Roux, 2001).

4.2.1.3. The Number of Soil Borings

To get a full picture of the subsurface soil condition throughout the project, the quantity of soil borings needs to be ascertained. An investigation site typically needs two soil borings, whereas a slope area typically needs three. Based on practical experience, an extra soil boring can be deployed for every 600 square meters, with the overall amount of soil borings being modifiable for sites larger than 6,000 square meters. Notably, the location of the variable ground profile necessitates further appropriate soil borings (Broere et al., 2007).

4.2.1.4. Others

A more thorough site investigation is necessary due to the importance of vital buildings, such as those located in public areas, including the safety of these structures. Should the site's geology be straightforward, consisting of consistent rock and gravel layers, the investigation might only need to meet the bare minimum requirements. However, when the geological condition is complicated, a more thorough investigation should be carried out (Hooimeijer & Tummers, 2017).

Table 4.1. Engineering Types and Investigation Methods

Investigation method		*Surface survey*		*Geophysical survey*		*Borehole drilling and sampling test*						*Groundwater investigation*				*Field test*	*Laboratory test*
Engineering types		*Data collection*	*Reconnaissance*	*Seismic survey*	*Electrical and electromagnetic survey*	*Drilling and sampling*	*Standard penetration test*	*Cone penetration test*	*Vane shear test*	*Borehole televiewer*	*Lateral load test*	*Observation well and piezometer*	*Field permeability test*	*Pumping test*	*Water logging test*	*Plate loading test*	
Building	Shallow foundation	○	–	–	–	○	○	○	–	–	–	○	–	–	–	○	○
	Pile foundation	○	–	–	–	○	○	△	–	△	△	○	–	–	–	–	○
	Foundation excavation	○	–	–	–	○	○	△	○	–	–	○	○	○	–	–	○
Underground structure	Buried structure	○	○	–	–	○	○	△	△	–	△	△	–	△	–	–	○
	Large-scale underground structure	○	○	○	○	○	–	–	–	–	○	△	○	–	△	–	○
Road or railway	Cut	○	○	○	○	○	○	–	–	–	–	△	–	–	–	○	○
	Fill	○	○	△	–	○	○	○	–	–	–	△	–	–	–	○	○
	Tunnel	○	○	○	○	○	○	–	–	△	–	○	○	○	△	–	○
	Bridge	○	○	–	–	○	○	–	–	–	○	△	△	–	–	○	○
Dam	Dam foundation	○	○	○	○	○	△	–	–	–	○	△	–	–	○	–	○
	Dam	○	○	–	–	○	–	–	–	–	–	–	–	–	–	–	○
Slope		○	○	○	○	○	○	–	–	△	△	–	○	○	△	○	–

Legend: ○, often used; △, sometimes used; –, rarely

4.2.2. Depth of Subsurface Exploration

A subsurface depth determination is necessary for an investigation plan. The guiding concept for the decision is that the depth under investigation needs to be able to offer enough details to assess the overall structure's safety as well as that of the construction projects (Zacny et al., 2008). In engineering, the min boring depth aims to reach where a net increase in the vertical effective stress σ_o under a foundation, with depth as illustrated in Figure 4.4, appears to have limited influence, may be determined by the rules as follows:

i. Under a foundation with depth, find a net increase in the effective stress i.e., σ_o

ii. Find an estimate variation of the vertical effective stress given by $\sigma_{v'}$

iii. Confirm the depth i.e., $D = D_1$, where the increase in effective stress equals one-tenth of q while q is the foundations net stress (estimated).

iv. Determine the depth ($D = D_2$) where $(\sigma_o' / \sigma_v') = 0.05$

v. As indicated in Figure 4.4, find the depth ($D = D_3$), which is the separation between the bedrock and the foundation's lower face (Shepard, 1935).

vi. The minimum depth of boring should be found by picking the shallowest of the depths (D_1, D_2, D_3) determined before (Mukherjee et al., 2006).

vii. For the pile foundation, delve at least three meters below the bearing layer, as indicated in Figure 4.4.

4.3. BOREHOLE EXPLORATION

One of the most popular techniques for conducting geological investigations is borehole drilling, which is used for three main objectives: establishing the profile, collecting soil samples, and installing monitoring systems. An experienced geotechnical engineer can perform the soil profile determination in the investigated area with a few bored holes after soil sampling. The purpose of the soil boring work, which uses samplers, rods, and boring machines, is to investigate various geological materials, including rocks and soils. On plains sites, the wash boring method in conjunction with thin-walled or split tube method are the soil boring techniques that are frequently used (Daniels & Dyck, 1984). The soils are sampled on slope sites using the rotary method. Furthermore, the wash boring method—which is covered in the following sections—is frequently used in conjunction with SPT field test method, which uses a split-spoon sampler. In order to provide readers with a better understanding of the purpose of the penetration methods and samplers, this chapter first defines borehole exploration followed by samplers in later sections. Readers can also consult the references (Hvorslev, 1949).

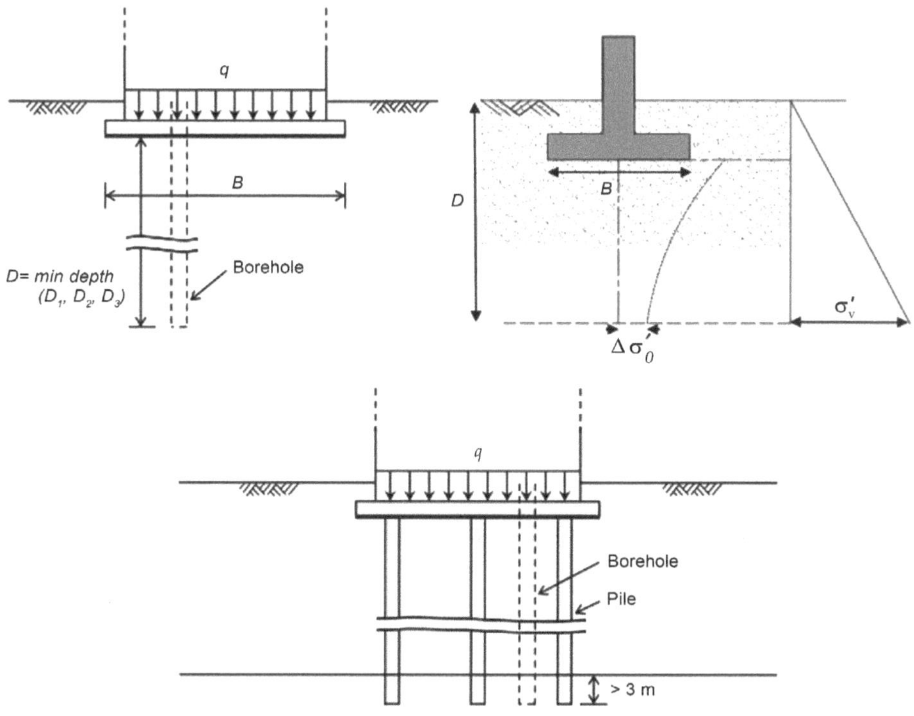

Figure 4.4. *Initial assessment of the depth of borehole.*

4.3.1. Wash Boring Method

When using the wash boring technique, a fishtail drill bit (Figure 4.4) is typically used to penetrate soil by washing and boring it down to the planned sampling borehole depth. The soil is then sampled for various purposes that will be covered in the upcoming sections using a thin-walled tube or a split spoon sampler. Figure 4.5 depicts a schematic diagram of the wash boring process (Selamat et al., 1892). It uses a sharp chiseled hollow drilled rod, and water is pressurized inside the drill rod and it is then controlled, rotated, raised and dropped from the ground. The drill rod is circulated with the soil water slurry driven to the surface. When applying sampling techniques, a horizontal water jet is typically utilized to prevent direct downward washing which could disturb the sample. Typically, the sampling interval remains one or two meters, but it shouldn't exceed two meters. Denser intervals are required for soil profiles that undergo notable changes (Starr & Ingleton, 1992).

4.3.2. Rotary Drilling Method

The process of rotary drilling involves using a sharp rotating drill bit that cuts or crushes rock layers or soil at a depth that varies depending on the hardness of the rocks, the soil formation, and the downward pressure being applied.

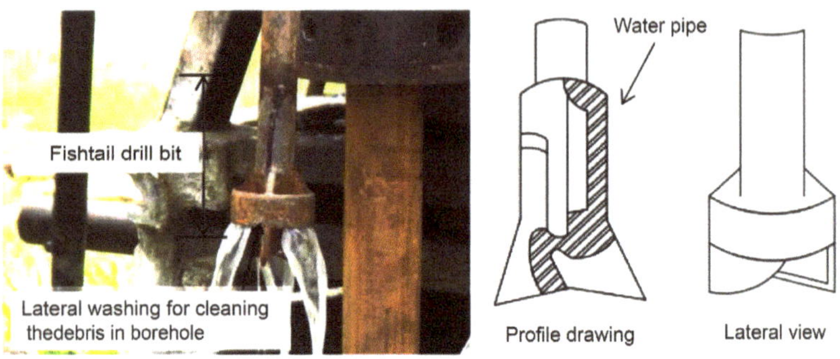

Figure 4.5. *Fishtail drill bit.*

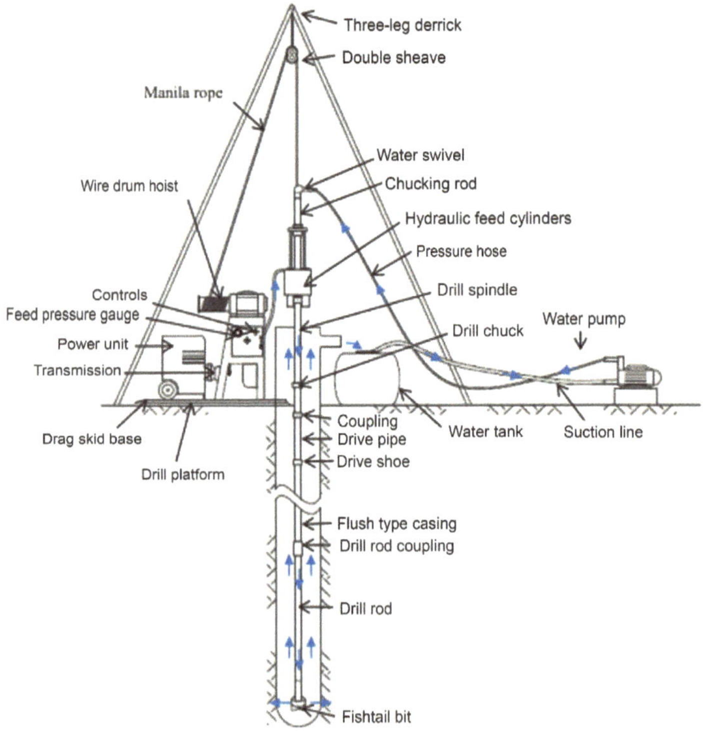

Figure 4.6. *Wash boring operation schematic.*

When rotating and penetrating, the drill bit's face discharge, as seen in Figure, must be kept sharp (Wijk, 1991). Since it becomes one of the main expenses of this work, the contractors and employees must exercise experience and wisdom in selecting the drill bit type that's compatible with the corresponding formation. Additionally, a fluid circulation is typically applied during this drilling work to keep the drill bit cool and lubricated and to keep the hole free of cutting and reclamation. Figure provides a schematic image of a rotary drilling technique (Sapińska et al., 2015).

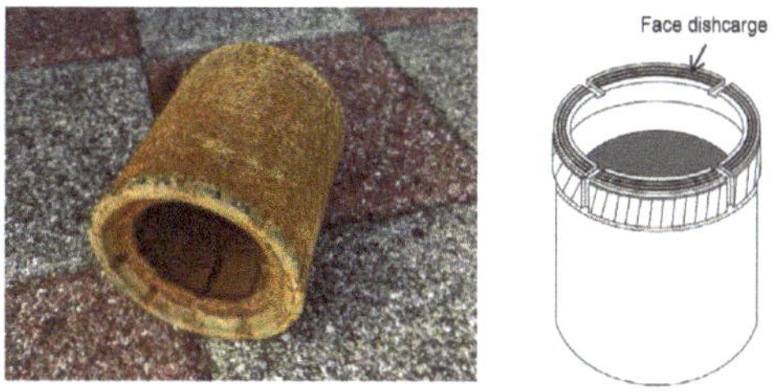

Figure 4.7. *Drill bit with diamond bits.*

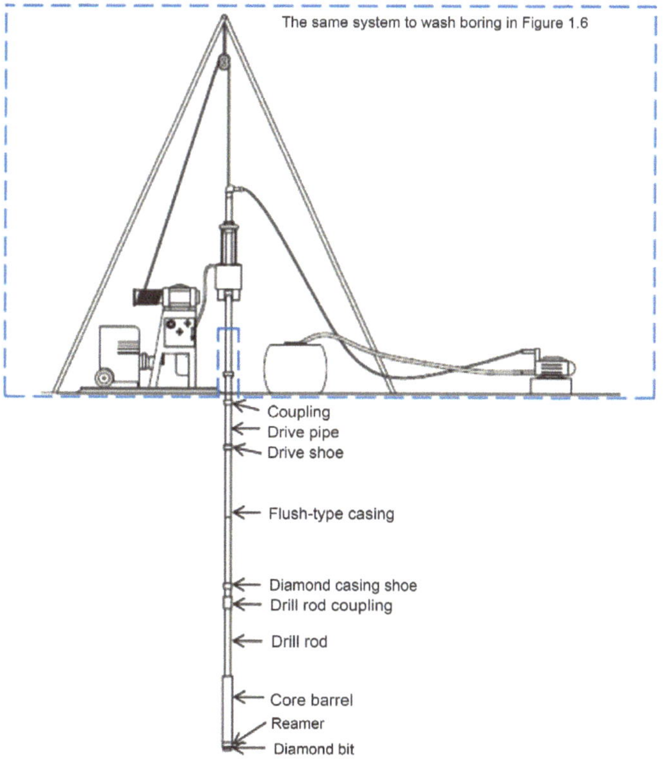

Figure 4.8. *An example of a wireline coring rig.*

4.4. SOIL SAMPLING METHODS

The following sampling techniques are needed to obtain soils in the ground for various design parameters. In general, an undisturbed soil sample is preferred for the soil strength and compressibility and strength properties; however, a disturbed soil sample is more frequently used for physical soil properties because it is less expensive. Three different types of sampling methods and their disturbance in the soils are described here (Cline, 1944).

4.4.1. Thin-Walled Tube Sampler

For testing the soil's consolidation behavior, compressibility, and strength, relatively undisturbed cohesive soil samples are typically obtained using the thin-walled tube sampler, which is primarily composed of stainless steel.

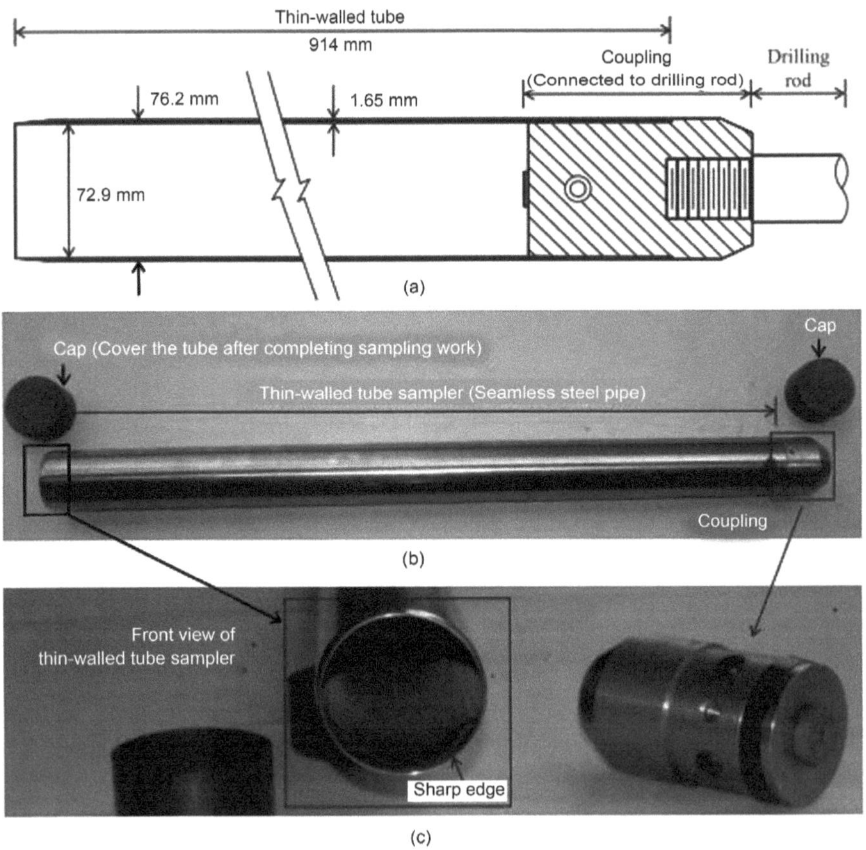

Figure 4.9. *A thin-walled tube sampler; (a) an illustration; (b) a picture; (c) a close-up shot.*

Figure 4.9 shows the sampler that has a sharpened opening that is commonly used (Bray et al., 2016). It has an outside diameter of 76.2 mm and an inside diameter of 72.9 mm, and it is usually between 700 mm and 900 mm long. To preserve the soils and the samplers' regular shape, the sampler ought to be gently pushed. Caps serve to seal the tubes and wax is used to seal moisture. When delivering the thin-walled tube to the lab, it should be vibratory motion-limited and kept in a freezer until testing is completed. Here, ASTM D1587/D1587M-15 (2015) should be used as a guide (Nia & Hamedani, 2010).

4.4.2. Split-Spoon Sampler

The thick-walled tube driven into soils with hammer drops is frequently used to sample disturbed soils for generic property tests since thin-walled tube samplers have trouble penetrating hard soils, like cemented, sandy and gravelly soils (Landon et al., 2007).

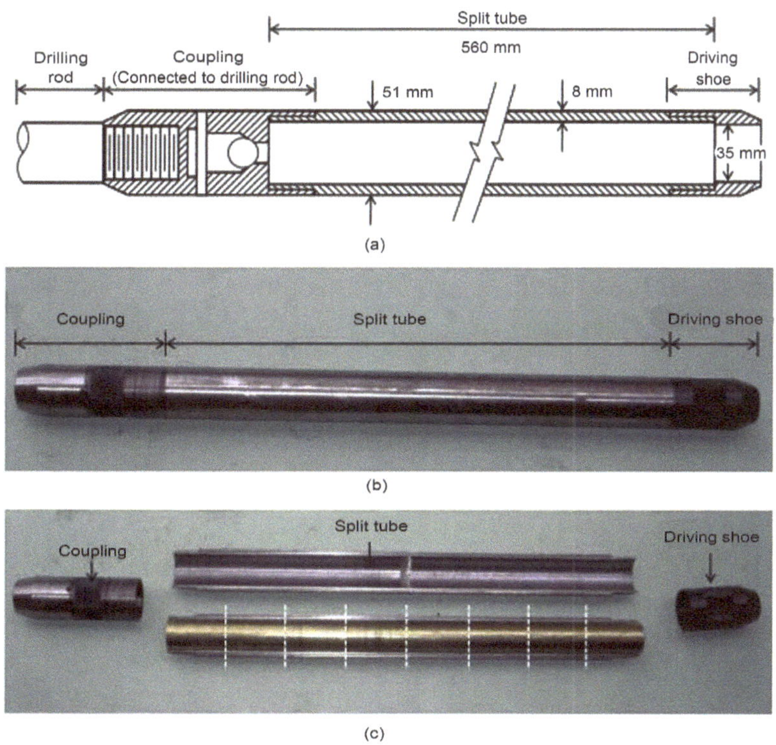

Figure 4.10. *A split-spoon sampler: an illustration (a); an assembled image (b); and, a disassembled parts (c).*

The typical sampler for the SPT is a 15 cm long, thick-walled, open-drive split barrel sampler with an inside diameter of 35 mm (Figure 4.10). It is made up of a coupling part that is attached to a drill rod, a driving shoe, and a split tube measuring 560 mm. The split tube contains eight small tubes, of which a sample is sent to the lab for examination, testing, and observation. The source (ASTM D3550–01, 2001) can be used as a guide (Mishra & Robinson, 2018).

4.4.3. Piston Sampler

A piston sampler, which is made up of a thin-walled tube with a piston connected to the length of rod that passes through head of the sampler and inside the hollow-boring rod, can be employed to sample very sensitive soils or soils with a high fines content. A piston at the tube's lower end is used to lower the sampler into the borehole during the sampling procedure (Wright, 1967). A clamping device located at the very top of the rods locks the piston and tube together, as illustrated in Figure 4.11. Because of this, the piston can both keep soil in the tube when it is being removed with a vacuum and stop loose soil or water from getting inside the tube during sampling. Although the tube dimensions of the various piston samplers vary, the basic principles of operation remain the same (Brown, 1956). As a result, the instrument dimensions are missing from Figure 4.11.

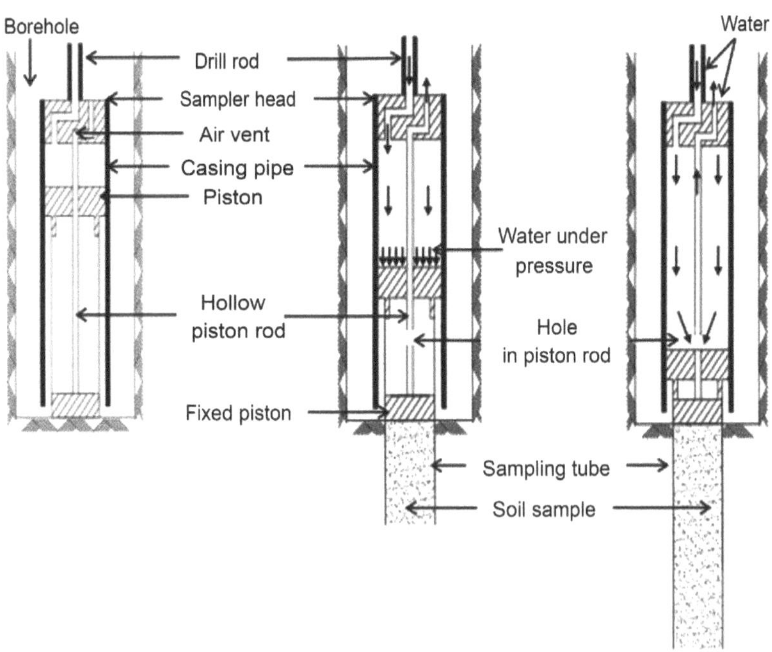

Figure 4.11. *Sampling process by piston sampler.*

4.4.4. Core Sampler

Soft rock, cementation containing sands, silts, plus stiff to hard clays, are typically sampled with a core sampler. Loose, cohesionless and loose soils; and gravelly soils are not good candidates for the sampler. As seen in Figure 4.12, triple-tube core barrels are the commonly used rotary samplers with a detachable thin-walled tube, called a liner, inside an outer tube equipped with a cutting bit (Mackereth, 1958).

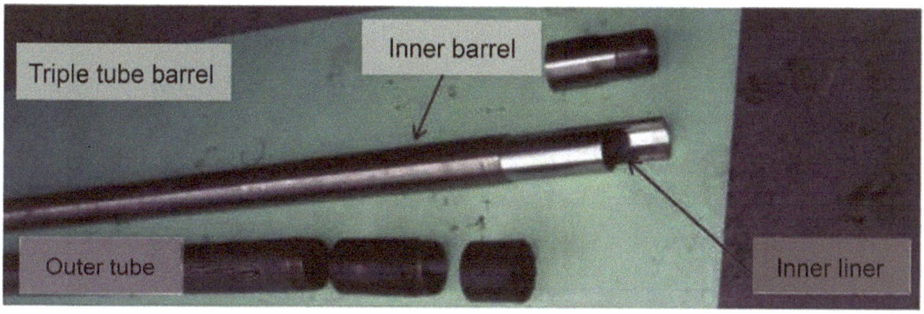

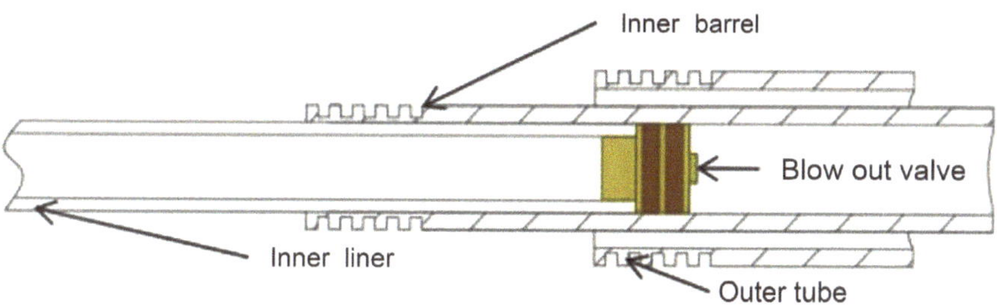

Figure 4.12. *Schematic diagram of triple-tube core barrel.*

The rotary sampler is 61 cm long and has two diameters: 63.5 mm for HQ, which is commonly used for slopes; and 50.8 mm for NX, which is commonly used for slopes and plains. After rotating and pushing the outer tube containing the cutting bit into the soil to the necessary depth, the sample fills the liner. After the outer rotating tube cuts the sample, it slides into the liner, which is the inner tube, or the liner, which is equipped with a smooth cutting shoe. Thus, the liner is used to collect the rock core sample (Jamison et al., 1950). For additional geological research, the rock core is set aside in the core box marked with the sample depth, date and the project name as indicated in Table 4.2.

Table 4.2. Common Borehole Sizes

Core barrel type	Hole diameter	Core diameter	Core sample (Sample preservation with labeled core box)
NX	75.8 mm	50.8 mm	Upper one 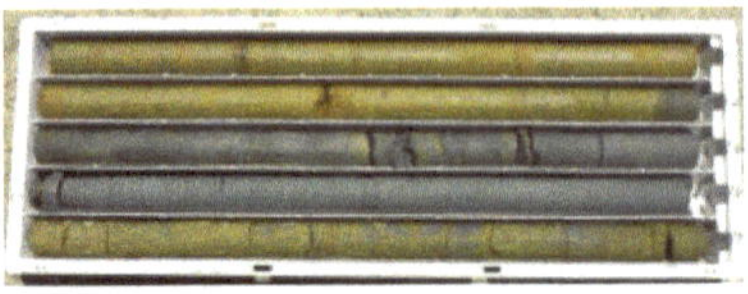
HQ	96.3 mm	63.5 mm	Lower one 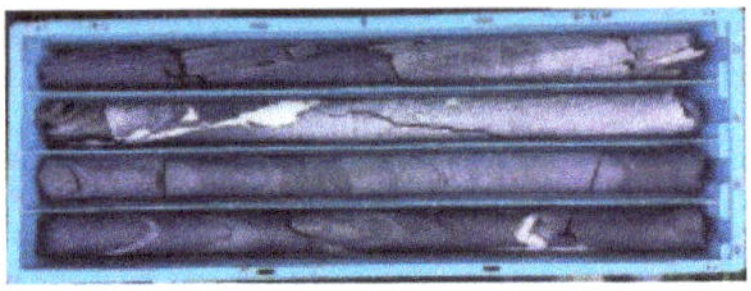

Following boring and sampling, the process known as "rock quality designation" (RQD) is typically carried out. RQD employs the core sample to assess the local geological status and determine the rock core quality extracted from a borehole. RQD is a percentage that represents the degree of fracture or joining in a rock mass (Pratje, 1950). As Table 4.3 illustrates, rocks with an RQD of 75% or higher are considered hard and high-quality, while rocks with an RQD of less than 50% are considered weathered and low-quality. It's important to note that RQD is the percentage of whole drill core pieces found during a full core run are at least 10 cm (Gillespie et al., 1985).

Table 4.3. RQD Classification Index

RQD (%)	Rock mass quality
<25%	Very poor
25–50%	Poor
50–75%	Fair
75–90%	Good
90–100%	Excellent

4.5. SAMPLING DISTURBANCE

It is inevitable that the soil samples will be disturbed when inserting the samplers described before into the ground to remove soil. Inside wall friction is one of the main reasons for sample disturbance while employing the thin-walled tube method (Hvorslev, 1949). One way to get eliminate or mitigate soil-sampler wall friction is to make the cutting edge's diameter a little smaller than the sampler tube's inside diameter (Clayton & Siddique, 1999). This gives the cutting edge some inside clearance. Here is one way to express the area ratio:

$$A_r = \frac{d_0^2 - d_i^2}{d_i^2} \tag{1.1}$$

Where d_0 = outside diameter of the sampler tube.

d_i = inside diameter of the sampler tube.
A_r = area ratio (if A_r is smaller than or equal to 10%, the samples are considered undisturbed).

4.6. LABORATORY TEST

To investigate the physical and mechanical characteristics of soils in the field, a laboratory test is necessary. Only then can engineers accurately estimate the subsurface soil's engineering performance. As a result, practitioners ought to utilize laboratory testing wisely (Lau & Chandler, 2004). The lab test results and their relevance are compiled in Table 4.5. The relevant specifications, which are outside the purview of this book, should be followed by test procedures that are critical to guaranteeing test quality. For additional reading, the following source can be consulted (Casagrande & Fadum, 1940).

Laboratory tests	*Soil property*
General physical property tests	
Water (moisture) content test	ω, γ_{sat}, γ_d
Specific gravity test	G_s
Atterberg limits test	*LL, PL, PI, SL* Soil classification
Sieves and hydrometer analysis	Soil particle size distribution curve Soil classification
Relative density test	γ_{dmax}, γ_{dmin}, D_r
Permeability tests	
Constant head permeability test	*k* (for coarse-grained soil)
Falling head permeability test	*k* (for fine-grained soil)
Triaxial permeability test	*k* (for fine-grained soil)
Compressibility tests	
Oedometer test	σ'_c, *OCR*, C_c, C_{ur}
Shear strength tests	
Direct shear test	c', ϕ'
Triaxial compression test	
Triaxial consolidated drained (CD)	c', ϕ'
Triaxial consolidated undrained (CU)	c', ϕ' (with pore water pressure measurements)
Triaxial unconsolidated undrained (UU)	s_u
Triaxial unconfined compression test (UC)	q_u, s_u

4.7. FIELD TEST

When conducting in situ tests during soil boring exploration, they are referred to as field tests. It consists of the VST, SPT, CPT, flat dilatometer test, pressuremeter test, and plate loading test. While CPT is able to continuously measure soil properties throughout penetration, SPT is able to remove samples for lab testing. One of the inherent benefits of in situ tests is their ability to prevent disturbances to soil sampling (Kraeuter et al., 2019). Careful sampling techniques can yield undisturbed soil samples, as discussed before; however, sample transport and lab test procedures require close attention. Numerous important details about soil behaviors may be obtained with lab tests, including soil strength both drained and undrained, the path of stress, and the reactions of the soils to stress and strain. Therefore, in order to investigate the exact site geological information, both lab and field testing are necessary (Stanford, 2007).

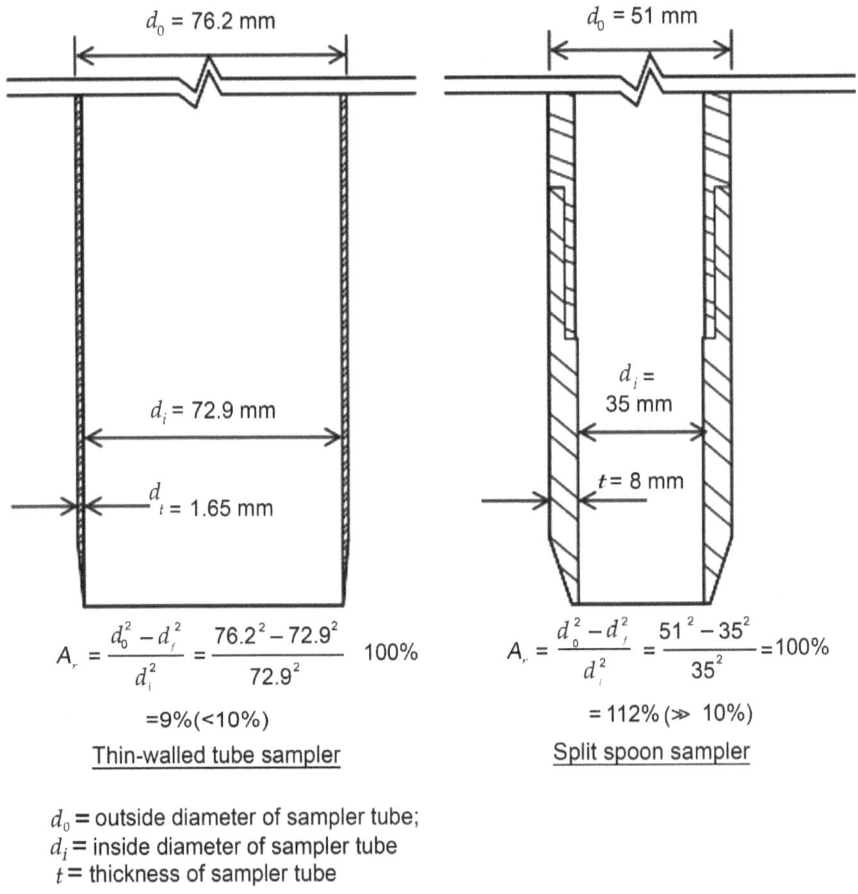

Figure 4.13. *Area ratio (A_r) calculation for thin-walled and split spoon samplers.*

While a field test can quantify the characteristics of in situ soils, it typically cannot directly determine the parameters (Yu et al., 2001) of the soil that are needed for a foundation designing foundations, like Young's modulus, friction angle, undrained shear strength, and so forth. Some empirical formulas or transformation equations can help calculate these essential factors (Kallio & Kaikkonen, 2005).

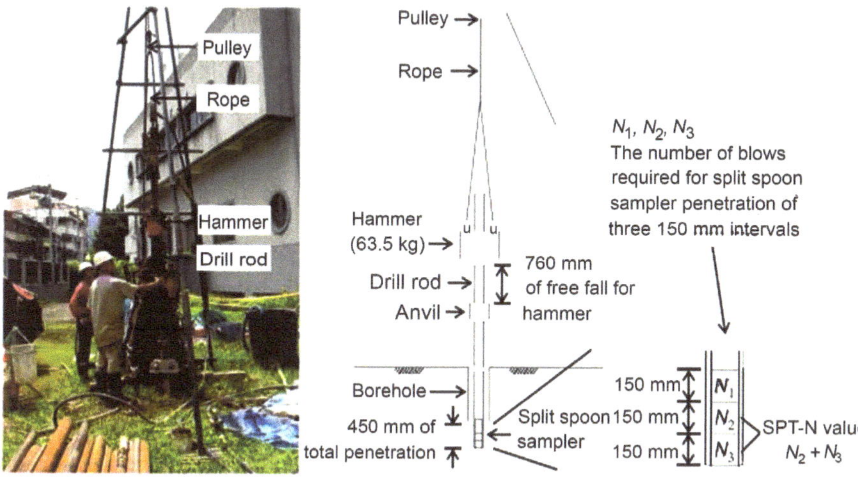

Figure 4.14. *The operation of SPT.*

4.7.1. Standard Penetration Test (SPT)

As illustrated in Figure 4.14, the SPT apparatus consists of a split tube equipped with the driving head and a solid cone point or a heavy-duty cutting shoe. The tube's head is threaded so that it can be connected to a hammer using a number of drill rods. The instrument is driven in the soil at the borehole's base using a trip hammer that weighs 63.5 kg and descends freely to 760 mm distance. The borehole should be plumb since the hammer and rods and hammer would be vertical (Fletcher, 1965).

The amount of blows needed to push it through the first 150 mm and the second 300 mm test zones is counted during the test (Al-Jabban, 2013). The first 150 mm are regarded as distributed material. It is becoming common practice to record the test zone's blow count, and it is highly advised. At the base of the borehole, a 63.5 kg free-falling trip hammer drops 760 mm and drives the device into the ground. The soil samples for the general physical property soil test are pushed into the split tube shown in Figure 4.10 during the penetration (Yusof Zabidi, 2018). A general relationship between the SPT-N value and the relative density of sand or the strength of clay is shown in Tables 4.4 and 4.5.

Table 4.4. Relationship of the Consistency of Clay, SPT-N Value, and Unconfined Compressive Strength

SPT-N value (blows)	Consistency	Unconfined compressive strength q_u (kPa)
<2	Very soft	<25
2~4	Soft	25~50
4~8	Medium	50~100
8~15	Stiff	100~200
15~30	Very stiff	200~400
>30	Hard	>400

Table 4.5. Relative Density of Sands as Per SPT-N Results

SPT-N value (blows)	Relative density
0~4	Very loose
4~10	Loose
10~30	Medium
30~50	Dense
>50	Very dense

The drop hammer's energy efficiency on the drop rods differs based on the region because of differences in local practices and equipment. Furthermore, there is a significant impact of the overburden stress on the SPT-N value. The obtained SPT-N values need to be adjusted using transformation and correction equations (Thorburn, 1986).

4.7.2. Cone Penetration Test (CPT)

In engineering practice, another often utilized field investigation technique is CPT. Similar to the SPT, soil testing is conducted with a steel cone that penetrates the ground. Data on pore pressure, side friction, and tip resistance, are collected during steady penetration to profile continuous soil layers along the depth and estimate the strength, pile bearing capacity, and overall classification of the soil. As illustrated in Figure 4.15, the steel cone configuration in the CPT is now standardized (David, 2018). It has an apex angle of 60° and a 10 cm^2 cross-sectional area of cone bottom (Robertson, 2016).

The CPT work procedures are shown in Figure 4.16, where the weight of the truck is used to produce a response. The majority of commercially accessible CPT rigs use electronic piezocone and friction cone penetrometers, the testing protocols for which are briefly laid out in this book. These devices continuously log and print out lithologic interpretations of every 2 cm interval, as well as computerized logs of tip and sleeve resistance, their ratio, induced pore pressure immediately behind the cone tip, and pore pressure ratio i.e., ratio of pore pressure to measured pressure at a 2 cm/s penetration speed) (Jefferies & Davies, 1991).

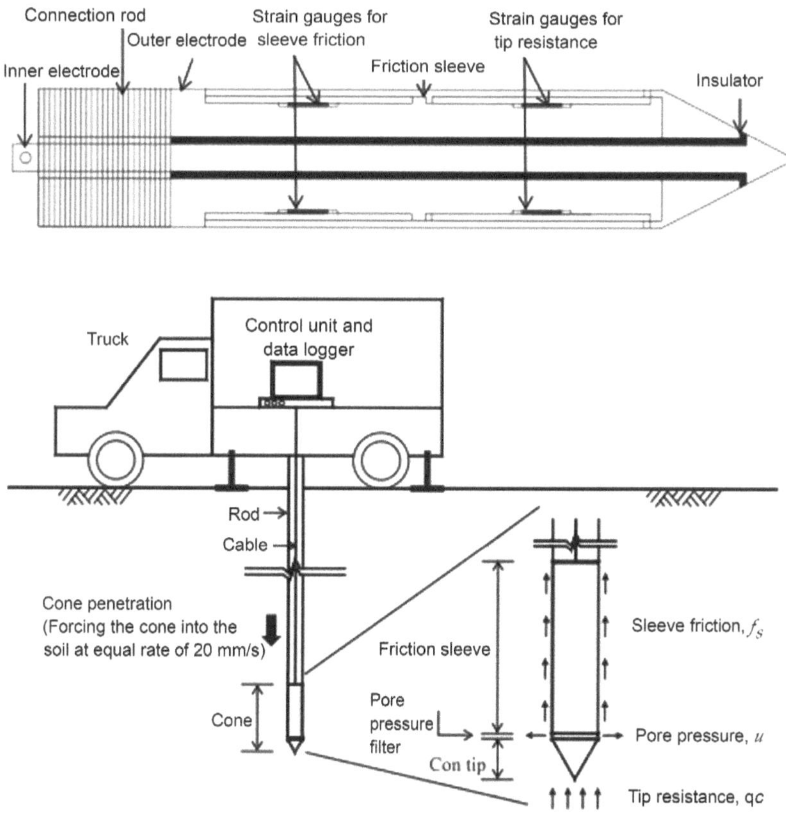

Figure 4.15. *Illustration and working of the CPT.*

A simple rule of thumb for classifying soil types is as follows: sands are defined as having qc > 5 MPa and u2 ≈ uo, while intact clays are more common if u2 > uo (hydro) and qc < 5 MPa. The pore water pressure magnitude can be used to identify intact clays, including hard clay (u2 ≈ 20·uo), stiff clay(u2 ≈ 8·u0), firm clay (u2 ≈ 4·u0), and soft clay (u2 ≈ 2·u0). Overconsolidated, fissured clays typically have negative u2 values, meaning that u2 is less than zero. The following is a list of all the symbols' meanings:

Basically, tip resistance (qc) is the ratio of the penetration force.

Sleeve friction (fs) is measured by the embedding of tension load cells as the probe advances through the soil.

The hydrostatic pressure is given by u0. Pore pressure (u2) is measured throughout the penetration (Niazi & Mayne, 2013). The main benefit of the CPT over the SPT is that it can collect continuous data on soils (refer to Figure 4.17); the drawback is that it cannot sample soils. Due to two primary factors, the SPT method is still mostly used in engineering procedures in many countries: the labor cost for SPT is typically less expensive compared to CPT trucks, and SPT data are more widely available. High pore pressure and low tip resistance ($u2 > uo$ and $qc < 5$ MPa) show clayey soils at depths of 2–6 m and 10–17 m in Figure 4.17. On the contrary, low pore pressure and high tip resistance ($u2 \approx uo$ and $qc > 5$ MPa) show sandy soils or sand mixtures at depths of 8–10 m (Xu et al., 2008).

(a)

(b)

Figure 4.16. *The CPT in action. Figure shows positioning the CPT truck (a); and pressing the cone into the ground at a 20 mm/s constant pace (b).*

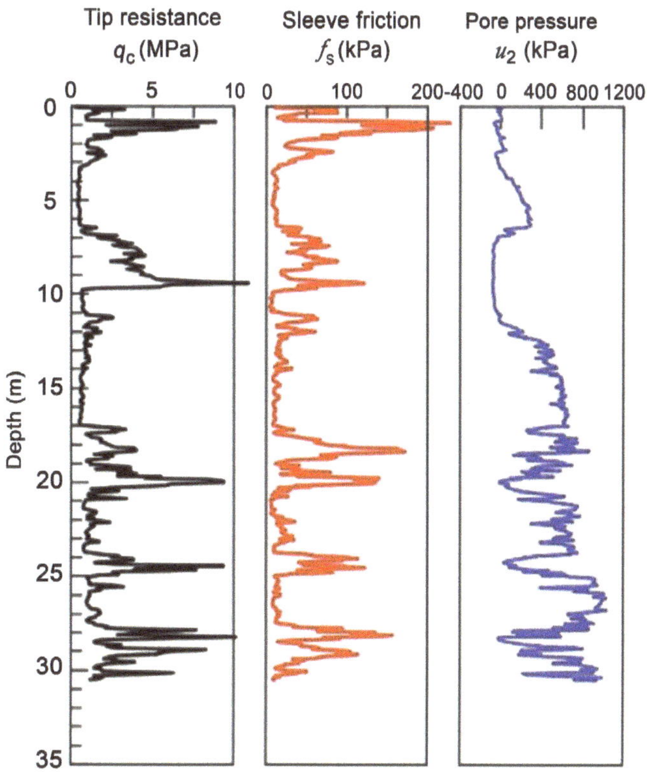

Figure 4.17. *Common CPT results.*

4.8. GROUNDWATER INVESTIGATION

Groundwater is a crucial component of field investigations since it is the primary cause of disasters in geotechnical engineering and it definitely affects the stability of construction. Groundwater seepage is complex because of geo-structural inhomogeneity, discontinuities, or fissures. Additionally, the properties of several aquifers discussed in detail before make field investigation challenging. To gain a deeper comprehension of the groundwater variation profile, integrated investigation techniques are required. This section provides an overview of commonly used techniques, such as in situ permeability testing, water pressure monitoring, and groundwater table observation (Walsh et al., 2013).

4.8.1. Groundwater Table and Pressure Measurement

The two techniques that are commonly used in engineering practice are introduced in the following paragraphs: piezometers to measure the water pressure in aquifer layers or sand at various depths and observation wells to monitor the groundwater table (Kalbus et al., 2006).

In completed boreholes, observation wells are typically built with a 50 mm diameter PVC opening with holes, screened with a filter pack or geotextile, and the groundwater table depth is measured in the PVC as can be seen in Figure. 4.18. Observation wells are typically used in a single aquifer because the groundwater table measured in the PVC of these wells usually reflects a mixed groundwater table across its depth of penetration (Xiao et al., 2020).

When installing an open standpipe piezometer, it is common practice to place a solid casing at the depth of interest plus a screened or slotted casing in the zone being measured for water pressure. As illustrated in Figure 4.19, both above and below the depth of interest, clay or bentonite is utilized to seal the water into the borehole and obstruct the supply of groundwater at other depths (Loheide et al., 2005).

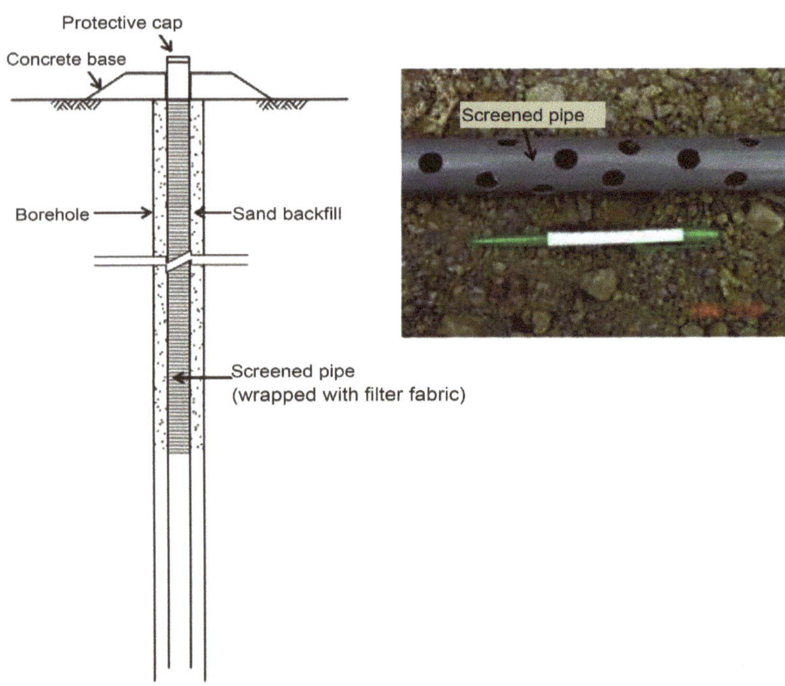

Figure 4.18. *Diagram and photograph of the pipe (which measures the water level in a single aquifer) and the water observation well.*

These days, data can be collected manually or automatically for monitoring the groundwater table in the observation well and the groundwater pressure in the piezometer. Pressure gages, also known as transducers, can operate as strain gages, vibrating wire, pneumatic, or other types of transducers that convert pressure into a voltage signal (Ho et al., 2021). These piezometers are connected to the surface so that portable readout units or data loggers can read them more quickly or frequently. This ability could be very important to a warning system. Crucially, though, in order to verify the automatic instrument's accuracy, a manual reading of the groundwater table is required (Motojima et al., 1981).

Piezometer wells are illustrated and a piezometer image is shown in Figure 4.19. Water from other soil layers was sealed off due to the bentonite seals in the borehole, as seen in the figure. One of bentonites' unique properties is that when mixed with water, they swell to form a gel-like semi-solid mass that can fill gaps and stop water from moving, creating barriers between water and soil (Niwa et al., 2012).

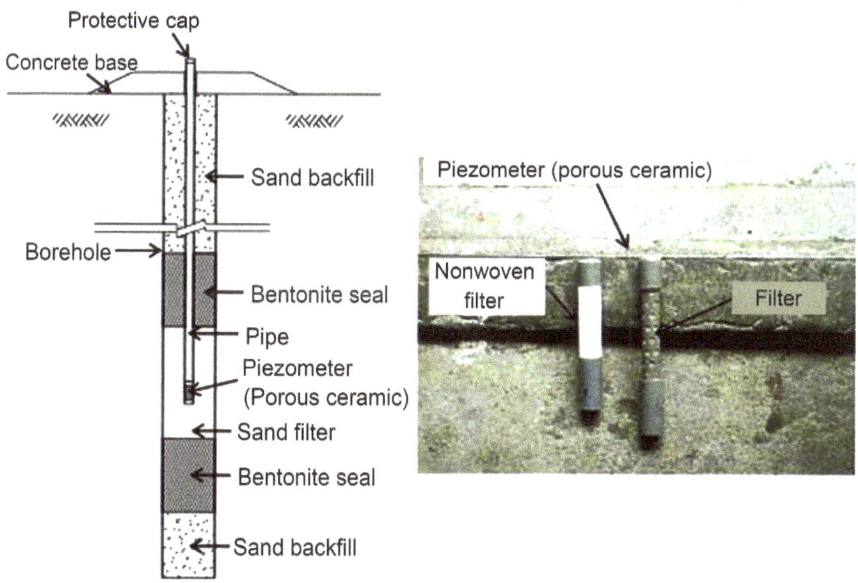

Figure 4.19. *Piezometer well and piezometer (used to measure the water level in any aquifer) illustration and photo.*

4.8.2. Field Permeability Test

In principle, undisturbed soil specimens sampled from the field can be used to measure the soil permeability, also known as hydraulic conductivity, of the soils using a variety of techniques, including constant and falling head laboratory tests; in practice, however, triaxial tests are used to indirectly determine the permeability of fine-grained soils in the lab testing (Elhakim, 2016). As an alternative, pumping tests, field falling or constant head tests can be used to measure permeability in the field. In engineering practice, the permeability can also be empirically determined for preliminary design purposes using the grain size distribution. Planning a construction site requires a solid understanding of soil properties in order to estimate seepage and design drainage systems (Day et al., 1985).

4.8.2.1. Permeameter Test

As seen in Figure 4.20, the permeameter test determines the permeability of soil at a borehole's bottom. In order to produce an unbalanced pressure head between the water inside the PVC and the soils at the bottom of the borehole—a constant head for sand, as shown in Figure 4.20(a), and a falling head for silt and clay, as shown in Figure 4.20(b)—the test temporarily fills the PVC with water in the borehole. Field permeability will be calculated using the relationship between filling water level and time in PVC (Jain et al., 2006). The in-field test's permeability equations of the falling and constant heads are comparable to laboratory tests found in soil mechanics textbooks, but they account for horizontal seepage. In Figure 4.20, hc is the constant head, h0 is the head at time t0, and h1 is the head at time t1.

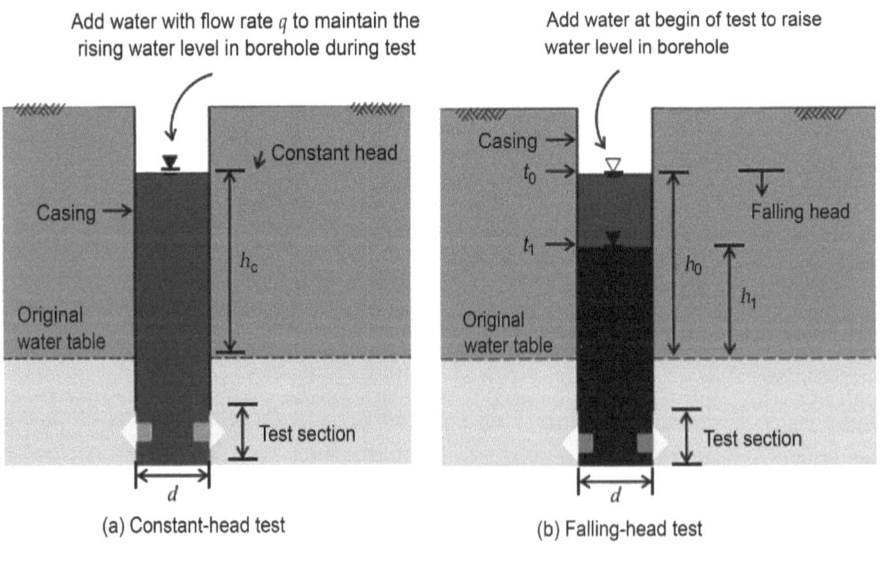

Figure 4.20. *Tests for borehole permeability illustrated: (a) constant; and (b) falling head test.*

Figure 4.20 further demonstrates that "small-scale" permeability values in the field are obtained from tests conducted in boreholes that only involve a small volume of rock or soil surrounding the test section. Such tests might not be indicative of the strata's mass permeability if the soil/rock is heterogeneous or has a substantial amount of fabric. Extensive experiments (like pumping tests) might yield superior outcomes. For further reading, the following reference is useful (Daniel et al., 1984).

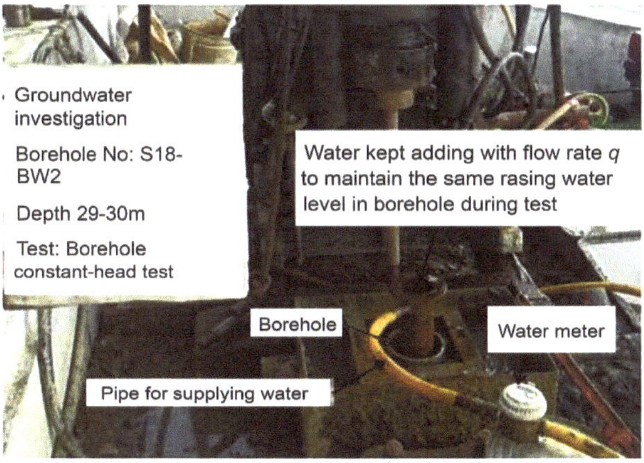

Figure 4.21. *Borehole constant head test in operation.*

4.8.2.2. Pumping Test

A large scale pumping test has both pumping and observation wells. Usually, a pumping well with at least 30 cm diameter and an observation well of at least 15 cm diameter are put in. The pumping test in Figure 4.22 works on the basis that when water is pumped from the well and the discharge (q) is noted, the drawdown and in-observation wells (h1 and h2) are measured, and the calculated pumping discharge and drawdown curve in the textbook of soil mechanics can be used to estimate the hydraulic conductivity (k) and other parameters (Butler & Healey, 1998).

This pumping test typically has an observation period of more than 24 hours. A workability assessment of subterranean drainage projects, such as catchment wells, vertical pumping wells, and horizontal drains, can be conducted using the results of the large-scale pumping test. Care must be taken that excessive pumping at the location could lead to neighboring buildings to settle and tilt. This can be prevented, though, if the information in this section is understood. In this instance, a pumping test limited to an unconfined layer is presented. People may find the test in the confined layer intriguing in relation to the confined aquifers' depressurization (MacDonald et al., 2008).

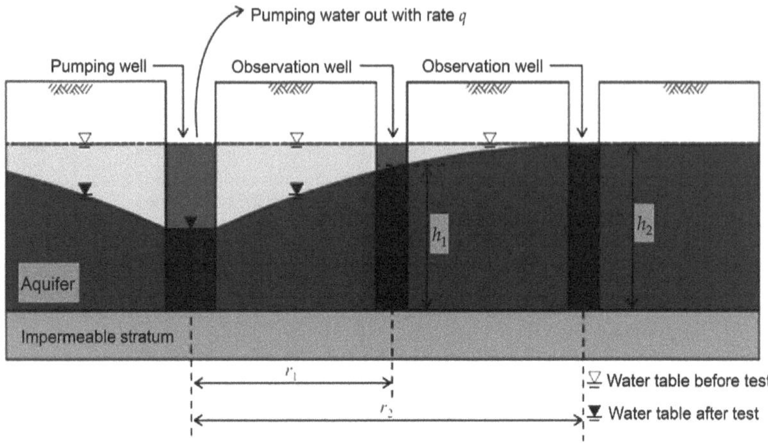

Figure 4.22. *Pumping test illustration for unconfined problem.*

4.9. GEOPHYSICAL SURVEY

In order to ascertain the geological, structural, physical, and mechanical properties of the foundation soil, geophysical surveys—which include airborne, marine, and land geophysical surveys as well as borehole surveys—implement geophysical methods, such as electromagnetic and electric, gravity, and seismic surveys; and radioactivity, and magnetic methods (Jordan, 2009). These techniques are employed to investigate the composition, characteristics, and thickness of individual layers beneath the surface of the terrain where a particular structure is to be built. The advantages of nondestructive testing techniques include their speed, reduced expense, continuous data logging, and ability to explore the site more thoroughly (Maurer et al., 2010). The techniques for seismic surveying and electromagnetic and electrical surveys will be covered in this section.

4.9.1. Electrical and Electromagnetic Survey

Based on conductivity differences of different minerals, cements, water contents, porosities, etc., of geological materials, electrical resistivity, which gages the layers' resistivity profile, has been widely used in surveying for bedrock features, buried objects, groundwater, and faults (Auken et al., 2006).

By using 2-dimensional electrical resistivity tomography (ERT), geological structures and groundwater distributions can be mapped based on material resistivity. The purpose of electrode array configurations is to collect data necessary for estimating ground resistivity variation in values, both vertically and laterally. Figure 4.23 displays an example of the outcomes of using 2D ERT. The detection depth range is about 20 m, and the ground resistivity detection line is about 50 m long. The image profile data for ground resistivity (in ohm-m) is shown to get the range of apparent resistivity (see Figure 4.24) (Xue et al., 2019). Two layers can be roughly distinguished in the survey area after interpreting the profile of ground resistivity. Layer 1: Approximately three meters below the surface of the earth, featuring a resistivity distribution that is roughly 10~80 ohm-m. Higher resistivity is the result of larger particles or a greater void ratio in the subsurface material (Clay, 2001). Layer 2: 3–20 meters below the surface of the earth, typically exhibiting a 5–12 ohm-m electrical resistivity, which is relatively low. Lower resistivity is caused by either a higher water content or finer-grained subsurface material.

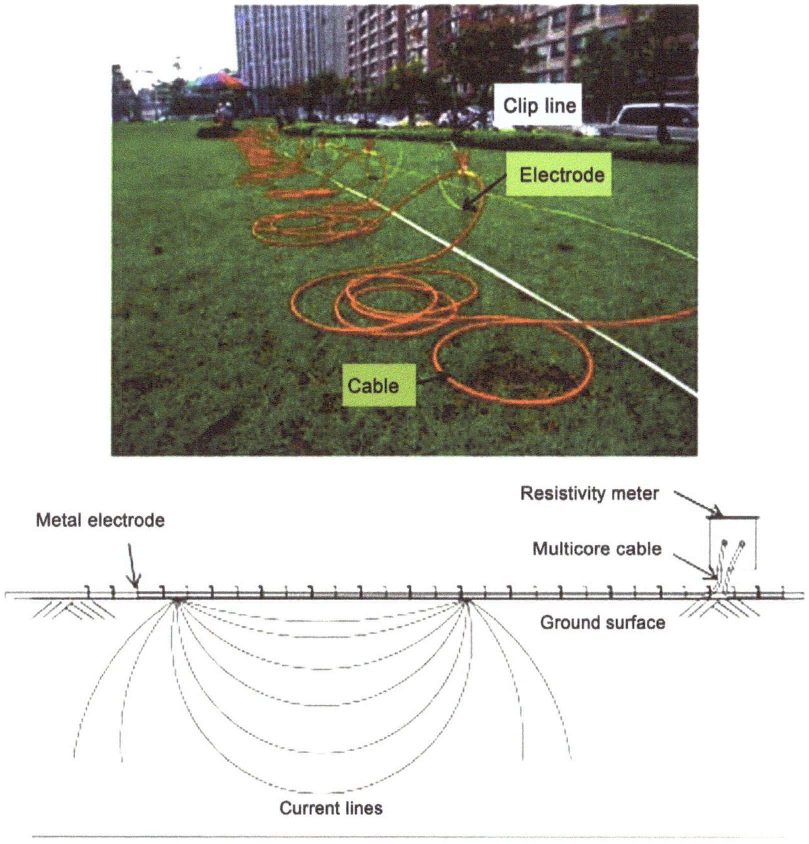

Figure 4.23. *Survey operation for electrical and electromagnetic fields.*

The distribution of soil layers can be accessed using a variety of survey data via ground resistivity detection. Crucially, since this nondestructive testing method determines electrical properties indirectly, the resistivity responses are still dependent on the characteristics of the stratum, including its composition, size, salinity, and water content (Ingeman & Baumgartner, 2006). To establish a connection between the physical properties and resistivity of those formation materials, core samples must be bored. When the ERT is calibrated against various direct geological investigation techniques, like the SPT and CPT, the investigation results become more accurate. Lastly, because droughts and rainfalls can have a significant impact on the resistivities of shallow layers, the rainfall records should be involved in the ERT projects (Godio & Bottino, 2001).

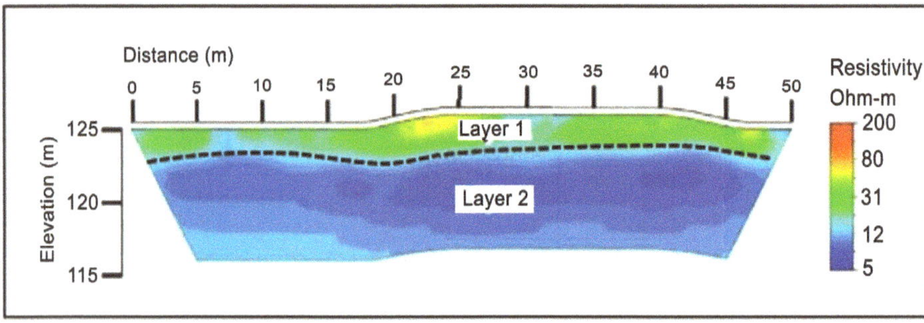

Figure 4.24. *An illustration of a 2D ERT survey outcome.*

4.9.2. Seismic Survey

Seismic surveys can be broadly classified as either refraction or reflection, based on the particular wave type that is being used. An impact source, like a sledgehammer, is used to produce active seismic waves (Nowacek et al., 2015). Different geological media have different porosities, fractures, densities and other characteristics that affect how the generated waves are transmitted. It's interesting to look at the reflection and refraction data collected by a group of sensors or geophones (McCauley et al., 2000). Following a back-analysis of the measured wave velocity utilized to represent various geological materials, Table 4.6 can be used to determine the geological conditions in the site of interest (Ashton et al., 1994).

An illustration of a seismic refraction method in action is provided in Figure 4.25. The detection depth range in this instance is about 50 m, and refraction seismic survey line stands about 150 m long. Following the recording of the seismic wave at each measuring point along the survey line and the computation of the seismic wave velocity, the profile distribution (velocity layered) is investigated. According to the refraction seismic survey, there are roughly five layers in the ground, each with a corresponding velocity wave. Layer 1: 0.4–0.6 km/sec seismic wave velocity, located about 0–3 m below the surface of the earth (Saenger et al., 2009).

Table 4.6. A Rough Range of P Wave Speeds (V_p) for Some Common Rock Types

Medium	*Velocity (km/s)*	
	Min	*Max*
Air	0.31	0.36
Weathered soil horizon	0.10	0.50
Gravel, dry sand	0.10	0.60
Loam	0.30	0.90
Wet sand	0.20	1.80
Clay	1.20	2.50
Water	1.43	1.59
Friable sandstone	1.50	2.50
Dense sandstone	1.80	4.00
Chalk	1.80	3.50
Limestone	2.50	6.00
Marl	2.00	3.50
Gypsum	4.50	6.50
Ice	3.10	4.20
Granite	4.00	5.70
Metamorphosed rock	4.50	6.80

Because of the significant energy loss experienced during seismic wave transmission, the surface soil layer may be porous or soft, resulting in a slow wave velocity transmission. Layer 2: the velocity of seismic wave is roughly 1.1 km/sec, and it is located about 1~11 m below the surface. Because of either a large-grained rock composition of this layer, or the relatively compact formation, there is little energy loss during wave propagation, and the wave velocity is high. Layers 3 and 4: the velocity of seismic wave is roughly 2.0 km/sec, and the depth is between 20 and 40 m. It may be a layer of rock that has collapsed, is full of muddy materials, or has multiple cracks in it. Seismic waves propagate with a comparatively large energy loss, which causes the wave velocity to be slower than the fifth layer (Gordon et al., 2003). Nevertheless, additional verification of this attribute necessitates core sampling from geological surveys. Layer 5: roughly 20–40 meters below the surface of the earth, with a seismic wave velocity of roughly 3.50 km/sec. This layer could possibly be formation of rock. With few mud inclusions and cracks, the rock mass remains largely intact. Because there is little energy lost during wave transmission, the wave velocity travels quickly. However, a reflection method, is also frequently employed in lieu of the seismic refraction method for determining the soil profile at shallow depths. In this method, the first and second layer's shear velocities, V1 and V2, are analyzed using signals captured by a surface geophone to determine the soil profile (Di Iorio & Clark, 2010).

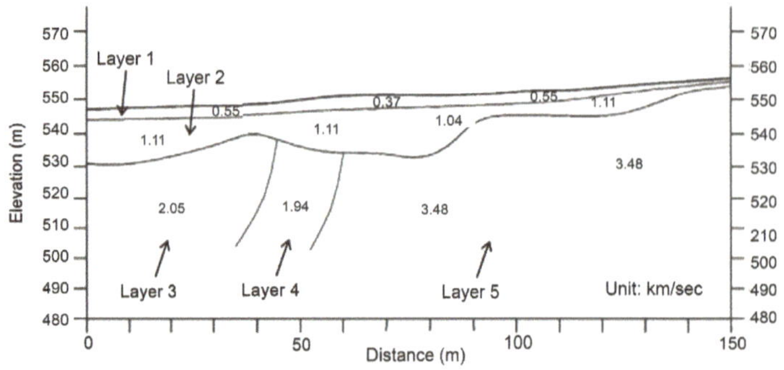

Figure 4.25. *Results of a seismic refraction survey.*

Basically, soil boring can give direct investigation results, but they can only look at a small area of the geological materials in the holes that are drilled in the site of interest. In conjunction with geophysical surveys, they enable the development of a 2D or 3D geological structure and offer comprehensive insights into subsurface geological conditions, which are particularly beneficial for intricate ground profile exploration (Greene & Richardson, 1988).

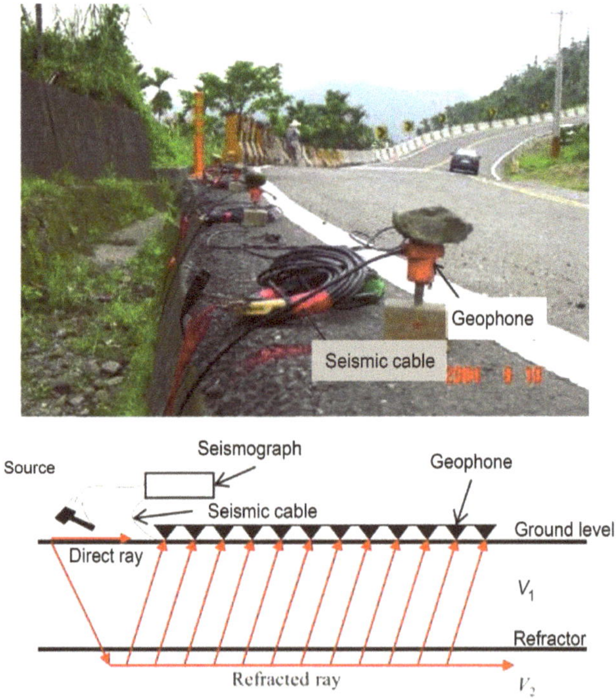

Figure 4.26. *Seismic survey operation.*

4.10. REPORTS AND INTERPRETATION

A site investigation report usually has two parts: a factual summary of the results of the investigation along with an engineering analysis, that contains suggestions for what should be done next. A trustworthy report should provide an accurate description of the project site's characteristics, the work accomplished, and results attained (Boutron et al., 2010).

The project's importance, scope, and purpose.

The site's fundamental characteristics, including its topography and hydraulics.

It is necessary to prepare a brief explanation and synopsis of the quantities, methods, and outcomes of various laboratory and field tests. It is recommended to present tables with lab and field test results, a condensed geological profile, and bore logs (Simmering et al., 2015).

The report should also include analyzes of the recommended foundation type, liquefaction potential, settlement, and/or allowable bearing capacity. Reports should include specific suggestions for more research if there are still questions that need to be answered because of how complicated the geology is at the site of project (Bauchner et al., 2019). The items in the geotechnical investigation report are listed in Table 4.7.

A building foundation construction project in Taipei City, north Taiwan, is used as an example to make it easier to talk about which excavation and foundation plan is used in that project. In order to incorporate the previously mentioned information, the investigative works that were utilized to examine the geological conditions have been outlined as follows (Kotrlik et al., 2011).

4.10.1. Abstract of Investigation Plan

Figure 4.27 displays the site building construction site plan that is presented in this chapter. The dimensions of the excavation area are roughly 3,200 m^2, 60–105 m in length, 40 m in width, and the shape is roughly trapezoid. It is anticipated that this structure will have five floors below ground and eighteen stories above (Manolis et al., 2011).

The excavation's design depth is roughly 19.7 meters below the surface of the ground. In order to comprehend the geological conditions and determine the properties of the soil, a sequence of investigations is carried out in order to plan and design a suitable foundation (Wimmer et al., 2014).

Table 4.7. List of Items in a Geotechnical Investigation Report

General description	Analysis	Other
1. The purpose and importance of the project. 2. Topographical features and hydrological conditions of the site. 3. A brief description of the various field and laboratory tests carried out. 4. Boring log/CPT. 5. Geophysical surveying. 6. Table listing the testing results.	1. Analyze and discuss the test results. 2. Determine soil profiles and geotechnical and hydraulic parameters. 3. Determine allowable bearing pressures, pile loads, etc. 4. Evaluate soil liquefaction potential. 5. Analyze excavation- induced influences.	1. Recommendations, such as dewatering, ground improvement, construction procedures, and further investigation.

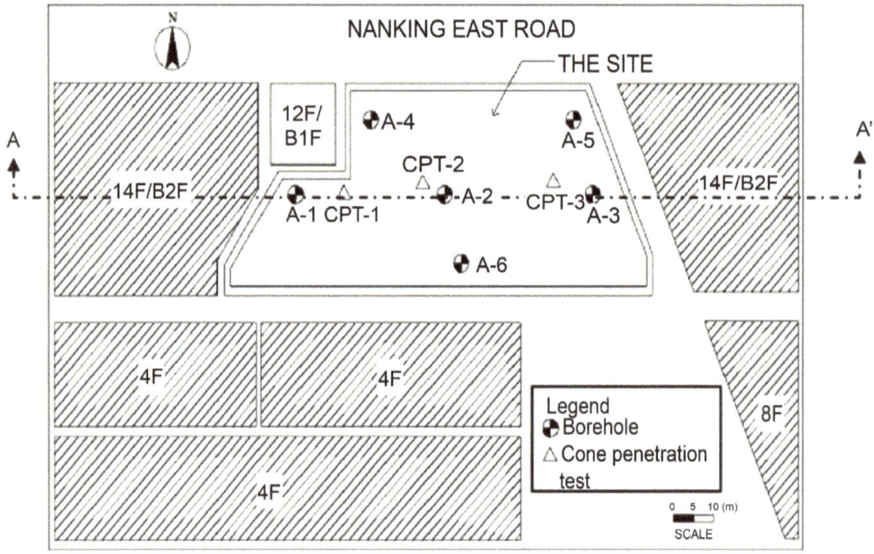

Figure 4.27. *The plan of site location and the layout of borings.*

4.10.2. Basic Information Collection

4.10.2.1. Location

As seen in Figure 4.27, the plan site is surrounded by 4- and 12-story buildings in addition to existing roads. Thus, throughout the design as well as construction stages, it is crucial to ensure the safety of surrounding buildings and residents as well as to reduce noise and vibration caused by construction (Gabszewicz & Thisse, 1992).

4.10.2.2. Regional Geology

This plan site is situated in the Taipei Basin, which is made up of the thick Sungshan Formation—a layer of alluvium that sits on top of the Chingmei Gravel Formation. There are six alternating layers of clay and sand in the Sungshan Formation. Silty clay (CL) makes up layers II, IV, and VI, while silty sand (SM) makes up layers I, III, and V. The primary layer influencing the study's excavation behavior is the Sungshan layer IV (25 m thick), which is categorized as soft to medium clay (Anderson et al., 1997).

4.10.2.3. Borings and Tests

The plan site is going to be located on the clay deposits, according to the basic data gathered. Because clay typically has a lower shear strength along with a poorer bearing capacity, consolidation settlement needs to be taken into account (Reese et al., 1975). This project site's subsurface exploration items comprise geological borehole drillings as well as lab and field tests (Galeazzi et al., 2010).

4.10.2.4. Layout and Depth of Borings

1. **Geological Borehole Drilling Volume:** the project site's base area is roughly 3,200 m^2, and the borehole configuration principle calls for at least one borehole for every 600 m^2. Consequently, as indicated in Figure 4.27, 3,200 m^2 divided by 600 m^2 = 5.3 boreholes are needed, and 6 boreholes should be distributed evenly throughout the plan site (Qi et al., 2016).
2. **Geological Borehole Drilling Depth:** the excavation is 19.7 meters deep, so the borehole drilling design must go deeper than the excavation to a hard stratum (Zhou et al., 2000).
3. **Sampler:** For this project, a thin-walled tube sampler is used to collect undisturbed soil for lab testing (Wakabayashi et al., 2010).

4.10.2.5. Field Test

The SPT and CPT are the two types of field tests used in this plan site investigation. In order to guarantee the accuracy of the test results, the clay's undrained shear strength is also ascertained directly using VST. Vane measurement of clay's shear strength will be very helpful in providing information for engineering design (Walsh & Cummins, 1976).

4.10.2.6. Laboratory Tests

To ascertain the mechanical and physical characteristics of the soil for this project, laboratory tests such as the Atterberg limit test, oedometer test, triaxial permeability test, and shear strength test are conducted. To obtain the necessary parameters that will be used in the foundation design, a variety of shear strength tests are conducted, such as the triaxial consolidated and unconsolidated undrained test (CU and UU), and direct shear test (Egner, 2000).

4.10.3. Investigation Results

Following the completion of geological borehole drilling, lab testing, and field testing at this location, a portion of a boring log is obtained, as an example is provided in Figure 4.28.

Borehole Log

Job title: Taipei survery | Location: Kuan Fu N. RD. | Borehole: A-3 Sheet 1 of 3
Depth: 52.5 m | Elevation: 6.07 m | G.W.L: 5.58 m | Date: 1998/10/30~11/01

Boring data					Testing										
Depth (m)	Sample no.	SPT-*N* blows	Log	Soil description	USCS	Particle size analysis(%)				Water content	Unit weight	Specific gravity	Void ratio	Liquid limit	Plasticity index
						G	S	M	C	∞(%)	(kN/m³)	G_i	*e*	*L.L*	*PI*
1				0.7											
	S-1	4		Sand and gravel	CL	0	11	50	39	34.5	17.85	2.67	0.201	50.5	23.7
2															
3	S-2	2			CL					30	18.44				
				Gray-black sility clay											
4	T-1					0	1	60	39	40.8	17.66	2.68	0.214	29	8.5
	S-3	1			CL					39.5	17.17				
5															
6	S-4	2		6.10	SM	0	3	68	29	42.2	17.56	2.68	0.217	30	8.8
7															
	S-5	3		Sand mixed sandy silt	SM					25.6	18.93				
8															
9				9.0											
10	S-6	2			CL					30.9	18.54				
11	S-7	3			CL					27.8	19.62				
12	T-2				CL-ML	0	19	71	10	29.5	18.74	2.71	0.187	24.6	5.3
	S-8	2			CL	0	5	70	25	29.3	18.93	2.74	0.187	23.6	7.4
13															
14	S-9	3			CL					36.8	18.25				
15				Gray-black sility clay											
	S-10	3			CL	0	1	48	51	41.1	17.85	2.74	0.217	40.9	18.5
16															
17	S-11	3			CL					38.5	18.05				
18															
	T-3					0	0	52	48	40.8	17.66	2.7	0.215	38.3	15.6
19	S-12	4			CL					34.8	17.95				
20															

Figure 4.28. *A section of a boring log.*

A soil profile, shown in Figure 4.29, can be ascertained based on the Boreholes A-1, A-2, and A-3 logs, which are labeled as Section A-A in Figure 4.29 (Bernardi, 1994). Keep in mind that other sections, such as the one across boreholes A-4, A-2, and A-6, should have their soil profiles determined if irregular or non-uniform layers of soil are warranted.

The silty clay layer (CL classified in USCS) that influences the behavior of this excavation has an SPT-N value of 2 to 8 blows, which is evaluated as soft to medium stiff soil. The excavation project will reach a depth of 19.7 meters. Figure 4.30 displays the SPT-N values distribution along the depth (MacCoun, 1998).

Because this plan site consists mainly of clay, the main points of the foundation design will be considered bearing capacity, consolidation settlement, and heave; therefore, a detailed analysis needs to be carried out in the report. For obtaining a more detailed examination on the soil strength of the clay soil underlying the foundation, undisturbed soil is obtained for UU test in the laboratory and VS test is applied in the field. According to the shear strength test results, s_u is approximately 63 kPa at a depth of 19.7 m, as shown in Figure 4.31.

Finally, a simplified subsoil profile judged by an experienced geotechnical engineer with the recommended parameters in Table 4.8 is to use for engineering design in the report.

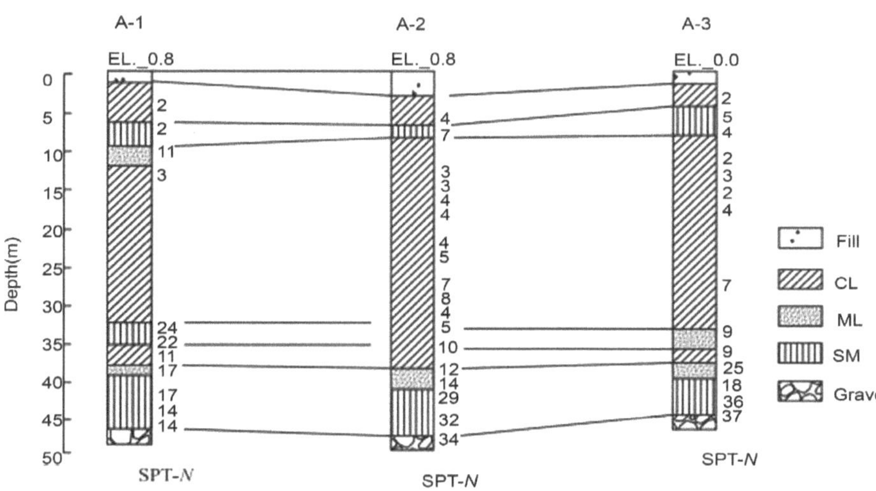

Figure 4.29. *Geological profile*

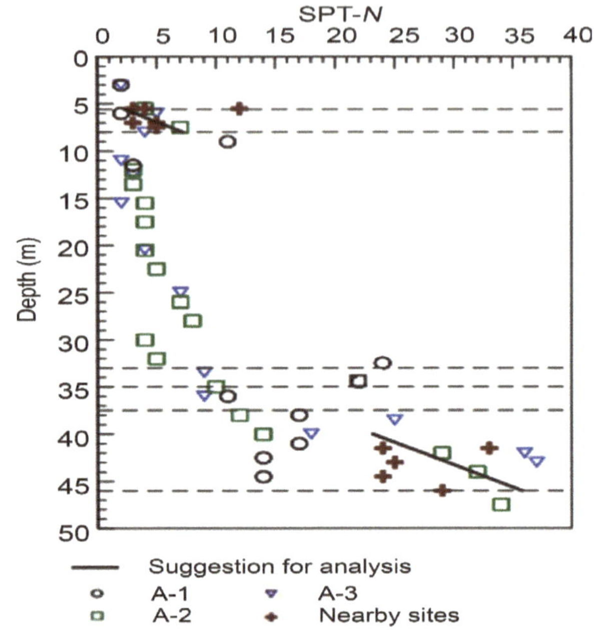

Figure 4.30. *The SPT-N values are spread out along the depth.*

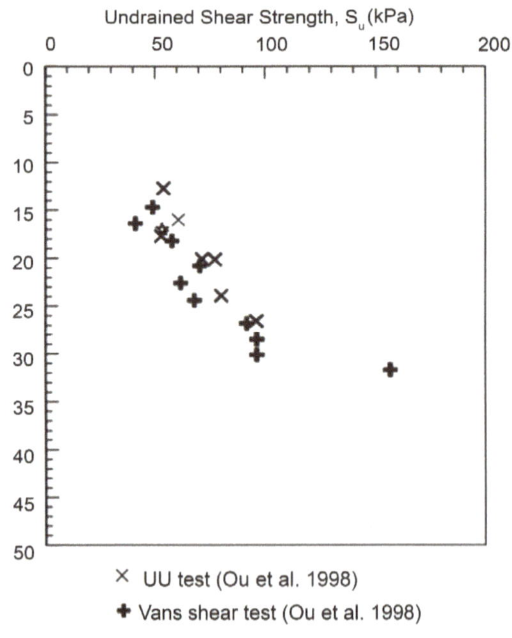

Figure 4.31. *Relationship between depth and undrained shear strength using variable tests.*

Table 4.8. Simplified Stratum Classification and Test Results

Layer	Depth (m)	USCS	SPT-N value	(kN/m³)	c' (kPa)	ϕ' (deg.)	ω (%)	LL	PI	s_u (kPa)
1	0~5.6	CL	2~4	18.25	–	30	33~38	33~36	13~16	-
2	5.6~8	SM	4~11	18.93	0	31	21~25	–	–	–
3	8~33	CL	2~8	18.15	0	29	32~40	29~39	9~23	40~100
4	33~35	SM	22~24	19.62	0	31	25~26	–	–	–
5	35~37.5	CL	9~11	19.13	–	29	28~32	37~41	15~19	150
6	37.5~46	ML or SM	14~37	19.62	0	32	23~24	–	–	–
7	>46	GP	>100	–	–	–	1	–	–	–

Note: γ_t = unit weight; ω = water content; *LL* = liquid limit; *PI* = plasticity Index.

SUMMARY

Various geological hazards, related to material properties and geological conditions, can affect foundation engineering throughout the construction and operation phase. These hazards include large settlements, bearing capacity failures, liquefaction caused by earthquakes, ground heaves, landslides and soil piping. Thus, surveying subsurface conditions and identifying potential geological hazards using a variety of investigation techniques and technologies are the goals of geotechnical site investigation. Furthermore, in order to obtain the proper parameters—physical, mechanical, and hydraulic—for an appropriate design that will prevent disasters and improve the safety and financial performance of construction, geotechnical site investigation becomes fundamental and essential.

A variety of geotechnical site investigation methods are frequently used during the investigation process in good engineering practice because each method or technique has a unique purpose, benefit, drawback and applicability. Additionally, not a single approach can simultaneously collect every aspect of the ground for design and analysis. Furthermore, a multitude of research techniques have been established and are still evolving globally. A few common and basic techniques are listed in this chapter, including electrical resistivity tomography, seismic surveys, groundwater investigations, lab testing, SPTs, and borehole drilling and sampling. This section gives readers an overview of geotechnical site investigation and its practical uses by condensing earlier techniques into a case study.

However, every research method or technique requires a certain level of background knowledge (like geology, geophysics, etc.). In applying those methods, practitioners occasionally come across challenging cases with unique and complex geological conditions. Since it is difficult for one person to be knowledgeable about all of the information and various investigation techniques, geotechnical site investigation work necessitates collaboration and group discussions in order to integrate the incomplete data from different investigation methods and team opinions in order to build an acceptable and precise geological model of the site that will support design at a later stage.

Accurate and trustworthy investigation results also depend on good planning and execution of the investigation project. However, the environment, time, and budget often make it hard to choose the investigation plan and the areas to look into. Geotechnical engineers and development units are constantly faced with the challenge of selecting the most practical and efficient approach. As a result, geotechnical engineers first need to have a clear understanding of the engineering project as well as the specifications needed for design and analysis. Then, after giving it careful thought, they must suggest a suitable investigation method. Furthermore, conducting research is a continuous process. If there are still certain uncertainties or unanswered questions, more research may be done. Once the reports have been meticulously and thoroughly assembled, they can offer valuable and significant data to support a feasible engineering design.

REFERENCES

1. Al-Jabban, W. J. M. (2013). Estimation of standard penetration test (SPT) of Hilla city-Iraq by using GPS coordination. *Jordan Journal of Civil Engineering, 7*(2), 133–145.
2. Anderson, S. P., Goeree, J. K., & Ramer, R. (1997). Location, location, location. *Journal of Economic Theory, 77*(1), 102–127.
3. Arnous, M. O. (2013). Geotechnical site investigations for possible urban extensions at Suez City, Egypt using GIS. *Arabian Journal of Geosciences, 6*(5), 1349–1369.
4. Ashton, C. P., Bacon, B., Mann, A., Moldoveanu, N., Ireson, D., Sinclair, T., & Redekop, G. (1994). 3D seismic survey design. *Oilfield Review (Netherlands), 6*(2), 33–115.
5. Auken, E., Pellerin, L., Christensen, N. B., & Sørensen, K. (2006). A survey of current trends in near-surface electrical and electromagnetic methods. *Geophysics, 71*(5), 249–260.
6. Bauchner, H., Golub, R. M., & Fontanarosa, P. B. (2019). Reporting and interpretation of randomized clinical trials. *JAMA, 322*(8), 732–735.
7. Bernardi, R. A. (1994). Validating research results when Cronbach's alpha is below .70: A methodological procedure. *Educational and Psychological Measurement, 54*(3), 766–775.
8. Boutron, I., Dutton, S., Ravaud, P., & Altman, D. G. (2010). Reporting and interpretation of randomized controlled trials with statistically nonsignificant results for primary outcomes. *JAMA, 303*(20), 2058–2064.
9. Bray, J. D., Beyzaei, C. Z., Cubrinovski, M., Riemer, M., Markham, C., Stringer, M. E., & Haycock, I. (2016). Recommendations for high-quality field sampling procedures: Using the Dames and Moore (DM) Hydraulic Fixed-Piston Thin-Walled Tube Sampler. *Journal of Geotechnical and Geoenvironmental Engineering, 2*(1), 114–355.
10. Broere, W., Faassen, T. F., Arends, G., & Van Tol, A. F. (2007). Modelling the boring of curves in (very) soft soils during micro tunnelling. *Tunnelling and Underground Space Technology, 22*(5–6), 600–609.
11. Brown, S. R. (1956). A piston sampler for surface sediments of lake deposits. *Ecology, 37*(3), 611–613.
12. Butler Jr, J. J., & Healey, J. M. (1998). Relationship between pumping-test and slug-test parameters: Scale effect or artifact? *Groundwater, 36*(2), 305–312.
13. Casagrande, A., & Fadum, R. E. (1940). Notes on soil testing for engineering purposes. *Geotechnical Testing Journal, 3*(2), 23–88.

14. Clay, R. B. (2001). Complementary geophysical survey techniques: Why two ways are always better than one. *Southeastern Archaeology, 5*(2), 31–43.
15. Clayton, C. R. I., & Siddique, A. (1999). Tube sampling disturbance—Forgotten truths and new perspectives. *Proceedings of the Institution of Civil Engineers-Geotechnical Engineering, 137*(3), 127–135.
16. Cline, M. G. (1944). Principles of soil sampling. *Soil Science, 58*(4), 275–288.
17. Crisp, M. P., Jaksa, M., & Kuo, Y. (2020). Optimal testing locations in geotechnical site investigations through the application of a genetic algorithm. *Geosciences, 10*(7), 2–65.
18. Daniel, D. E., Trautwein, S. J., Boynton, S. S., & Foreman, D. E. (1984). Permeability testing with flexible-wall permeameters. *Geotechnical Testing Journal, 7*(3), 113–122.
19. Daniels, J. J., & Dyck, A. V. (1984). Borehole resistivity and electromagnetic methods applied to mineral exploration. *IEEE Transactions on Geoscience and Remote Sensing, GE-22*(1), 80-87.
20. Day, S. R., & Daniel, D. E. (1985). Field permeability test for clay liners. *Hydraulic Barriers in Soil and Rock, ASTM STP, 874*, 276–288.
21. Di Iorio, L., & Clark, C. W. (2010). Exposure to seismic survey alters blue whale acoustic communication. *Biology Letters, 6*(1), 51–54.
22. Diez Roux, A. V. (2001). Investigating neighborhood and area effects on health. *American Journal of Public Health, 91*(11), 1783–1789.
23. Egner, W. (2000). The use of laboratory tests in the diagnosis of SLE. *Journal of Clinical Pathology, 53*(6), 424–432.
24. Elhakim, A. F. (2016). Estimation of soil permeability. *Alexandria Engineering Journal, 55*(3), 2631–2638.
25. Fletcher, G. F. (1965). Standard penetration test: Its uses and abuses. *Journal of the Soil Mechanics and Foundations Division, 91*(4), 67–75.
26. Forster, J. C. (1995). Soil sampling, handling, storage and analysis. In *Methods in Applied Soil Microbiology and Biochemistry* (Vol. 1, pp. 49–121).
27. Friedman, L. S., & Richter, E. D. (2004). Relationship between conflicts of interest and research results. *Journal of General Internal Medicine, 19*, 51–56.
28. Furgale, P., Barfoot, T., & Ghafoor, N. (2010). Rover-based surface and subsurface modeling for planetary exploration. In *Field and Service Robotics: Results of the 7th International Conference* (Vol. 2, pp. 499–508).

29. Gabszewicz, J. J., & Thisse, J. F. (1992). Location. *Handbook of Game Theory with Economic Applications, 1*, 281–304.
30. Galeazzi, S., Point, O., Haddadi, N., Mather, J., & Druesne, D. (2010). Regional geology and petroleum systems of the Illizi–Berkine area of the Algerian Saharan Platform: An overview. *Marine and Petroleum Geology, 27*(1), 143–178.
31. Gibson Jr, R. L., & Tzimeas, C. (2002). Quantitative measures of image resolution for seismic survey design. *Geophysics, 67*(6), 1844–1852.
32. Gillespie, D. M., Stites, D. L., & Benke, A. C. (1985). An inexpensive core sampler for use in sandy substrata. *Freshwater Invertebrate Biology, 4*(3), 147–151.
33. Godio, A., & Bottino, G. (2001). Electrical and electromagnetic investigation for landslide characterisation. *Physics and Chemistry of the Earth, Part C: Solar, Terrestrial & Planetary Science, 26*(9), 705–710.
34. Gordon, J., Gillespie, D., Potter, J., Frantzis, A., Simmonds, M. P., Swift, R., & Thompson, D. (2003). A review of the effects of seismic surveys on marine mammals. *Marine Technology Society Journal, 37*(4), 16–34.
35. Greene Jr, C. R., & Richardson, W. J. (1988). Characteristics of marine seismic survey sounds in the Beaufort Sea. *The Journal of the Acoustical Society of America, 83*(6), 2246–2254.
36. Guan, Z., & Wang, Y. (2020). Statistical charts for determining sample size at various levels of accuracy and confidence in geotechnical site investigation. *Géotechnique, 70*(12), 1145–1159.
37. Ho, Y. T., Wang, Y. L., Chang, L. C., Wang, T. P., & Tsai, J. P. (2021). Optical system for monitoring groundwater pressure and temperature using fiber Bragg gratings. *Optics Express, 29*(11), 16032–16045.
38. Hooimeijer, F., & Tummers, L. (2017). Integrating subsurface management into spatial planning in the Netherlands, Sweden and Flanders. *Proceedings of the Institution of Civil Engineers-Urban Design and Planning, 170*(4), 161–172.
39. Hornik, J., Cherian, J., Madansky, M., & Narayana, C. (1995). Determinants of recycling behavior: A synthesis of research results. *The Journal of Socio-Economics, 24*(1), 105–127.
40. Hvorslev, M. J. (1949). Subsurface exploration and sampling of soils for civil engineering purposes. *Geotechnical Testing Journal, 2*(1), 56–117.
41. Ingeman-Nielsen, T., & Baumgartner, F. (2006). CR1Dmod: A Matlab program to model 1D complex resistivity effects in electrical and electromagnetic surveys. *Computers & Geosciences, 32*(9), 1411–1419.

42. Jain, P., Powell, J., Townsend, T. G., & Reinhart, D. R. (2006). Estimating the hydraulic conductivity of landfilled municipal solid waste using the borehole permeameter test. *Journal of Environmental Engineering, 132*(6), 645–652.
43. Jaksa, M. B., Kaggwa, W. S., Fenton, G. A., & Poulos, H. G. (2003). A framework for quantifying the reliability of geotechnical investigations. In *9th International Conference on the Application of Statistics and Probability in Civil Engineering* (Vol. 4, pp. 1285–1291).
44. Jamison, V. C., Weaver, H. A., & Reed, I. F. (1950). A hammer-driven soil-core sampler. *Soil Science, 69*(6), 487–496.
45. Jefferies, M. G., & Davies, M. P. (1991). Soil classification by the cone penetration test: Discussion. *Canadian Geotechnical Journal, 28*(1), 173–176.
46. Jordan, D. (2009). How effective is geophysical survey? A regional review. *Archaeological Prospection, 16*(2), 77–90.
47. Kalbus, E., Reinstorf, F., & Schirmer, M. (2006). Measuring methods for groundwater–surface water interactions: A review. *Hydrology and Earth System Sciences, 10*(6), 873–887.
48. Kallio, T., & Kaikkonen, A. (2005). Usability testing of mobile applications: A comparison between laboratory and field testing. *Journal of Usability Studies, 1*(4–16), 23–28.
49. Kotrlik, J. W., Williams, H. A., & Jabor, M. K. (2011). Reporting and interpreting effect size in quantitative agricultural education research. *Journal of Agricultural Education, 52*(1), 132–142.
50. Kraeuter, A. K., Guest, P. C., & Sarnyai, Z. (2019). The open field test for measuring locomotor activity and anxiety-like behavior. *Pre-clinical Models: Techniques and Protocols, 2*(1), 99–103.
51. Landon, M. M., DeGroot, D. J., & Sheahan, T. C. (2007). Nondestructive sample quality assessment of a soft clay using shear wave velocity. *Journal of Geotechnical and Geoenvironmental Engineering, 133*(4), 424–432.
52. Lau, J. S. O., & Chandler, N. A. (2004). Innovative laboratory testing. *International Journal of Rock Mechanics and Mining Sciences, 41*(8), 1427–1445.
53. Loheide, S. P., Butler Jr, J. J., & Gorelick, S. M. (2005). Estimation of groundwater consumption by phreatophytes using diurnal water table fluctuations: A saturated unsaturated flow assessment. *Water Resources Research, 41*(7), 55–119.
54. MacCoun, R. J. (1998). Biases in the interpretation and use of research results. *Annual Review of Psychology, 49*(1), 259–287.

55. MacDonald, A. M., Barker, J. A., & Davies, J. (2008). The bailer test: A simple effective pumping test for assessing borehole success. *Hydrogeology Journal, 16*, 1065–1075.
56. Mackereth, F. J. H. (1958). A portable core sampler for lake deposits. *Limnology and Oceanography, 3*(2), 181–191.
57. Mahmood, M. A. (1987). System development methods—A comparative investigation. *MIS Quarterly, 2*(1), 293–311.
58. Manolis, E., Osman, T. E., Herold, R., Koenig, F., Tomasi, P., Vamvakas, S., & Raymond, A. S. (2011). Role of modeling and simulation in pediatric investigation plans. *Pediatric Anesthesia, 21*(3), 214–221.
59. Maurer, H., Curtis, A., & Boerner, D. E. (2010). Recent advances in optimized geophysical survey design. *Geophysics, 75*(5), 177–194.
60. McCauley, R. D., Fewtrell, J., Duncan, A. J., Jenner, C., Jenner, M. N., Penrose, J. D., & McCabe, K. (2000). Marine seismic surveys—A study of environmental implications. *The APPEA Journal, 40*(1), 692–708.
61. Mishra, S., & Robinson, R. G. (2018). A combined split-spoon sampler and spherical penetrometer: Laboratory trials. *Géotechnique Letters, 8*(2), 118–123.
62. Moh, Z. C. (2004). Site investigation and geotechnical failures. In *Proceedings of International Conference on Structural and Foundation Failures* (Vol. 4, No. 3, pp. 2–4).
63. Motojima, I., Shimada, J., Fujiwara, S., & Ohta, K. (1981). Proposal for the groundwater pressure measurement to investigate the groundwater movement in the bedrock. In *ISRM International Symposium* (Vol. 3, pp. 44–110). ISRM.
64. Mukherjee, S., Bartlett, P., Glass, B., Guerrero, J., & Stanley, S. (2006). Technologies for exploring the Martian subsurface. In *2006 IEEE Aerospace Conference* (Vol. 6, No. 3, pp. 11–99).
65. Nia, A. A., & Hamedani, J. H. (2010). Comparative analysis of energy absorption and deformations of thin-walled tubes with various section geometries. *Thin-Walled Structures, 48*(12), 946-954.
66. Niazi, F. S., & Mayne, P. W. (2013). Cone penetration test based direct methods for evaluating static axial capacity of single piles. *Geotechnical and Geological Engineering, 31*(4), 979–1009.
67. Niwa, M., Takeuchi, R., Onoe, H., Tsuyuguchi, K., Asamori, K., Umeda, K., & Sugihara, K. (2012). Groundwater pressure changes in Central Japan induced by the 2011 off the Pacific coast of Tohoku Earthquake. *Geochemistry, Geophysics, Geosystems, 13*(5), 34–88.
68. Nowacek, D. P., Clark, C. W., Mann, D., Miller, P. J., Rosenbaum, H. C.,

Golden, J. S., & Southall, B. L. (2015). Marine seismic surveys and ocean noise: Time for coordinated and prudent planning. *Frontiers in Ecology and the Environment, 13*(7), 378–386.

69. Park, C. (2013). MASW for geotechnical site investigation. *The Leading Edge, 32*(6), 656–662.
70. Parsons, R. L., & Frost, J. D. (2002). Evaluating site investigation quality using GIS and geostatistics. *Journal of Geotechnical and Geoenvironmental Engineering, 128*(6), 451–461.
71. Pratje, O. (1950). A new core sampler and its first test. *Deutsche Hydrografische Zeitschrift, 3*, 100–107.
72. Purdy, C. M., Raymond, A. J., DeJong, J. T., Kendall, A., Krage, C., & Sharp, J. (2022). Life-cycle sustainability assessment of geotechnical site investigation. *Canadian Geotechnical Journal, 59*(6), 863–877.
73. Qi, G., Zhengying, W., & Hao, M. (2016). An experimental research on the rock cutting process of the gage cutters for rock tunnel boring machine (TBM). *Tunnelling and Underground Space Technology, 52*, 182–191.
74. Reese, L. C., Cox, W. R., & Koop, F. D. (1975). Field testing and analysis of laterally loaded piles in stiff clay. In *Offshore Technology Conference* (Vol. 4, No. 2, pp. 12–23).
75. Robertson, P. K. (2016). Cone penetration test (CPT)-based soil behavior type (SBT) classification system—An update. *Canadian Geotechnical Journal, 53*(12), 1910–1927.
76. Roy, A., & Apparao, A. (1971). Depth of investigation in direct current methods. *Geophysics, 36*(5), 943–959.
77. Saenger, E. H., Schmalholz, S. M., Lambert, M. A., Nguyen, T. T., Torres, A., Metzger, S., & Méndez-Hernández, E. (2009). A passive seismic survey over a gas field: Analysis of low-frequency anomalies. *Geophysics, 74*(2), 29–40.
78. Sapińska-Śliwa, A., Wiśniowski, R., Korzec, M., Gajdosz, A., & Śliwa, T. (2015). Rotary-percussion drilling method: Historical review and current possibilities of application. *AGH Drilling, Oil, Gas, 32*(2), 313–323.
79. Selamat, M. R., Rosli, R. N., Ramli, M. H., Azmi, M., Kumaravelu, P., & Govindasamy, D. (2017). Comparative characterization of the Quaternary deposits of Northwest Peninsular Malaysia by wash boring and CPTU site investigation methods. In *AIP Conference Proceedings* (Vol. 30008, No. 1, pp. 1–200).
80. Shepard, E. R. (1935). Subsurface exploration by earth resistivity and seismic methods. *Eos, Transactions American Geophysical Union, 16*(1), 78–91.

81. Simmering, M. J., Fuller, C. M., Richardson, H. A., Ocal, Y., & Atinc, G. M. (2015). Marker variable choice, reporting, and interpretation in the detection of common method variance: A review and demonstration. *Organizational Research Methods, 18*(3), 473–511.

82. Simons, N., Menzies, B., & Matthews, M. (2002). *A Short Course in Geotechnical Site Investigation* (Vol. 5). Thomas Telford Publishing.

83. Singh, K., & Chauhan, R. (2002). Exploration of subsurface strata conditions for a limestone mining area in India with ground-penetrating radar. *Environmental Geology, 41*(8), 966–971.

84. Sklet, S. (2004). Comparison of some selected methods for accident investigation. *Journal of Hazardous Materials, 111*(1–3), 29–37.

85. Stanford, S. C. (2007). The open field test: Reinventing the wheel. *Journal of Psychopharmacology, 21*(2), 134–135.

86. Starr, R. C., & Ingleton, R. A. (1992). A new method for collecting core samples without a drilling rig. *Groundwater Monitoring & Remediation, 12*(1), 91–95.

87. Tam, C. M., Tong, T. K., Leung, A. W., & Chiu, G. W. (2002). Site layout planning using nonstructural fuzzy decision support system. *Journal of Construction Engineering and Management, 128*(3), 220–231.

88. Thorburn, S. (1986). Field testing: The standard penetration test. *Geological Society, London, Engineering Geology Special Publications, 2*(1), 21–26.

89. Uzielli, M., Vannucchi, G., & Phoon, K. K. (2005). Random field characterization of stress-normalized cone penetration testing parameters. *Géotechnique, 55*(1), 3–20.

90. Wakabayashi, J., Ghatak, A., & Basu, A. R. (2010). Suprasubduction-zone ophiolite generation, emplacement, and initiation of subduction: A perspective from geochemistry, metamorphism, geochronology, and regional geology. *Bulletin, 122*(9–10), 1548–1568.

91. Walsh, D., Turner, P., Grunewald, E., Zhang, H., Butler Jr., J. J., Reboulet, E., & Fitzpatrick, A. (2013). A small diameter NMR logging tool for groundwater investigations. *Groundwater, 51*(6), 914–926.

92. Walsh, R. N., & Cummins, R. A. (1976). The open-field test: A critical review. *Psychological Bulletin, 83*(3), 482–503.

93. Wang, Z. Z., & Jiang, S. H. (2023). Characterizing geotechnical site investigation data: A comparative study using a novel distribution model. *Acta Geotechnica, 18*(4), 1821–1839.

94. Wijk, G. (1991). Rotary drilling prediction. In *International Journal of*

Rock Mechanics and Mining Sciences & Geomechanics Abstracts (Vol. 28, No. 1, pp. 35–42).

95. Wimmer, S., Rascher, W., McCarthy, S., & Neubert, A. (2014). The EU paediatric regulation: Still a large discrepancy between therapeutic needs and approved paediatric investigation plans. *Pediatric Drugs, 16*(5), 397–406.
96. Wright, H. E. (1967). A square-rod piston sampler for lake sediments. *Journal of Sedimentary Research, 37*(3), 975–976.
97. Xiao, X., Gutiérrez, F., & Guerrero, J. (2020). The impact of groundwater drawdown and vacuum pressure on sinkhole development: Physical laboratory models. *Engineering Geology, 279*, 105894.
98. Xu, X., Schneider, J. A., & Lehane, B. M. (2008). Cone penetration test (CPT) methods for end-bearing assessment of open-and closed-ended driven piles in siliceous sand. *Canadian Geotechnical Journal, 45*(8), 1130–1141.
99. Xue, G., Chen, W., Cheng, J., Liu, S., Yu, J., Lei, K., & Feng, X. (2019). A review of electrical and electromagnetic methods for coal mine exploration in China. *IEEE Access, 7*, 177332–177341.
100. Yang, R., Huang, J., Griffiths, D. V., Meng, J., & Fenton, G. A. (2019). Optimal geotechnical site investigations for slope design. *Computers and Geotechnics, 114*, 103–111.
101. Yu, S. L., Kuo, J. T., Fassman, E. A., & Pan, H. (2001). Field test of grassed-swale performance in removing runoff pollution. *Journal of Water Resources Planning and Management, 127*(3), 168–171.
102. Yusof, N. Q. A. M., & Zabidi, H. (2018). Reliability of using standard penetration test (SPT) in predicting properties of soil. In *Journal of Physics: Conference Series* (Vol. 1082, No. 1, pp. 012–094).
103. Zacny, K., Bar-Cohen, Y., Brennan, M., Briggs, G., Cooper, G., Davis, K., & Stoker, C. (2008). Drilling systems for extraterrestrial subsurface exploration. *Astrobiology, 8*(3), 665–706.
104. Zhou, W., Beck, B. F., & Stephenson, J. B. (2000). Reliability of dipole-dipole electrical resistivity tomography for defining depth to bedrock in covered karst terranes. *Environmental Geology, 39*(7), 760–766.

CHAPTER 5

SPECIAL GROUND IMPROVEMENT METHODS

Contents

5.1. INTRODUCTION

The need for 'ground improvement' as a geotechnical activity has come to stay, since the need is ever present to improve poorer soils to adapt them for their intended uses in geotechnical and foundation engineering. Such a need arises due to the pressure, rather 'squeeze,' on land created by the ever increasing tempo of developmental activities thanks to the escalation in the construction of infrastructural projects of diverse kinds, the world over (Kurian, 2005). This chapter aims to introduce a few of the relatively recent, noteworthy, and successful ground improvement techniques that are currently in use and have a lot of potential for widespread application across the globe (Mishra, 2016).

5.2. VIBRO-COMPACTION OF SOILS

Trying to make the soil better so that regular foundations can be built on it is one way to deal with unfavorable foundation soil. "Soil/ground improvement methods" are generally defined as techniques intended to improve the quality of the soil. One such geotechnical technique for compacting extremely compressible and loose granular fills and deposits is the use of "vibroflot." (It is well known that vibrating loose, cohesionless soils is the most effective way to compact them.) "Vibro-compaction" is the general term for this compaction technique. In the 1930s, the technique was initially developed in Germany (Baumann & Bauer, 1974).

5.2.1. Vibroflotation

The vibroflot (Figure 5.1) is made up of a powerful vibrating machine with a heavy vibratory unit connected to it. The machine is jetted into its position by releasing the bottom jet, which causes the flow to go upward. The vibrating unit is activated upon reaching the required depth, maintaining the upper jet open and the bottom jet closed. This is how soil around a vibroflot is vibrocompacted; the cylinder size depends on the vibroflot's capacity, and the upper jet's downward flow facilitates the process (Brown, 1977). A cone-shaped depression forms on the surface as an outcome of this compaction, and it is then filled in. The unit is gradually pulled upward, compaction is applied in the same way at each stage, and the surface depression is filled in until the job is finished and then the unit is removed from the ground. Now that every point has been covered, the unit is moved to the next closest one, and the procedure is repeated until the entire area is covered with overlapping cylinders of the compacted soil (as seen in Figure 5.2). It is clear that just the foundation line requires going through the aforementioned compaction process if a strip-footing foundation is being considered (Zhi-guang et., 2020).

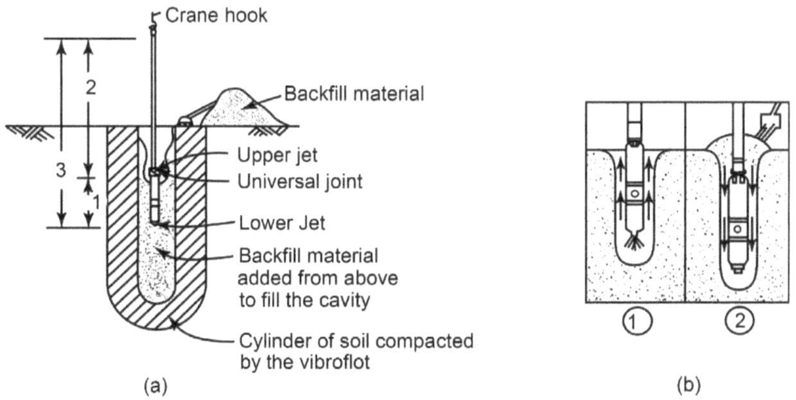

Figure 5.1. *Granular Soil Compaction with Vibroflot (a) Vibroflotation-based soil compaction with horizontally vibrating unit (1); a follow-up pipe (2); a vibroflot probe (3. (b) Vibroflotation Procedures 1. Probe is lowered by jetting; flow is upward and the bottom jet is open (1); Backfill material was shoved in when the upper jet opened and the flow descended (2) (Brown, 1997).*

Brown (1977) dwells at length on the various aspects of vibrocompaction in cohesionless soils. Even though vibroflotation is normally confined to cohesionless granular soils, Wong (1975) has shown its effect on weak cohesive soils. The holes left by the vibrating mandrels in the case of cohesive soils are backfilled with compacted sand or gravel

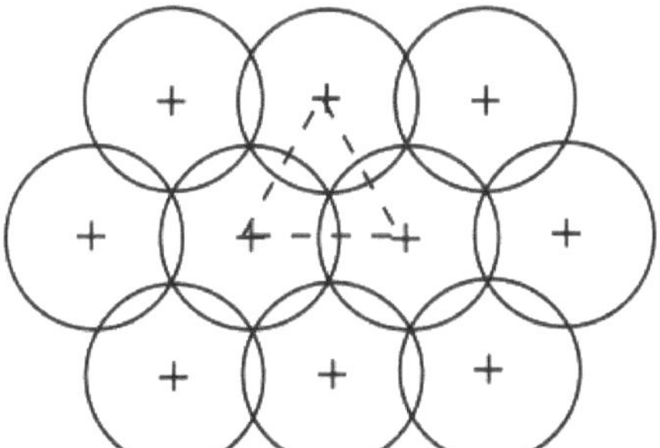

Figure 5.2. *For compacting large areas, vibro-flotation pattern are shown (Brown, 1997).*

5.2.1.1. Pile Driving by Vibration

Although unrelated to vibro-compaction, this subject is related. Techniques have been developed whereby driving of piles – both sheet and ordinary – is effected by vibration instead of hammer impact that is the most popular way of fixing precast displacement piles (Kurian, 2005). Nonetheless, attempts to do this are more frequently made when dealing with thin piles of steel sheet where the vibrating unit is positioned at the top. This method of driving steel sheet piles for the diaphragm-type cellular cofferdam is described by (D'Appolonia et al., 1955).

Densification occurs when a pile is driven into cohesive-less, loose soils using the vibratory technique. This is an additional benefit because it increases the soil's bearing capacity and decreases settlement. It would be prudent to take this into consideration when designing the pile. It should be mentioned in this regard that some vibration is produced during the traditional hammer impact method of driving piles, which has associated advantages (Webb & Hall, 1969).

5.2.2. Vibro-Techniques to Improve Ground

5.2.2.1. Vibro-Flotation (VFD)

The treated granular soil can be compacted by vibro-plotting to relative densities of up to 80%, resulting in a modulus compressibility of 80–150 N/mm^2, which can increase to 150–200 N/mm^2 in especially favorable soil conditions (Triantafyllidis & Kimmig, 2019).

The movement caused by the flushing action of water combined with high frequency vibration lessens the interfacial friction among the soil particles, shaking them into optional packing and density positions.

5.2.2.2. Vibro-Displacement (VDD)

Vibro-displacement is a method that can be used to improve cohesive soils that are not amenable to vibro-compaction because of the cohesive bond between the particles, which can't be broken by vibration (Triantafyllidis & Kimmig, 2019). This method involves lowering a vibrator that has been specially equipped to the appropriate depth. The surrounding soil is compressed locally as a result of the vibrator's action. A pipe integrated into the rig is used to drop crushed rock or gravel into the hollow space beneath the vibrator tip as it is being withdrawn. After that, the fill material is repeatedly raised and lowered on the vibrator, which densifies it and forces some of it to penetrate the surrounding soil to remain firmly attached (Ivanove & Ksenofontov, 2014).

5.2.2.3. Vibro Cast-in-Place Concrete Columns (VCIPC)

This is the VDD method's expansion. The technique involves pumping concrete down a tremie in order to replace the stone column with an unreinforced concrete column, similar to what happens in VDD (El Kamash et al., 2022). Following the placement of a charge, the vibrator is progressively lowered and retracted, desizing and enlarging the diameter. This creates a large, dense concrete column that is locally compressed and well-bonded to the adjacent soil, increasing the column's load carrying capacity (Okumus et al., 2015).

The above-mentioned piles are examples of unreinforced concrete piles that can be used to transfer loads to deeper soils. Low bearing capacity organic and cohesive soils are said to respond especially well to the VCIPC method's economy and efficiency (Fonseca et al., 2004).

5.2.3. Stone Columns

A relatively new method of ground improvement known as "stone columns" basically entails replacing some of the site's soil. The goal is to create closely spaced, well-compacted columns of graded and clean stone at the foundation site, usually with the aid of a vibroflot. It's not completely clear how the stone column actually works, but it's possible to imagine them as bearing piles (Hughes et al., 1976). This is because they create areas or pockets of granular material that has greater shear strength compared to the surrounding soil, which increases the soil's ability to withstand general shear failure, which is the same thing as its bearing capacity. However, creating columns requires a lot of stone, and in soft soil, they might only be slightly less expensive than traditional pilings—especially when taking into account the latter's greater working load capacity (Hughes et al., 1974).

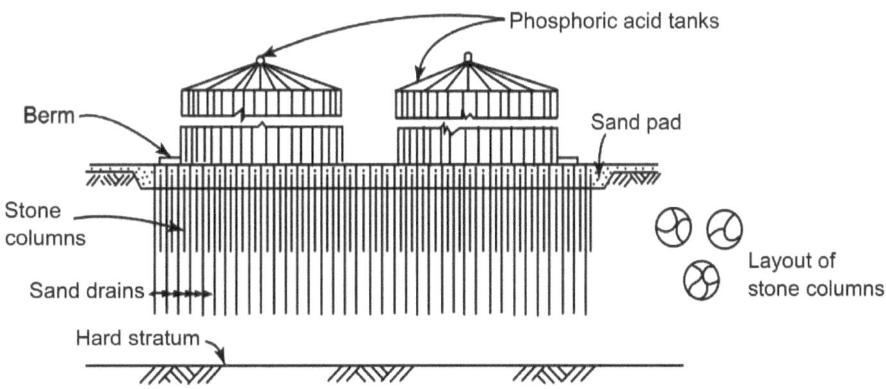

Figure 5.3. *Stone columns for the foundation of a storage tank (Datye & Nagaraju, 1977).*

Serious questions have been raised about the effectiveness of the stone column design method proposed by (Hughes & Withers (1974) that is further supported by field trials (Penman, 1977).

The process described above for creating stone columns entails moving soil and then filling the empty space with stones in the shape of columns. It is an add-on or byproduct of vibroflotation (Note how this resembles a "displacement pile" (Ambily & Gandhi, 2007). Another method for creating stone columns is the "replacement" technique, which involves removing the in-situ soil and replacing it with broken stones. The term "rammed stone columns" describes this kind. However, because of the reduced lateral bulging under load caused by the increased side soil compaction from vibroflotation, the former is more effective.

The construction of 2000 rammed stone columns at the Kandla Fertilizer Project in Gujarat, India, and their field testing are detailed by (Datye & Nagaraju, 1975). Instead of utilizing a vibroflot, these stone columns were created using standard bored piling machinery. This method involved driving an open-ended casing pipe, clearing the bore, inserting a charge of crushed sand and stone, and compacting it with a falling weight hammer, and pulling the pipe back a short distance at the same time. Step-by-step, the procedure was repeated until the whole column was built. (As previously mentioned, this is essentially a "replacement" technique, compared to the "displacement" method of the vibroflot method.) The load-settlement diagrams that were produced during the testing produced very positive outcomes. Figure 5.3 displays the plan for installing stone columns at Kandla for a twin chemical storage tank (Ambily & Gandhi, 2007).

- **Soil-Cement Piles:** The process of filling boreholes with soil cement and compacting it is described for the construction of soil cement piles (Hughes & Withers, 1974).

5.3. GROUND IMPROVEMENT BY KELLER

Founded in 1860 in Offenbach, Germany, Keller Grundbau GmbH is a prominent global provider of specialized ground improvement and piling services. Their primary area of business is ground improvement, where they are arguably the world's most well-versed and fully-equipped contractors. They have a manufacturing wing that supplies equipment for projects requiring such equipments, just like BAUER, and they have strong internal R&D support that helps with the ongoing improvement and upgrading of these equipments. Through its branch offices and affiliated businesses, Keller provides its extensive range of services in more than 20 countries worldwide (Egan & Slocombe, 2010).

The information that follows is an attempt to provide specifics about Keller's numerous geotechnical engineering endeavors, taken from their technical publications (Arulrajah et al., 2009). More than 75 years ago, Keller was the first to create and employ "depth vibrators" (Figure 5.4) for compacting deep pockets of loose sand. A traditional geotechnical technique called compaction via transmitting vibration is used to directly densify granular soils, thereby solving a variety of static, dynamic, and seismic foundation issues (Keller et al., 2013).

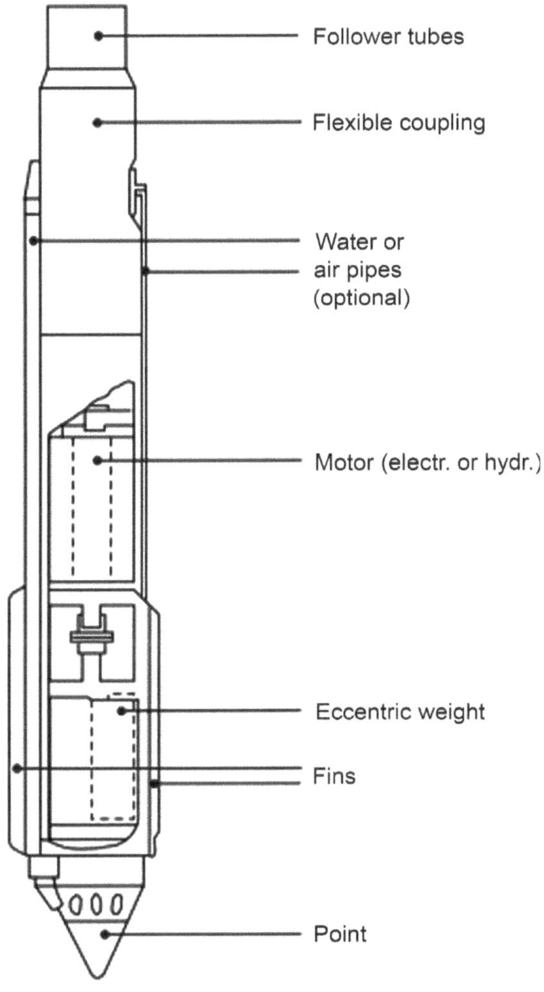

Figure 5.4. *A depth vibrator is shown (Mosely & Priebe 1997.*

5.3.1. Vibro-Compaction

Reaching depths of up to 35 meters, the goal of vibro-compaction is to densify coarse-grained soils, which range from coarse gravel with up to 15% silt to fine sands. Keller was the one who first presented the method in 1936.

The process's efficacy stems from vibration's ability to cause soil particles to realign themselves into the configurations that result in their ideal density (Figure 5.5a). Soil settlements ranging from 5–15% of the depth of treatment are the consequence of this. Figure 5.5b illustrates how the vibration is caused by the eccentric weight rotating (Baumann & Bauer, 1974).

The stages involved in the vibro-compaction process are shown in Figure 5.6. The method has the potential to save footing design costs and has been found to raise the soil's safe bearing pressures to 1000 kN/m^2 (1 MN/m^2). Shearing resistance increases significantly when there is a 5–8° increase in the internal friction angle (φ). A significant decrease in foundation settlements results from an increase in the modulus of deformation, or alternatively, the Young's modulus, E, to 100 MN/m^2. The soil, the amount of time required for the procedure, and the probe spacing—which ranges from 1.5 to 4 meters—all affect the aforementioned data (Duan et al., 2019).

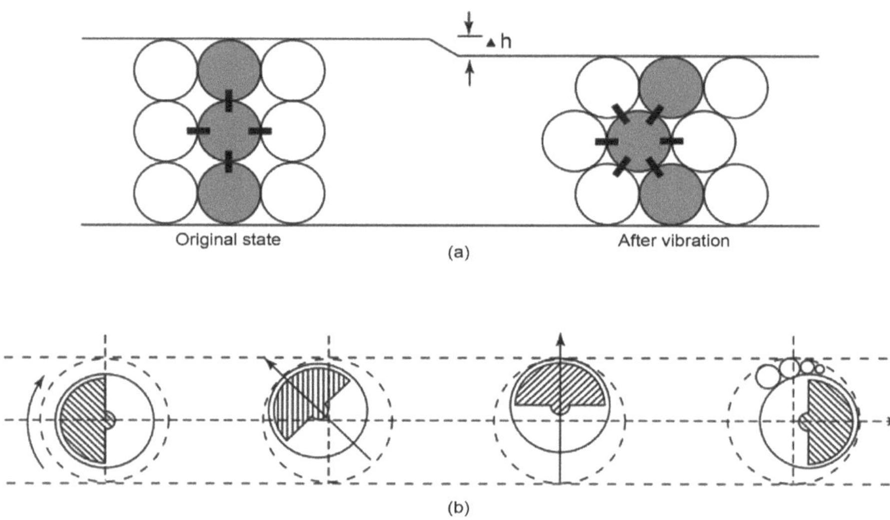

Figure 5.5. *Vibro-compaction results (a); and eccentric mass rotation (b).*

5.3.2. Vibro-Replacement

Vibro-replacement, a technique for building stone columns in the weak fills and soils composed of cohesive soils that do not respond to vibro-compaction, was first developed by Keller in 1957. Figure 5.7. Shows the steps that are involved in creating stone columns by vibro-replacement. The stone fill must be introduced and compacted gradually by the controlled vibrator up and down movement itself in order to create the required keying effect and improve the interaction between the surrounding soil and the stone columns (Renton-

Rose et al., 2000). The ideal is to make the soil-stone mixture homogeneous so that the two behave as a single material as a foundation support with low compressibility and better bearing capacity. One of the stone column's many benefits is that it acts as a vertical drainage path and a radial drainage face, which speeds up the consolidation of cohesive soil and causes settlements to occur more quickly. In general, they are spaced between 1.2 and 2.3 m apart below the main load-bearing foundations as well as much as three meters apart below the floor slabs (Sexton et al., 2017).

5.3.3. Design Principles

The ranges of particle sizes that vibro-replacement and compaction are effective over are depicted in Figure 5.8.

In cases where the weak materials are deeply buried, the treatment can be stopped at a depth where the new structures' added stress is negligible (bulb of pressure) (Fu et al., 2016). In fact, vibro-replacement has made it possible to economically develop a number of locations that are underlain by thick layers of weaker natural soils or artificial ground like fill (Sarnecki et al., 1993).

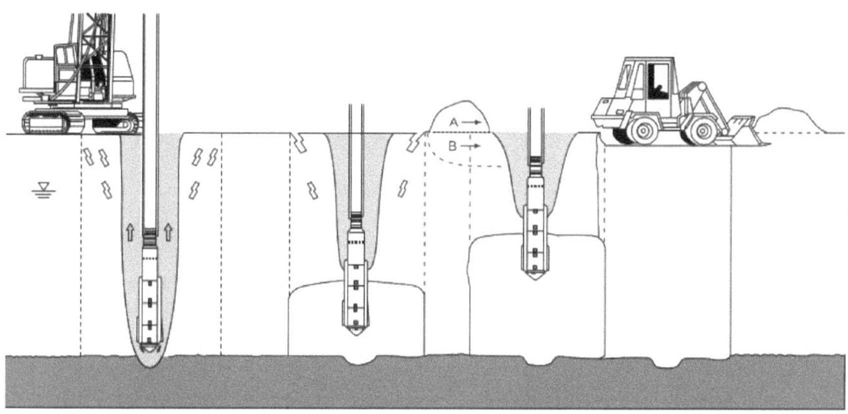

The process	1 Penetration:	2 Compaction	3 Refilling	4 Finishing
STEPS	At full water pressure the vibrator penetrates to design depth and is surged up and down as necessary to agitate sand, remove fines and form an annular gap around the vibrator. Once at depth the water flow is stopped or reduced.	Under the action of induced horizontal forces the soil particles surrounding the base of the vibrator are rearranged to a denser state of compaction. The vibrator is raised in steps after the required compaction is achieved.	During compaction sand fill is added which is either imported (A) or from the existing soil (B). The surface of the area being compacted may either settle up to 15% of the treated volume or the equivalent volume has to be imported.	With an economical layout of the compaction probes, an optimum improvement can be achieved. The surface of the compacted area has to be relevelled and, if required recompacted with a surface vibrator.

Figure 5.6. *Steps involved in the vibro-compaction process.*

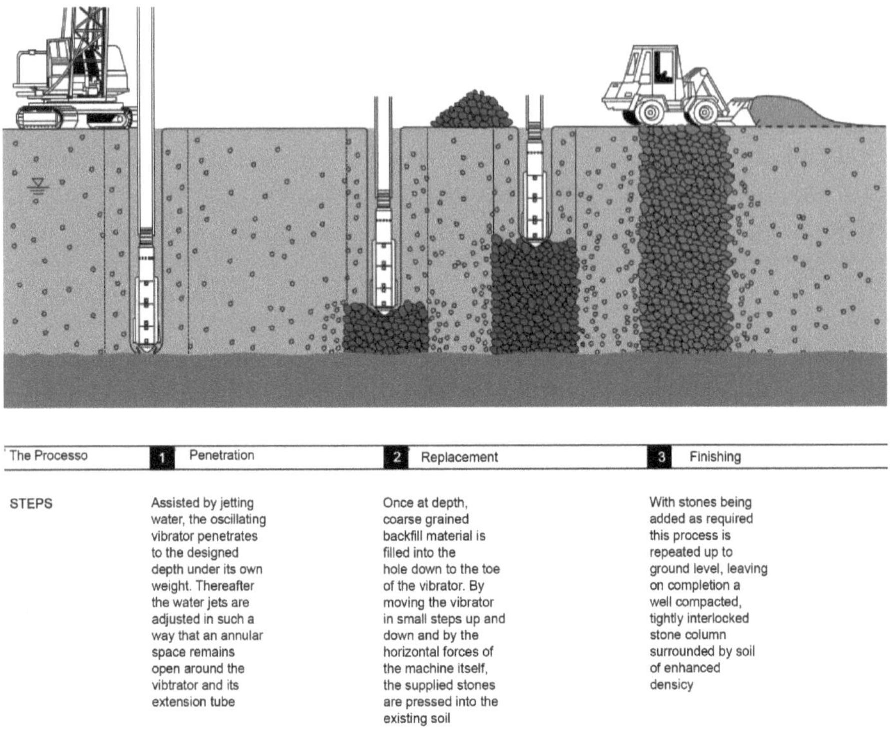

Figure 5.7. *Stone column by vibro-replacement – steps involved in the vibro-replacement of stone columns.*

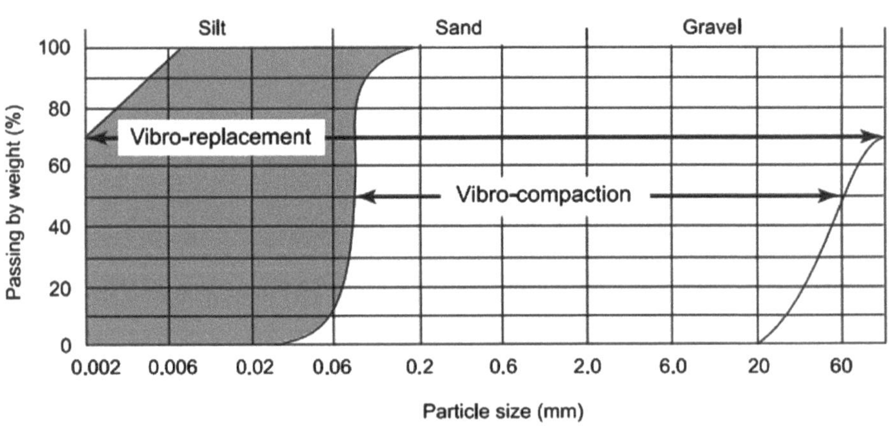

Figure 5.8. *Particle size obtained after vibro-replacement and compaction.*

The resistance of granular soil to liquefaction under seismic loading depends on its relative density. When it comes to earth retaining issues, active earth pressure falls and passive earth pressure rises with relative density. This provides benefits for retaining structure design when it comes to the load from the retained earth (Pettersson, 2010).

5.3.3.1. Tests to Ascertain Ground Improvement

Tests for penetration, like the Static CPT and the SPT, provide a practical way to determine how much vibro-compaction or vibro-replacement has improved the ground, especially the original ground between the stone columns (Kimball, 2013). Load tests, like the Plate Bearing Test, are possibly the most effective way to determine whether soil quality has improved in cohesive soils since, in contrast to cohesionless soils, the results of penetration tests are not consistently related to engineering properties (Blair-Early & Zender, 2008).

5.3.4. Examples of Vibro-Replacement

In order to meet the demands of Germany's "high speed railway," tracks are increasingly being laid in R.C. on rigid slabs, or pavements, rather than the traditional flexible tracks on a ballast. As a result, a stable track is laid that requires less maintenance and has a long operational life.

When installing stone columns in extremely soft ground or soils that have organic layers that are insufficiently supportive laterally, it is possible to create a grouted body by combining cement suspension with the column material. A more advanced version of this method installs premixed materials (gravel and grout) using a bottom-feed vibrator (Priebe, 1995). In actuality, a vibrated concrete column is a very effective substitute for a grouted column. For this, pumped concrete is utilized, creating a foundation that resembles a pile.

The lateral support provided by the surrounding soil determines the load bearing capacity in stone columns. Low support—for example, from soft organic layers—will cause lateral deformation, or bulging, which will result in large settlements. These types of soils require conventional stone columns to be set in firm soil, with the portion that passes through weak soils filled in with grout or using pre-mixed materials. "Partially grouted columns" is the term used for these (Ausilio & Conte, 2007). Even at extremely low water tables, stone columns can improve the ground and meet the geotechnical specifications needed to build the kind of rigid pavements needed to support trains. It offers a quick, affordable, and adaptable solution that can be quickly adjusted to the specific ground conditions (Gazdek et al., 2011).

5.3.4.1. Examples from Germany

Figure 5.10 shows how Keller improved the ground for an embankment with two tracks onto the Hamburg-Berlin line by using stone columns (De Koninck,

1979). There were layers of peat and organic soil on the ground. Measurements showed that during the installation of the stone columns, the predicted soil deformations brought on by track vibration from the rolling stock and engine movement had already taken place (Spallek et al., 2010).

5.3.4.2. Examples from Malaysia

Raju et al. (1997) discuss how stone columns were used under the Shah Alam Expressway in Malaysia, which is close to Kuala Lumpur. These columns were used in very weak soils like coastal marine deposits and tin mine tailings with cu values less than 10 kN/m^2. The first soil is mostly made up of alluvium with deposits of tin mine tailings scattered throughout. The majority of the soils left over from tin mining operations are slime and loose sands, which are typically very soft clayey silts. However, (Bhopal & Rowley, 2014). The marine soils near the coast are primarily made up of continental and quarternary marine deposits of silt and clay, occasionally combined with peat. The clays are typically consolidated and extremely soft. The ground improvement and foundations accounted for 38% of the project's total cost. All 900,000 linear meters of stone columns were involved (De Koninck, 1979).

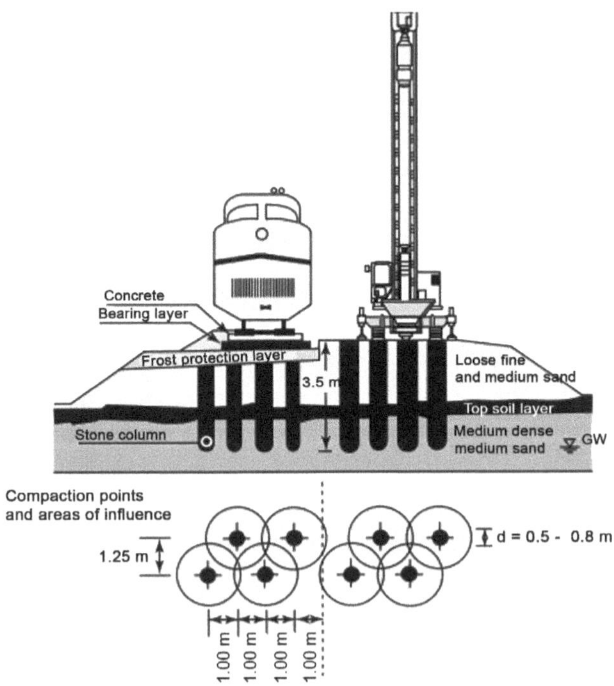

Figure 5.9. *Stone columns designed for the railway embankments in Germany (Sondermann, 1996).*

5.3.4.2.1. Quality Control

A computerized quality control system created by Keller, which outputs a full time history of the vibration's movement and, consequently, the entire installation process, is at the center of the installation procedure. This is represented by two graphs are generated to represent the process: 1) time versus depth, and 2) time versus current usage. In this way, each stone column is tracked, evaluated, and guaranteed to be of a certain quality before moving on to the next (Stiles, 1993).

Figure 5.11 depicts a section through Shah Alam's marine clay, where a blanket of compacted sand at the top connects the stone columns. The soft soil's overall compressibility has decreased by up to 5.13 improvement factors.

Further information about vibro-replacement employed at the location of the tin mine tailings mentioned above is provided by (Raju & Hoffmann (1996). Because the soil's cu values were so low, testing techniques were required to determine which of its properties would disturb the soil the least. It was determined that the Dutch electric piezocone penetrometer (EPP) test proved the most appropriate. Because of the deposits' artificial origins, it was discovered that the soil's depth and layering were highly variable, necessitating a lot of testing. It was discovered that the dynamic cone penetrometer test, DPT, in accordance with DIN 4094 is a quick and affordable way to determine the soil's stratification (Westgard, 2003).

The highway embankments had side slopes of 1V:2H and varied in height from 2 emd10 m above the current ground level. Reinforced earth walls were employed where available space was limited (Figure 5.12). On a square grid, the stone columns' spacing varied from 1.5 emd2.5 meters. The column had a diameter of 1 m, which increased to 1.2 m in very soft soils.

The columns that are over 100 times stiffer than the soft soil improve its overall stiffness, which makes settlements much less likely. The friction angle (φ) also gets better overall because the columns of compacted stone have high values compared to the soft soils, which means they have a higher bearing capacity. One important function of the stone columns is their function as sizable vertical drains, which helps the soft soil to consolidate quickly (Calvin, 1983). The sand blanket placed atop the columns helps with this process even more (see Figure 5.11).

Settlement monitoring showed that, in contrast to the design assumption, the rates of settlement tended to stabilize within two months. The tests were carried out on four and single columns at specific locations (Figure 5.13). As seen, loads were transferred using foundation blocks to the column along with the surrounding soil. As an additional quality control measure, load tests were conducted on blocks that ranged in size from 10 — 50 mm against a design load between 300 — 710 kN per column, 24 hours after the design loads were placed (Rodbard et al., 1968).

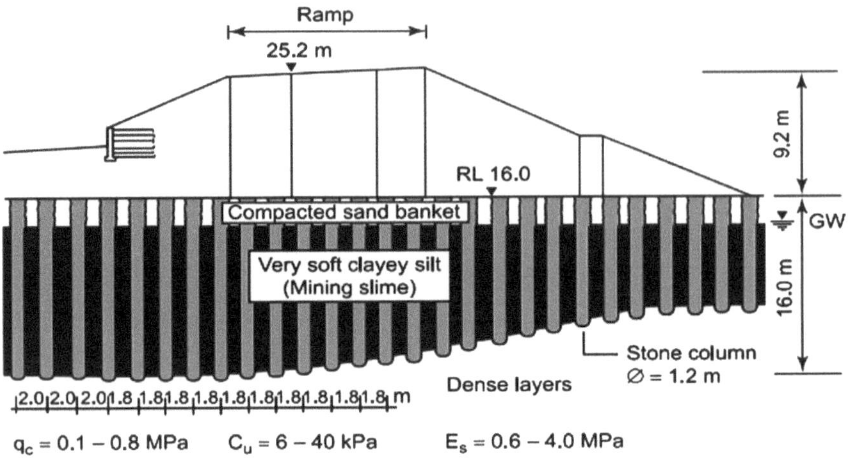

Figure 5.10. *Stone columns in Malaysia that are being passed through mining slime (Raju et al., 1997).*

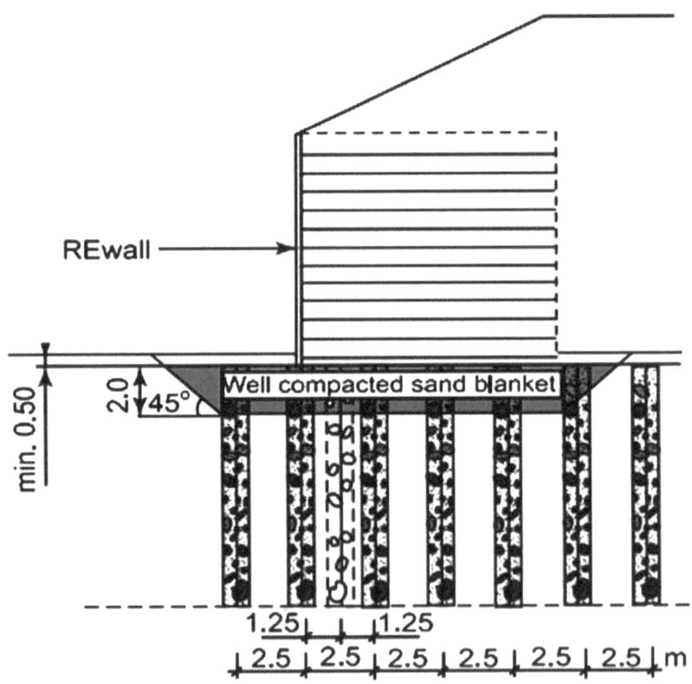

Figure 5.11. *The reinforced earth embankment on the stone columns is shown (Raju & Hoffmann, 1996).*

5.3.5. Additional Aspects of the Vibro-Technique

5.3.5.1. Vibro-Compaction

The initial method for compacting non-cohesive soils on a field scale was called "vibro-compaction," which involved using vibratory probes to cause the soil to vibrate. The 1938 process has been successfully implemented at many sites worldwide. The development of "stone columns"—dense stone columns tightly interlocked with the surrounding soil—was facilitated by its offshoot, the vibro-replacement technique. These columns allowed for the safe and cost-effective construction of warehouses, residential developments, industrial structures, etc., in a variety of weak soils and artificial fill materials (Baumann & Bauer, 1974) This technique has actually been selected for a number of significant structures in the USA and Europe, demonstrating its worth and safeguarding and promoting it as a cost-effective and safe ground improvement method for acting as foundations for a wide range of structures and soil types.

The Prime Minister's Office and other government departments were housed on a vast stretch of deserted land near Kuala Lumpur, Malaysia's capital, which was dubbed "Putra-Jaya" in the late 1990s. This impressive example of the use of stone columns was achieved by the German firm Keller, considered to be the industry leaders in this field, and the process totaled thousands of columns.

In the vibro-technique, when applied to noncohesive soils with natural densities lower than their max dry density, vibrations changes the structure of the grains, which lowers the void ratio. This decrease depends on the grain shape, size, distribution, and the level of vibration. The intergranular forces resulting from mechanical contact among the grains within a noncohesive soil momentarily disappear under the impact of induced vibrations, and the grains begin to move toward their maximum density position (Triantafyllidis & Kimmig, 2019).

In order to achieve vibro-compaction in non-cohesive soil to a predetermined depth, the vibrator is started and sunk as far as needed with the help of jetting at the tip. A compacted soil cylinder forms around the probe as a result of the vibration continuing in that position, which causes horizontal vibrations in the soil patricides. After the necessary number of steps, the probe rises to its next higher position, and the process is repeated until the compacted cylinder forms all the way to the bottom. After that, the probe was moved to the nearby location, and the operation proceeded. A nearly homogeneous compaction could be accomplished throughout the site by positioning the compaction points within appropriate patterns (Dobrescu & Brăguță, 2018). The soil surrounding the vibrator moves downward as the density increases, creating a cone-shaped depression on its surface (see Figure 5.6) that needs to be filled in all the time. At the conclusion of the work, the initial soil surface would have decreased if the site's material itself had been used for this purpose. Granular fill material that has been imported can be used in this way to avoid using it again. In terms of compaction, it is beneficial to use a fill material whose grain size differs from that of the construction site soil, especially if it is absent from the site soil (Duan et al., 2019).

Because of the strong cohesive bond between the particles in very fine-grained soils, like clays, vibro-compaction is unable to disperse the particles for reorganization as it can in non-cohesive soils. In this instance, the vibrator moves the dirt in a radial pattern, filling the empty space with granular material then compacting it. In this manner, granular material columns are created, stabilizing the cohesive soil around them. This method is referred to as vibro-replacement (Ausilio & Conte, 2007).

Stone columns can be arranged in groups or in lines beneath strip footings, and they can be positioned beneath rectangular or square spread footings, depending on their size.

As a stabilization technique, vibro-compaction/vibro-replacement lowers the soil's compressibility and subsequently settlements while raising its shear strength and subsequently bearing capacity.

Vibroflotation is the process of employing the vibrator, known as the vibro-compactor, which is used for stabilization. It is made up of four fundamental components (Figure 5.4): The vibrating unit with water/air jetting provisions, suspended from the extension tubes, the moving base unit with the crane that suspends and hoists the extension tube and vibrator, the stone delivery system used in vibro-replacement, and the control and verification devices are the first four components. The vibrator is the primary element among the aforementioned. Its dimensions are 3.0–4.9 m in length and 300–460 mm in diameter. Close to the base of the vibrator, vibrations are caused by the fast rotation of eccentric weights on a shaft (Figure 5.5), which is powered by a motor in the vibrator casing's upper part (Cai et al., 2017).

The motor has two power sources: hydraulic and electrical. Initially, there were 30 to 80 Horse Power (HP) motors in use [745.7 watt (W) = 1 HP]. The machines in use today have much more horsepower than the standard of 165 HP in late 1990s. These days, centrifugal forces are far greater than those of the late 1990s, when they reached 200 kN between the frequencies of 20 and 50 Hz. An elastic coupling separates the vibrator from extension tubes so that vibrations don't go up the tubes and past them. This makes it possible for the vibrations to reach the intended level in the soil and to operate at their fullest potential (Muro et al., 2001).

Flexible hoses carry air or water to the very top of extension tubes, where it is then forced through the tube to the vibrator tip, or nose, where it releases at high pressure, helping the vibrator penetrate the soil (Nagy et al., 2018).

Additionally, companies like Keller manufacture and supply sophisticated vibrator assemblies that offer a continuous record of every compaction point. Depth, power, and stone consumption measurements are time-stamped and produced as a computer printout in an instant for documentation. This effectively means that every stone column on the site, regardless of how many there are, is quality controlled during the construction itself, which does a great deal to boost client confidence (Muro et al., 1998).

The cone-shaped depression that forms as part of the vibro-compaction process, around the extension tube and vibrator assembly is filled with granular material that is either imported from elsewhere or is derived from the natural deposits on the site (see Figure 5.6). It is known as the "top fill system."

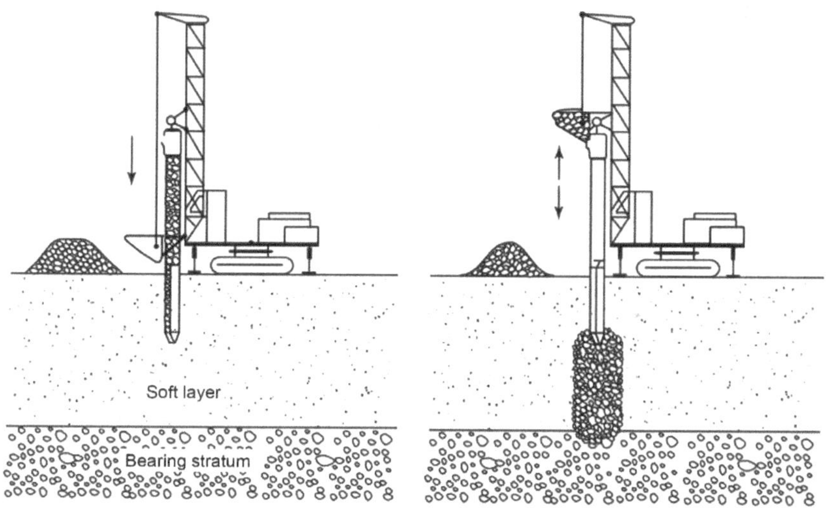

Figure 5.12. *Shows the bottom feed system for vibro-replacement. (Moseley & Priebe, 1997).*

Large amounts of loose stone must be imported and handled in order to create the stone columns used in the vibro-replacement process. The "bottom feed system" feeds the loose stone straight to the point of compaction. This system feeds stone into a skip that is mounted on vertical leaders and moves up and down (Figure 5.14). The stone is charged by the skip into the pipes that are connected to the vibrator and extension tubes. The stone is released at the nose of the vibrator and continuously fed until it is compacted (Yuan & Song, 2013). Following penetration, the vibrator in the vibro-replacement process is partially or completely removed, and the hole it made is then filled with a charge of stone—typically enough to fill the hole to a minimum depth of one meter. The stone is moved both downward and radially into the surrounding soil by the vibrator's nose cone, which is lowered into the stone. In this manner, a compact stone column with high integrity is built up to the ground's surface, carefully interlocking with the surrounding ground (Jeong, et al., 2004).

The soil's grain size distribution largely determines how effective vibro-compaction is. The ranges of grain sizes that can work for stone columns (vibro-replacement) and vibro-compaction are shown in Figure 5.15. A characteristic of vibro-compaction is the use of a CPT for pre- and post-compaction testing (SPT is a suitable substitute), which is done in accordance with a geometrical grid, to determine the efficacy of compaction (El Ouni et al., 2009). A typical compaction grid with pre- and post-penetration testing points is shown in Figure 5.2. Typically, each probe treats an effective area of 3 — 10 m^2.

In conclusion, vibro-replacement strengthens weak soils whereas vibro-compaction densifies the clean granular soils. Both methods result in less differential and total foundation settlement and an increase in the soil's carrying capacity (Liong, 2022). In addition to the above, it should be noted that using both approaches together lowers the chance of soil liquefaction in earthquakes (Kurian, 2005). In this regard, the function of stone columns in forming drainage channels for the purpose of releasing the excess pore pressures caused by earthquakes is especially noteworthy. According to, vibro-replacement has been engineered to reduce the potential for soil liquefaction in a number of instances from around the globe (Sanogo et al., 2021).

(Raju et al., 1997) provide an account of the process of building stone columns (via vibro-replacement) for the Shah Alam Expressway in Malaysia. The Shah Alam Expressway is a modern 6-lane highway that links Kuala Lumpur, the capital of Malaysian Federation, and Shah Alam, the coastal city of Klang and the capital of the Federal State of Selangor. Between 1996 and 1997, the work was finished in two packages.

Out of all of the above, the first section was mostly alluvium with tin mine tailings scattered throughout, located in the Kuala Lumpur region. For the most part, the latter consisted of very soft clayey silt and loose sand and slime. The remaining area, which was near the coast, was primarily made up of continental and marine quaternary deposits of clays, silts, and some peat. The clay were very soft and normally consolidated.

The nature of the soil was such that about 40% of the project cost was spent on foundations and ground improvement activities. The work involved over 900,000 lin. m. of stone columns covering 7 interchanges and overpasses (Al-Homoud & Wehr, 2006).

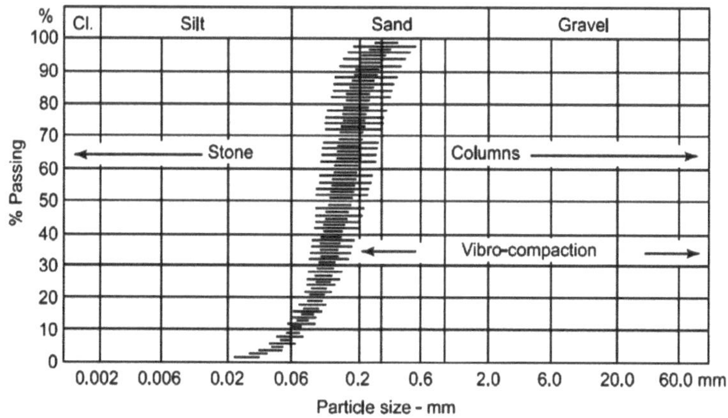

Figure 5.13. *Feasible particle size range for both vibro-compaction and stone columns is shown.*

5.4. SOILCRETE

Developed by Keller and first introduced in Germany in 1979, "Soilcrete" is a trade name for a jet grouting-based soil stabilization process. The word "crete" from concrete refers to the product's quality. Originally employed for small-scale underpinning projects, it is now utilized for a variety of purposes when soil stabilization and improvement are required.

Jet grouting, which produces the product known as "soilcrete," allows columns of soil cement to be formed in-situ in any geometrical pattern required for the task at hand (Put another way, it's a cement-based method of stabilizing soil. It makes use of water/cement suspension cutting jets under high pressure, with nozzle exit velocities greater than 100 m/s.

Depending on the soil treated, the jetting fluid, and the jet speed, the high speed jet can loosen the soil up to radial distances (up to 2.5 m). The cement suspension is combined with the loosen soil and rearranged. A portion of the soil-cement mix is flushed to the topmost part of the borehole through the space between the hole itself and the grout rods (Kolovos et al., 2013).

5.4.1. Uses

The fact that soilcrete can be used to stabilize and seal almost any type of soil, from clay to loose sediments, is by far its greatest benefit. It is applicable to soils with shifting layers, organic, and non-homogeneous soils. Keller has used this technique to treat even soft rock formations like sandstone with a weaker grain texture (Helson et al., 2017). Soilcrete functions as a sealing and/or stabilizing structure. In contemporary constructions, both roles are becoming more and more necessary (Friedman, 2008) The compressive strength of soilcrete can have compressive strength between 2 — 25 N/mm^2, depending on the soil type and cement content; the latter value is higher than that of regular structural concrete.

By using an appropriate grout suspension and then, if required, bentonite addition, soilcrete can be sealed against water infiltration. Numerous uses call for soilcrete's dual strengthening and sealing qualities, which makes the mix's careful design essential (Ashlock, 1974).

5.4.2. Production

Soilcrete is made in three different methods. The geometric configuration chosen, soil condition and the amount of soilcrete elements needed determine which method will be used for a given project. Below is an explanation of these techniques (Peterson, & Anand, 2004).

5.4.2.1. Single Direct Process (Soilcrete S)

Without an air shroud, it cuts and mixes soil simultaneously using a grout jet with an exit velocity of at least 100 m/s. small to medium-sized jet grout columns are the applications for it (Spagnoli et al., 2022).

5.4.2.2. Double direct process (Soilcrete D)

It uses a grout jet with an exit velocity of at least 100 m/s to cut and mix the soil at the same time. Using a specially designed airjet nozzle, air shrouding is used to increase the grout jet's cutting power and range. This process is primarily applied to sealing, panel walls, and underpinning projects (Arson & Juge, 2012).

5.4.2.3. Triple Separation Process (Soilcrete T)

It uses a water jet that is covered in air and has an exit velocity of at least 100 m/s to cut soil. Another nozzle situated beneath the water jet is used to simultaneously inject grout. The grout pump has pressures over 15 bar, or 1.5 MN/m^2. The method is applied to sealing, cut-off wall, and underpinning projects (Khoshsirat et al., 2019).

5.4.3. Soilcrete by Jet Grouting

Soilcrete is the term used to describe soil columns combined with the cement slurry in water through the use of high pressure grouting. This highly technological method uses cement to be an admixture to stabilize in-situ soil.

It's basically soil cement, but instead of being added to air-dried and ground-up soil, thoroughly mixed with water, and then left to cure like regular soil cement, this type of cement is delivered into the soil via a high-pressure pump (Fang et al., 1994).

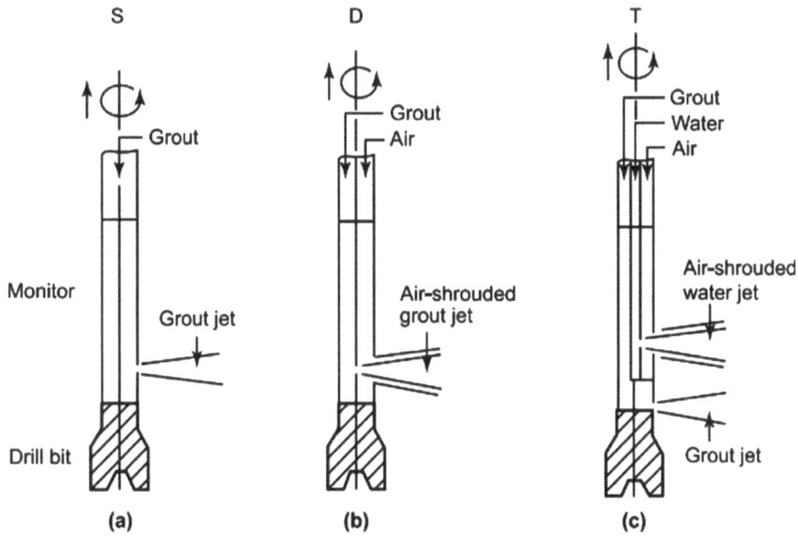

Figure 5.14. *The methods for producing soilcrete are depicted as Single system (a); double system (b); and Triple system (c) (Chambosse & Kirsch, 1995).*

Bore hole drilling is the first step in the creation of soilcrete. Grouting is done upward and from the bottom up. After drilling the neighboring bore hole, the soilcrete pillar is constructed in the same way as previously described (Wang et al., 2012). In order to form a thick column or continuous wall of soilcrete in the soil, the columns are designed to overlap as seen in Figure 5.16.

The soil particle sizes range that can be effectively covered by soilcreting is depicted in Figure and is observed to include silt, sand, and gravel. In underpinning, for example, soilcrete is used. An old load-bearing masonry wall is shown in Figure 5.17 resting on a thick wall of soilcrete built beneath it. In reality, soilcrete is the application of cement-based admixture stabilization to in-situ circumstances. German company GKN KELLER in Offenbach, is one of the top companies in the world for ground engineering, with operations across multiple nations, and they invented this technique. Section 5.2 has already discussed their leadership in this field of stone columns (Nikbakhtan et al., 2010).

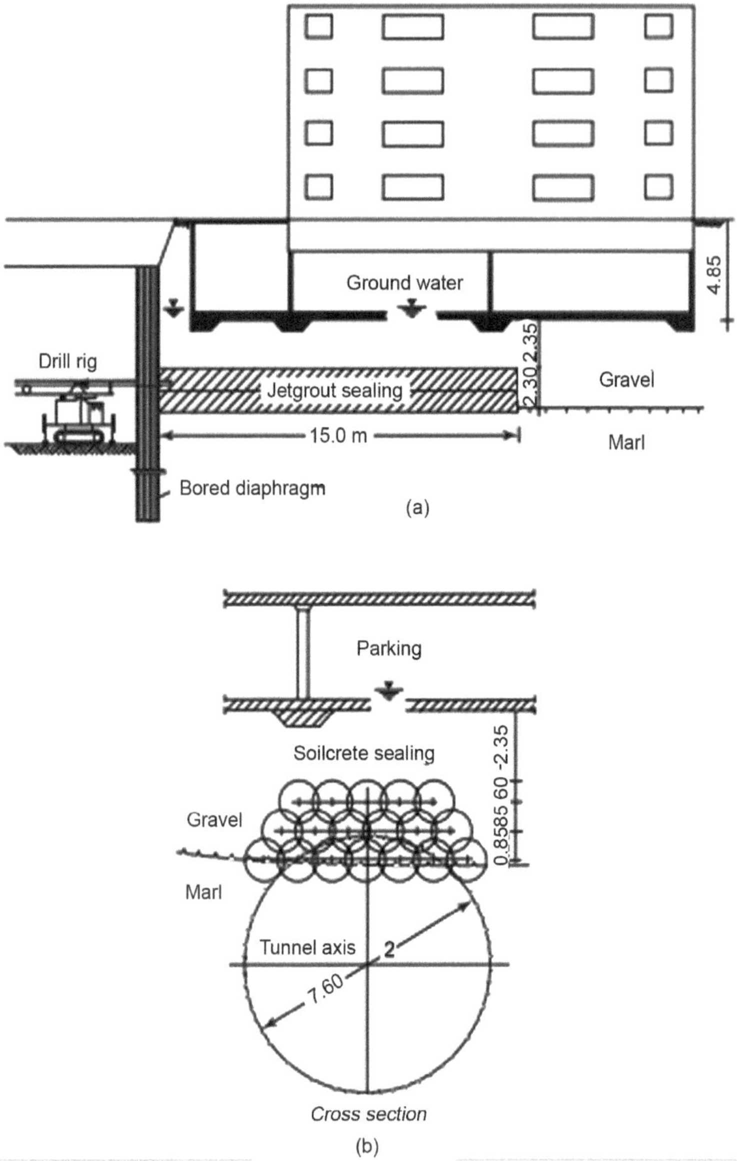

Figure 5.15. *(a) Soilcrete sealing horizontally beneath an apartment building; (b) A section displaying the tunnel beneath the soilcrete sealing.*

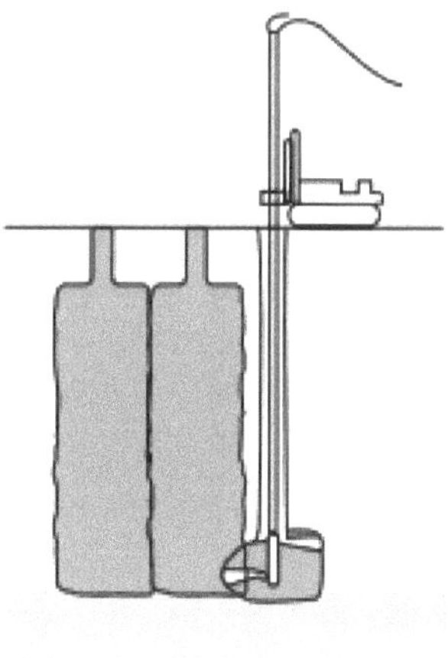

Figure 5.16. *Continuously overlapping soilcrete columns.*

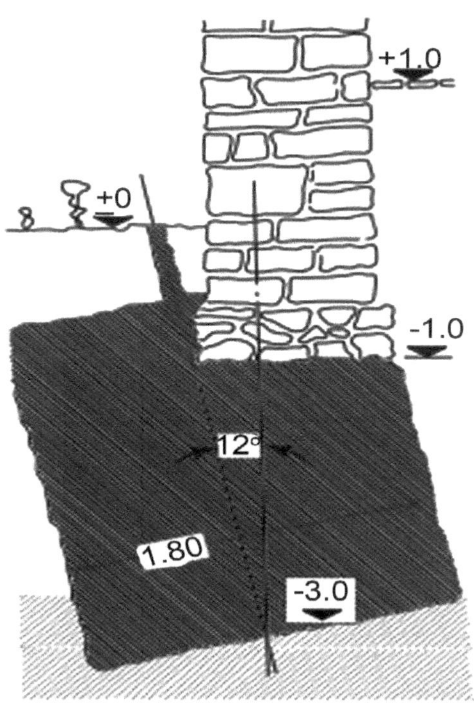

Figure 5.17. *The soilcrete wall underpinning the old masonry wall.*

5.4.4. Additional Features of Soilcreting

In geotechnical terms, grouting is the process of strengthening, stiffening, or waterproofing naturally occurring rock or soil formations. Cementitious grouts can be used in low pressure permeation grouting to treat cohesionless soils up to medium sands. Chemical grouts must be used for permeation in finer soils; this allows them to be used to cover silt sizes. They may, however, be more expensive and present toxicity issues (Emad et al., 2021). In an attempt to overcome these limitations, alternative techniques for grout infiltration into the ground have been explored (Helson et al., 2017). One such technique that fractures the ground is high pressure injection grouting, also known as "soilfrac grouting." Low slump, high viscosity grouts can also be expanded under high pressure into bulbs that compact and fortify the ground. This is known as "compaction grouting."

Jet grouting, a novel technology, and uses a completely different method called erosion to allow the grout to pass through a variety of soil types. In this process, the soil structure is broken down by high-pressure erosive water or grout jets, which also remove soil particles and substitute them with grouts made of cement that are applied throughout the erosive operation. The residual soil particles combine with the grout to create a treated mass that the Keller group—some of the industry's pioneers—names "soilcrete." The early 1970s saw the invention of this technique in Japan, and later that decade, it was introduced to Europe. Its ability to treat soil with basic cement grout, especially irrespective of its structure and grading, has brought it widespread recognition today (Spagnoli et al., 2022).

- **The Jet Grouting Process:** To insert the jetting pipes, also known as the "monitor," the first step is to drill a borehole with a diameter of 100 to 150 mm to the full treatment depth that is suggested. Depending on the most effective way to access the area that needs to be treated, it could be inclined or vertical. Where required, bentonite suspension can be used to stabilize the borehole (Shen et al., 2013).

The next step is to use jetting to erode the ground in preparation for grout slurry placement. Three different systems can be used for this purpose. The simplest of the three systems, called Single system (S), involves inserting a hollow grout pipe with one or more 2–4 mm diameter nozzles into the borehole such that the nozzles are set up at the exact treatment depths (Figure 5.15a). With pressures that reach as much as 600 bars (60 N/mm^2), grout is pumped into the ground via nozzles, eroding the soil as it goes and mixing it with the loosen soil from erosion. The monitor is raised and rotated at a constant rate to continue the jetting process, which aids in the creation of the soilcrete—a column of treated soil. Under pressure, extraneous suspension rises to the surface for disposal or collection by ascending the annular space between the borehole's sides and the grout pipe, known as annulus (Nikbakhtan & Osanloo, 2009)

The more sophisticated Double system (D) (Figure 5.15b) introduces a cover of compressed air at 2–15 bars, increasing the treated volume and significantly enhancing the erosive effect produced by the grout jet.

In the Triple system (T) (Figure 5.15c), which uses grout, water, and air, the erosive jet is highly efficient due to air-shrouded water jetting, and the grout is placed separately by injection nozzles. As can be seen in the schematic, the T-system uses 3 concentric pipes as opposed to the D-system's. The pressures involved are as follows (in bars): water: 500, grout: 5–30, and air: 2–15. The rotational speed and lift speed are approximately 10 rpm and 150 mm/min, respectively (Modoni & Bzówka, 2012).

Jet grouting works well on cohesive soils, but it works best on cohesionless soils, such as sands. In actuality, loose, poorly graded sands—the opposite of well-graded sands—form the largest column diameters. According to the T method, the usual ranges of column dia (in m) for various soils are as follows: silty sand: 1.4–1.6, sandy gravel: 2.0–2.4, and clayey silt: 0.8–1.0.

The typical option is cementitious grouts. Relative to strength or stiffness, water-to-cement ratios as low as 0.6 are utilized. To reduce costs, small amounts of bentonite are being added as anti-bleed agents, and fuel ash that has been ground up has been partially substituted for cement. When permeability is a top priority, cement-bentonite mixtures with a betonite content of up to 10% by cement weight are typically used, with water/cement ratios as high as 2.0 (Ochmański et al., 2015).

Based on unconfined compression tests, the compressive strength of soilcrete measurements have yielded following values (in N/mm^2) for various soil types: 5–30 for gravels, 5–25 for sands, 4–18 for silts, and 0.5–8 for clays. In every instance, the average coefficient of permeability is between 10–4 and 10–7 mm/s.

Groundwater control and underpinning are among the main uses for jet grouting.

5.4.4.1. Groundwater Control

The installation of permanent cut-off curtains, or walls, beneath dams (Figure 5.18), where durability and low permeability are important, is a prominent example of groundwater control. These can be created by putting two or more overlapping rows of columns together (Figure 5.19a, b). These have been used in leachate isolation, landfills, new dam construction, dam remediation, water flow prevention in tunnels and excavations, etc. The best way to seal gaps in the tangent pile walls is to use jet grout (Shaqour et al., 2008).

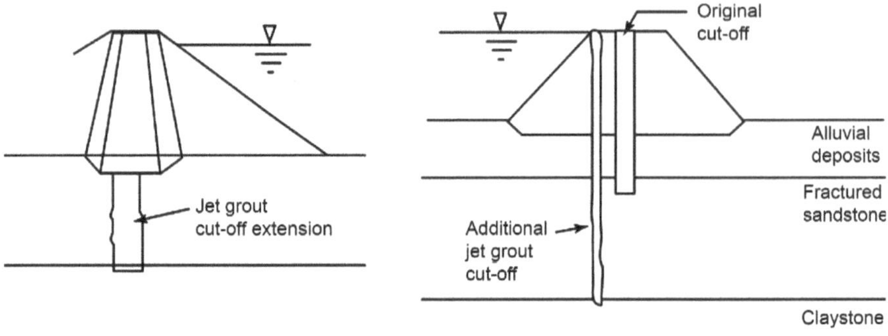

Figure 5.18. *Example of soilcrete cut-off wall under a Dam (Bell, 1997).*

5.4.4.2. Underpinning

The field of underpinning has heavily favored the jet grouting operation due to its geometric flexibility, comparative independence from soil variability, and ease of access without requiring the need to expose existing foundations. Wherever necessary, inclined drilling can be used to gain access to the desired zone of treatment (Figure 5.19). Grouting could be accomplished from inside damaged buildings with headroom as low as three meters. One such instance is depicted in Figure 5.20, wherein old timber pile foundations that are beginning to show signs of deterioration are covered in soilcrete. Figure 5.21 shows multiple instances of grouting the soil beneath existing buildings to allow for tunnel excavation without causing undue stress to the buildings (McClements et al., 2019).

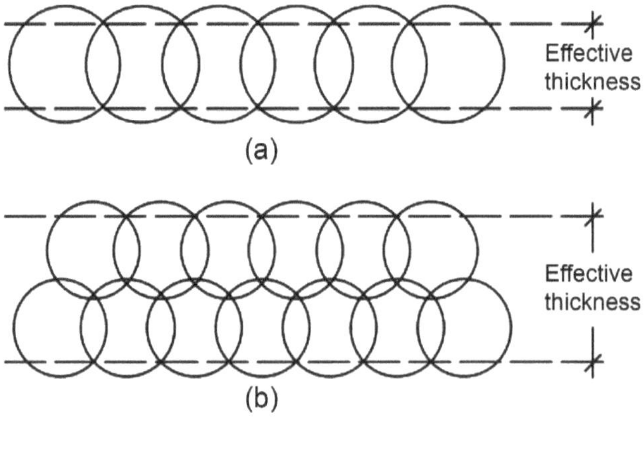

Figure 5.19. *Cut-off columns that overlap in (a) a single row; and (b) double row Bell (1997).*

Jet grouting, also known as soilcreting, is one more instance of how technology is driving the development of contemporary practice in geotechnical engineering.

Californian grouting contractors started experimenting with low slump (usually 25 mm at most) mortar type grout in the early 1950s. They discovered that loose soil formations beneath damaged structures could be densified by injecting the material with high pressure. They dubbed it "compaction grouting" because it helped to compact the soil. Instead of penetrating soil pores, this grout stays in the form of a homogenous mass which compacts loose soils through controlled displacement. In conditions of voided and loose soil, the compaction grout mass can be nearly cylindrical or spherical, and bulbs with a diameter of one meter are not unusual (Yarkoni et al., 2010).

Compaction grouting was conceived as a radical alternative to the widely used permeation techniques at the point of its inception. In order to fill in the joints and interstitial voids, this usually required injecting cementitious materials or chemicals into soils or rocks.

Since then, compaction grouting has been widely accepted as the best way to densify soils in the United States, with the industry recording a turnover of $10–20 million by the middle of the 1990s. In the United States, compaction grouting was given new life in 1977 when it was discovered that it could be used to control settlement brought on by loosening and ground loss associated with urban tunneling (Figure 5.34). In the US, compaction grouting is now a standard when building subways. It initially appeared as a ground (site) improvement method for new construction in the early 1980s. Around the middle of the 1980s, compaction grouting started to be used a lot in northeastern United States to fix sinkholes in areas that were prone to them (Chai et al., 2005).

Compaction grouting is mostly used to fix man-made or natural problems with the compaction of different types of soil. However, it can also be used to level out structures that have settled, and it can also be used as a pre-treatment method to make the site better before fresh structures are built.

Grout is typically placed in a quasi-continuous cylinder across the injection pipe's axis as it is withdrawn or inserted in the most popular construction method (Soukup et al., 2008).

The creation of compaction grout piles or pedestals that function as structural components for load transfer when coupled with the densified surrounding soil is another area in which the technology is put to use (El-Kelesh, et al., 2001).

Loose granular soils with average SPT N values between 0–15 or 20 that are either either above or underneath the ground water table are the best candidates for compaction grouting. The N-values are improved by densification by at least ten points. It is also possible to treat cohesive fills that are poorly placed and with N -values around 0 to 10. Nevertheless, it is not advisable to attempt treating thick clay strata or saturated silt. In this instance, the pore pressure created by compaction grouting increases dramatically and does not dissipate quickly enough. The outcome is an abrupt decrease in the soil's shear strength (Alberi et al., 2018).

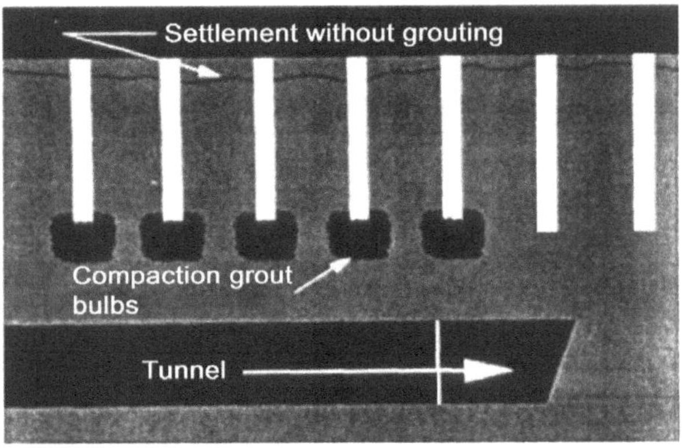

Figure 5.20. *The use of grouting to compact rock for urban tunneling.*

5.4.5. Design

From a design perspective, the pipe's inner diameter should be at least 50 mm, and depending on the circumstances, pipe spacing should be between one and three meters. Rather than a continuous injection, initial and subsequent, two-stage densification is made possible by primary and secondary sequencing, with the latter occurring inside the region covered by the former. It is preferable to use "upstage" grouting, which involves injecting grout from the bottom up, rather than "downstage" grouting, which involves injecting grout from the top down (Gemperle et al., 1998). This involves driving the pipe—which has a sacrificial cap on one end—all the way down, injecting it, and then forming a bulb at the tip. Once the first filling is done, the pipe is moved to the next step above and the grout is injected. The process is repeated and once at the top, this process ends (Li et al., 2019). The equipment used for compaction grouting must be able to produce pressures at the tip of about 40 bars. Regarding the injection rate, one may adhere to the following recommendations– The following recommendations can be used to determine the injection rate: 0.015–0.03 m^3/min for poorly draining soils near the surface; 0.030–0.120 m^3/min for free draining or dry soils with sufficient cover; and 0.120–0.360 m^3/min for loose or voided situations with substantial cover (Lowekamp et al., 2013).

5.4.6. Grout Material

The goal is to create a mix with just the right amount of granular components to generate internal friction and fines to allow for water retention and flow plasticity. Cement is not necessary in the grout mix if the goal is densification of the surrounding soil instead of forming a structural element in the ground. Typically, moist, appropriately graded silty sand suffices to accomplish the densification goal. Sand or sandy soil, either on its own or in combination with cement used as an additive, is the primary ingredient in the compaction grout mix (Li et al., 2019). A typical ratio of sand to cement is 3:1. Cement can be substituted with fly ash up to 50%. To improve pump ability, bentonite is commonly added in small amounts (about 3% of the total weight of fly ash and cement). Compaction grouting exhibits self-proofing properties when executed with precision. At the point of minimum resistance and maximum need, the compactive effort is at its highest. CPT or SPT are the two types of in-situ quality assurance testing (Hamed et al., 2017).

The majority of compaction grouting was done with modified, cutting-edge concrete pumps prior to the advent of sophisticated modern plants. When digging holes up to 7.5 m deep in fairly soft ground, hand-held percussion rotary machinery is usually enough to insert the grout pipe. Pneumatic or hydraulic driving equipment might be required for medium soils. In every situation, a tight fit between the soil and the grout pipe must be guaranteed (Kılıc et al., 2002). Consequently, the grout pipe may need to be extracted using specialized equipment. The most widely used equipment includes hydraulic cylinders set up as extraction gears. In the United States, Two examples of compaction grouting are given by one is from Michigan, where remedial compaction grouting was employed in a building site in which the dewatering system triggered the migration of huge amounts of fines, resulting in ground subsidence, and the other is from Maryland, for the Baltimore metro system to inhibit the settlement of older masonry buildings adjacent to the tunnel sections (Erol & François, 2014).

5.5. SOIL FRACTURING – SOILFRAC

The oil industry first employed the "soil fracturing" technique to make channels for oil to move toward oil wells (Smiley, 2001). Keller engineers modified this method in the 1960s in an effort to find answers for geotechnical issues. Similar to "soilcrete," "soilfrac" method is also a Registered Trademark of Keller.

Structures that have significantly settled for a variety of reasons can be raised by heights measured in decimeters using this technique. Since it uses grout injection, it can be applied to the restoration of leaning structures like the well-known—or notorious—Leaning Tower of Pisa.

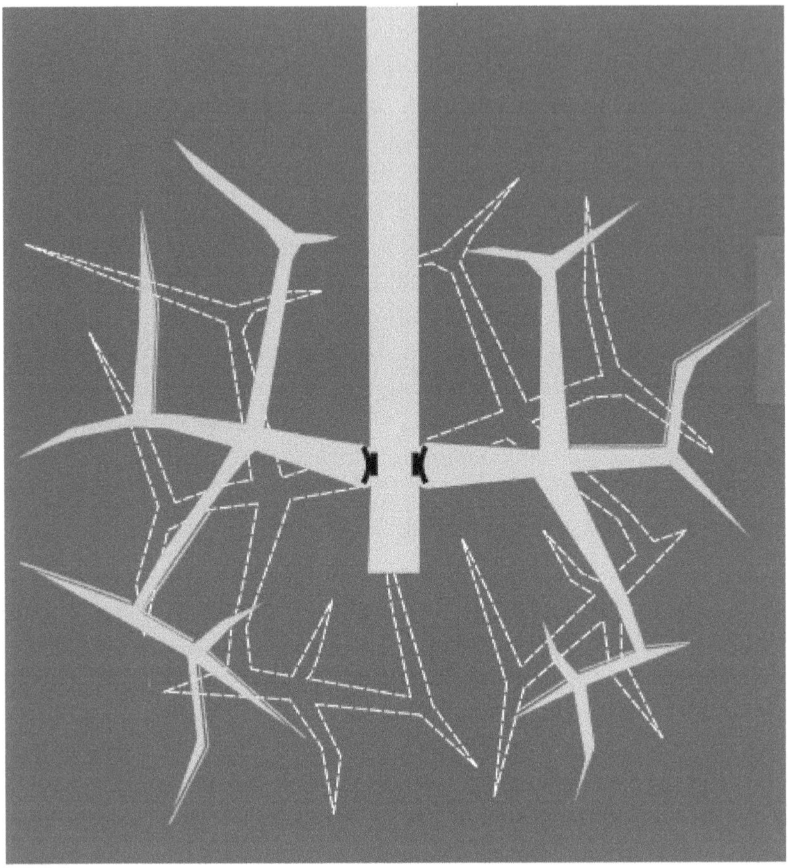

Figure 5.21. *The schematic diagram of soilfrac process.*

The procedure basically entails causing soil fractures and filling in the resulting void spaces with grout (Figure 5.22). For the aforementioned, sleeve pipes are buried in the ground and sealed with a hard cement-bentonite mixture in the annular space separating the sleeve pipe and the borehole wall (Figure 5.23). A grout hose with a "double packer" (Figure 5.23) on the bottom is pushed into the sleeve pipe in order to inject the soilfrac suspension. By sealing the sleeve pipe on both sides, the double packer makes it possible to grout over the section that needs to be treated (Ding et al., 1999).

Soilfrac has a broad range of applications, such as protecting and raising structures as well as restoring foundations. Regarding lifting, accurate differential lifting within 1 mm has been integrated and contributed to a net lifting in the decimeter range without compromising the structure's integrity or limiting its uses (Shirlaw, 1995).

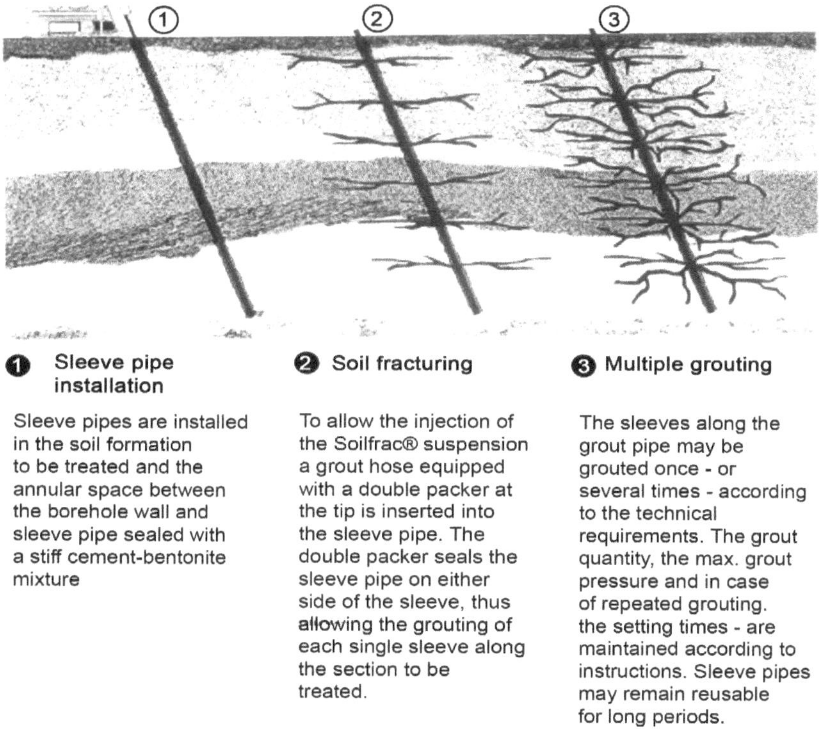

Figure 5.22. *Three steps in the soilfrac process.*

In their exciting historic work, Keller's Site Report 61–29E details how they used soilfrac grouting to provide active settlement compensation while tunneling beneath Antwerp, Belgium's main railway station. The scope of work under the historic structure is depicted in Figure 5.23. With the help of compensatory grouting, settlement was reduced from an estimated 60–120 mm to 5 mm, and tilt was reduced to 1 in 2000 (Gottardi et al., 2015).

Following contact grouting, the pipes that comprised the tunnel's roof were installed using pipe-jacking. A maximum of 4 mm of settlement was noted. After each pipe jacking operation, compensation grouting was performed. In this way, loose zones were quickly stabilized, and active settlement compensation was used to raise the footings above their initial level. Constructed between 1899 and 1905, the imposing historical structure was not damaged or disturbed during the entire tunneling project (Stelte & Paßlick, 2009).

5.5.1. Soilfrac – Additional Aspects

As the name suggests, soil-fracturing is the process of applying high pressure to break up the soil and fill the void with cement grout, which is left in the shape of lenses after the cement sets. Compaction and strength gain increase the bearing capacity of loose and soft soils, and the volume increase is employed to raise structures whose settlements have exceeded allowable limits.

The method is actually based on mineral oil technology, which has been modified to meet geotechnical engineering specifications (Baldock et al., 1992). An increasing number of cases involving the need to stop settlements if they are going too far and tilt buildings back to their original position during construction, if needed, are a result of the growth of new structures along with the widespread use of unfavorable sites for their location. This is done without ultimately compromising the building's intended functionality (George & Dasaka, 2023).

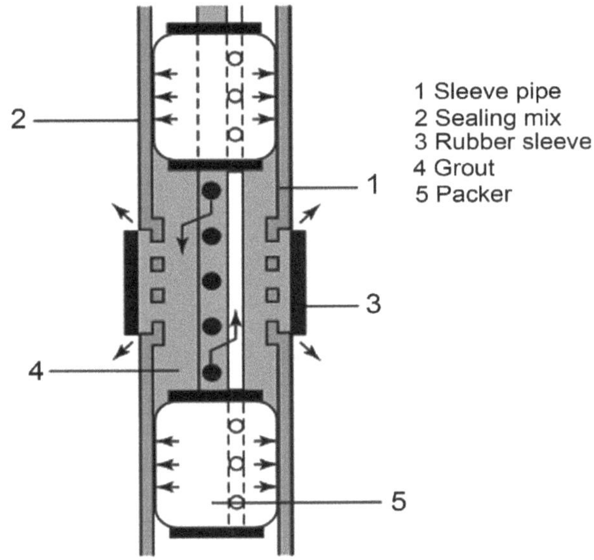

Figure 5.23. *A double packer assembly is depicted.*

5.5.2. Technology

Through the use of grout injection, the soil fracture occurs in a controlled and localized manner in this process. In order to achieve this, a network of injection pipes is inserted into the soil so that the 300–1000 mm sleeve openings are uniformly distributed throughout the soil layer that has to be treated. A suitable single or multiple packer system is used to serve each individual sleeve (Burke, 2000).

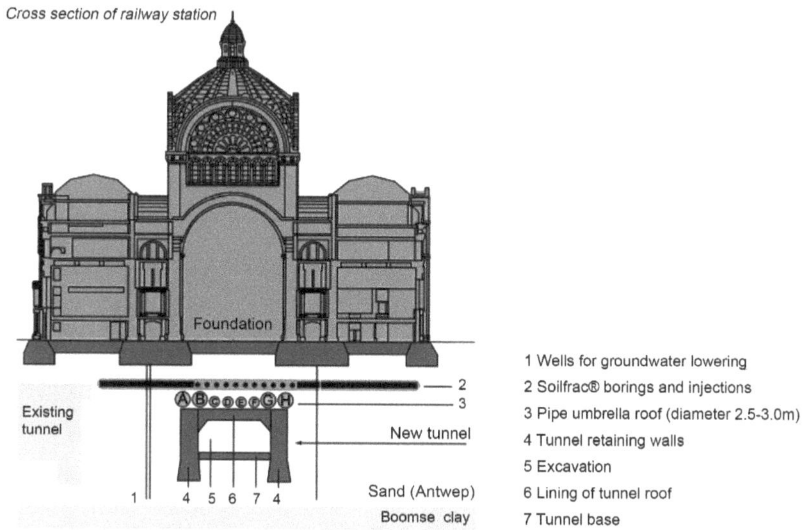

Figure 5.24. *Using soilfrac grouting for settlement compensation in Antwerp, Belgium.*

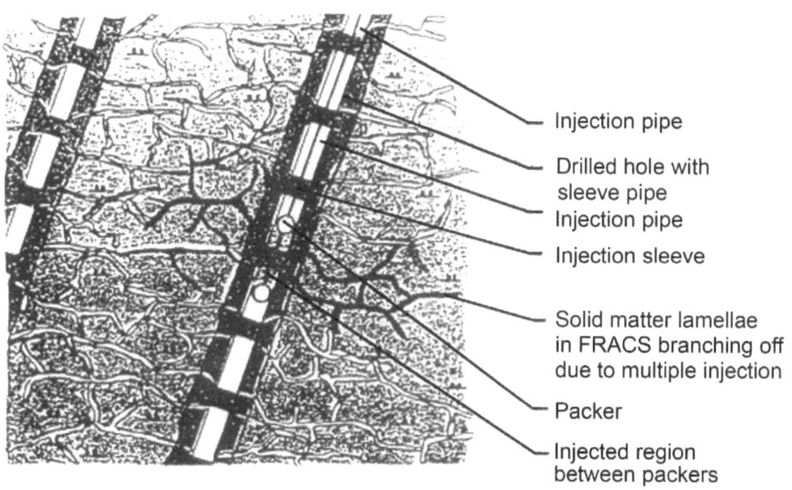

Figure 5.25. *Details of the soil fracturing pipes used (Raabe & Esters, 1997).*

A solid matter skeleton with many individual lenses branching off, as seen in the figure can be created by injecting different agents at appropriate gel times. Steel or plastic are used for the injection pipes. The process results in the hardened cement lenses creating a "force-locking" link to the surrounding soil, which strengthens the soil macroscopically (Hooper & Rieber, 1995).

5.5.2.1. Raising Buildings

In order to elevate buildings through soil-fracturing, consultation with a knowledgeable geotechnical engineer to decide on the injection plan's layout and the layer to be elevated by injection, which is also the layer that causes settlement. Heave that can be produced by soil-fracturing can range from a few to several hundred millimeters.

Raabe & Esters, (1997) give the example of using soil-fracturing to stabilize the tower of St. Gertrude's Church in Bochum-Wadenschied, Germany. The tower weighs about 3 million tons and is 65 meters high. As of 1975, the tower's spire had shifted by about one meter. The primary cause of the damage was variations in the water table brought on by surface sealing. Also, the foundation soil on top of the sandy marl was made up of loess and loose glacial deposits. The purpose of soil-fracturing was to stop the tower from further deflecting, not to move it back to its original position. 1500 meters of injection lines were installed during the project to connect the marl to the foundation's lower face. Grout was injected at 250 mm intervals to accomplish the goal, beginning with the outer rows (Koep et al., 2003).

5.6. DYNAMIC COMPACTION

Similar to vibroflotation, dynamics' effective compaction of cohesionless soils has been used as a straightforward "falling weight" technique to accomplish this goal. The most cost-effective method for enhancing sizable regions covered by such soils is "dynamic compaction," which is the term given to the resulting process. It entails the free fall of a large, raised weight in an established geometric pattern onto the soil surface by a crane (Figure 5.26). The surface depression serves as a gage for the soil below's level of compaction. As is evident, both the technology and the underlying principle of the dynamic compaction technique are straightforward (Seedorf et al., 1998).

The technique can be applied to landfill sites like spoil tips or waste dumps, as well as all kinds of mixed or non-cohesive soils, or soils with little cohesive material. Dynamic compaction has also been used to treat industrial waste dumps and refuse in numerous cases with success. However, the technique is not suited to mostly clayey soils and is less suitable for high silt content soils (Griesbeck et al., 2009).

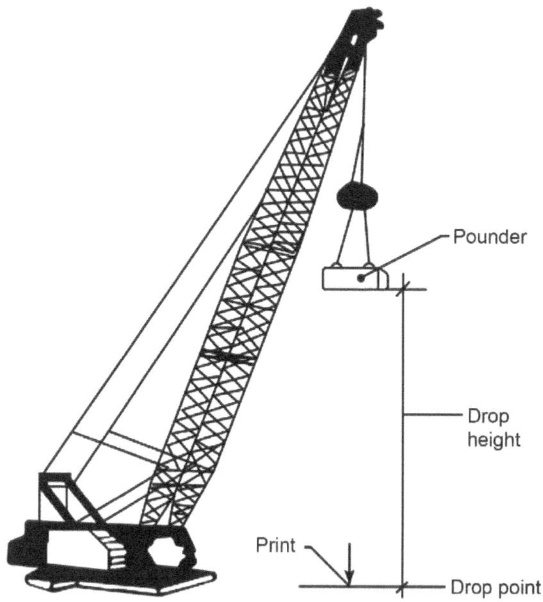

Figure 5.26. *Schematic diagram of a dynamic compaction (Yee, 1997).*

5.7. VACUUM CONSOLIDATION

In vacuum consolidation, a layer of saturated soil covered with an airtight membrane is compressed beneath a vacuum produced by suction pumps, using atmospheric pressure (that is, 0.1 N/mm^2 = 100 kN/m^2) in place of soil surcharge load (Figure 5.27).

The soil block to be vacuum-consolidated is first covered with a fill layer and a drainage layer. The process entails setting up vacuum pipes, excavating peripheral trenches, and then installing an impermeable, airtight membrane that extends to the bentonite-filled peripheral trench. The vacuum pumps are now operational, signaling the beginning of the consolidation process. Drains positioned horizontally in the drainage layer are used to accomplish dewatering. The degree of vacuum can be adjusted to control the consolidation settlement's rate and size (Long et al., 2015).

One of the method's many benefits is that it saves a lot of time compared to other traditional techniques like sand drains (Kosaka et al., 2016).

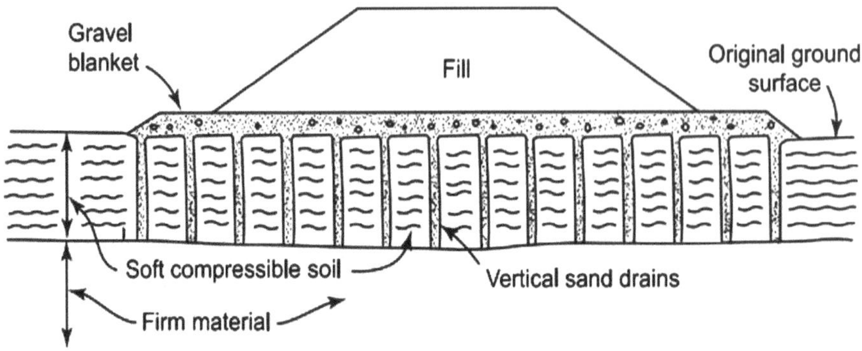

Figure 5.27. *Sand–drain – schematic of the sand drain (Peck et al., 1959).*

5.7.1. Vacuum Preconsolidation – Features

The pore water removal from a soil that is under load is known as consolidation. It is a time-dependent process that works with low-permeability cohesive soils (Due to its high permeability, it occurs instantly in cohesive soil.) The following three conditions must be met in order for consolidation to occur: 1) The load application (such as from new construction); 2) high water table, which causes the pore space in the soil below to become saturated; and 3) provision for pore water drainage. (Even with load, consolidation is impossible in a "sealed" system.)

Pre-consolidation is an attempt made prior to the start of construction to prevent large-scale post-construction settlements resulting from consolidation in a highly compressible soft soil. In order to lower the ensuing costs of maintenance and remedial actions and maintenance, this is required (Ni et al., 2019).

Pre-consolidation is required in order to minimize or manage post-construction settlement. Construction cannot be done on soft soils prior to pre-consolidation because of their extremely low shear strength and potential for a bearing capacity or shear failure. Pre-consolidation increases the soil's shear strength because it increases the parameters of shear strength i.e., φ and c. This allows for the development of a reliable and cost-effective design that has sufficient factor safety against a failure of the bearing capacity and settlement that stays within allowable bounds (Woo et al., 1989).

The earliest technique for pre-consolidation involves surcharge loading (made available following World War II) in conjunction with "sand drains," which improve drainage capacity. When an embankment is built, for example, to raise a highway or railway on a soft subgrade, the embankment acts as the surcharge since it is built at short heights to accommodate the soil's gradual bearing capacity and shear strength increase. The soil has the ability to flow towards the sand drains horizontally and vertically through them into the top drainage blanket. The water that has been drained in this way is directed into side ditches where it can enter natural water courses or sumps. With the development of geosynthetics in geotechnical engineering, the sand drain technique has seen numerous modifications, including "wick drains," "band drains," and others. In essence, these advancements are limited to the drain medium.

One novel and cutting-edge method of pre-consolidation is vacuum consolidation. This process involves creating vacuum in the soil, which in turn accelerates drainage. In this instance, surcharge is not required because the vacuum serves as a substitute for it. Stated differently, the vacuum acts as a "virtual" surcharge that results in consolidation (Seah et al., 2016).

5.7.2. IFCO's Vacuum Preconsolidation

The Dutch developer and instrument company, IFCO have a vacuum consolidation method that entails creating a ground vacuum in order to pre-consolidate cohesive soils with low permeability beneath the water table. At one site, they claim to have accomplished an amazing settlement of two meters in just 30 days (Fan et al., 2023).

The soil's shear strength increases as a result of the vacuum's creation of a decrease in pore-water pressure and an increase in effective stress. Three meters of sand fill at ground level is the same as a surcharge (load) of 0.5 bar (50 kN/m^2). According to reports, IFCO was able to reach a 95% vacuum.

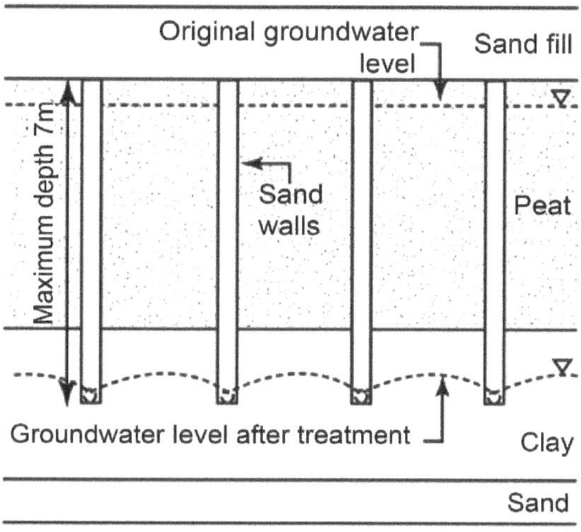

Figure 5.28. *Vacuum preconsolidation by IFCO (Vac. packed, 1998).*

The method makes use of a network of parallel, sand-filled trenches spaced a few meters apart throughout the site (Figure. 5.28) Generally, the trenches are a few hundred meters long, 250 mm wide, and 7 m deep, depending on how deep the soil needs to be covered. A deep trench cutting machine that can form up to 1 km of trench per day is used to excavate them.

The walls of sand serve as drains. The water that is drained into the sand wall from the side soil is collected by a pump at one end of the drain and guided down a perforated drain tube that is positioned at the trench bottom. Because clay has permeability that is five to ten times higher horizontally than vertically, the entire system is much more effective than vertical drainage systems (Kianfar et al., 2013).

The soil needs to be sealed off from the external atmosphere in order to create vacuum. For this purpose, airtight plastic sheeting is used to seal the trenches at ground level. Levees surrounding the area and the ability to pump extracted groundwater into this enclosed area above the sheets could act as a physical surcharge, hastening the settlement process. At the conclusion of pre-consolidation, when the vacuum system is disassembled, the soil will remain in its settled position while the water table rises to its initial level (Baral et al., 2021).

For roads, IFCO advises that wall drains be installed parallel to the axis, and for runways, perpendicular to the axis.

5.7.3. Additional Features of Menard Vacuum Consolidation

Vertical sand drains have been used for a long time to improve saturated and soft clayey soils by adding more drainage faces and paths and preloading or surcharging to create the necessary loading to cause consolidation. "Vacuum consolidation" is an alternative method that can get the same result more quickly and more effectively while also making the structure more stable against circular rotational slope failures and lateral deformations (Mesri & Khan, 2012).

The method involves covering the area that has to be treated with an airtight membrane (a PVC or HDPE geomembrane), creating a vacuum underneath the membrane with the help of dual venturi and vacuum pump. This vacuum corresponds to a preload fill of approximately 4 to 4.5 meters in height. If the method is combined with a physical surcharge, a greater preload effect may be obtained. To speed up drainage, vertical sand drains have been used, just like in the traditional surcharge method. For the first time, the technology was effectively deployed on a significant magnitude in France (Indraratna et al., 2011).

5.7.3.1. Historical Background

W. Kjellman introduced the concept of "vacuum preloading" at the Royal Geological University in Sweden in 1952. In 1957, this technique, with a vacuum pressure below 50 kN/m^2, was used in connection with the construction of a runway extension at Philadelphia Airport in U.S.A.

Menard carried out the first full scale trial in France in 1980 under the technical supervision of Professor J.M. Cognon. Further field trials followed in China and USA (Riley et al., 2000).

5.7.3.2. Conventional Method

The term "vacuum preloading" was first used in 1952 at the Royal Geological University of Sweden by W. Kjellman. This method—which uses a vacuum pressure of less than 50 kN/m^2—was applied in 1957 when Philadelphia Airport in the United States was building a new runway (Ajai et al., 2013).

In 1980, Menard conducted the first large-scale trial in France, working under Professor J.M. Cognon's technical guidance. There were more field tests in China and the USA after that.

5.7.3.3. General Description of the Vacuum Method

Vacuum consolidation was first proposed by Louis Menard in the early 1970s. Continuing this research, Cognon discovered in 1980 that vacuum pressure could be kept in the field for several months with constant vacuum pumping (de Oliveira et al., 2013).

The vacuum method involves covering the area that needs to be treated with an airtight, impermeable membrane. To prevent leakage, the membrane needs to be securely fastened and sealed into the peripheral trenches encircling the treatment area. One bar of vacuum pressure, or 100 kN/m^2, is employed. Following the deduction of leakage losses, a pressure range of 70–80 kN/m^2 is reasonable. It is important to take precautions to guarantee that the membrane's apertures for different tube connections are airtight (Somiya et al., 1964).

Vacuum consolidation is generally isotropic and approximately 30% more efficient than the preloading method. It produces a more uniform stress pattern across the treated soil area. The tiny "gas bags" in clay layer are diminished by the vacuum pressure, improving permeability and speeding up the consolidation process. The fill material is always present but submerged over the horizontal drains but below the membrane. In contrast to preload fill, where each meter of submerged fill loses a load intensity of 10 kN/m^2 due to submergence—more than half that resulting from the original dry weight—the surcharge load intensity does not decrease as a result! Preload or surcharge are frequently added to the vacuum method, either by adding more fill material or by applying water to the impermeable membrane (Badalian et al., 2006).

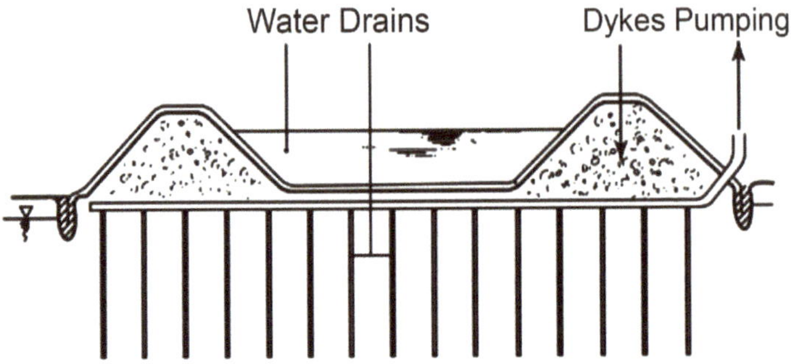

Figure 5.29. *Vacuum method with preloading (Yee, 1997).*

In the early projects, extensive instrumentation and routine monitoring were conducted as part of an R&D effort. The range of vacuum pressures was 80–90 kN/m^2. Over a few months, there was a mere 2 percent variation in vacuum pressure. The field experiences from multiple early project sites are compiled in Table 5.1 and attest to the effectiveness of this method.

Table 5.1. Vacuum Consolidation – Performance of Vacuum Consolidation at the Selected Sites in France (Yee, 1997).

Site No.	Duration of vacuum consolidation (months)	Spacing of cylindrical vertical drains (m)	$\left(\frac{C_c}{1+e_o}\right)$ (average)	Average settlement ratio* (%)
1	4.0	1.5	0.21	12.5
2	4.5	1.8	0.17	8.5
3	3.0	1.5	0.22	9.5
4	2.5	1.3	0.27	13.5
5	3.5	1.5	0.18	7.5
6	3.5	1.5	0.20	10.5
7	3.0	1.4	0.56	18.0

Settlement ratio is the proportion of the vertical drain length within the saturated clay layer that needs to be treated to the ground surface settlement (Terzis & Christodoulakis, 2009).

SUMMARY

In the chapter "Special Ground Improvement Methods," we delve into advanced techniques aimed at enhancing soil stability and load-bearing capacity for construction projects. This section explores innovative methods such as deep soil mixing, dynamic compaction, and stone columns. Deep soil mixing involves blending soil with cementitious material to form columns or walls, reinforcing the ground. Dynamic compaction employs heavy machinery to compact soil layers, reducing settlement and increasing density. Stone columns involve the insertion of compacted stone into the ground, improving drainage and supporting structures. Through these specialized approaches, engineers can address challenging soil conditions and ensure the durability and safety of infrastructure projects.

REFERENCES

1. Ajai, J. T., Imoko, B. I., & O'kwu, E. I. (2013). Comparison of the learning effectiveness of problem-based learning (PBL) and conventional method of teaching algebra. *Journal of Education and Practice, 4*(1), 131–135.
2. Alberi, K., Nardelli, M. B., Zakutayev, A., Mitas, L., Curtarolo, S., Jain, A., & Perkins, J. (2018). The 2019 materials by design roadmap. *Journal of Physics D: Applied Physics, 52*(1), 013–100.
3. Al-Homoud, A. S., & Wehr, W. (2006). Experience of vibrocompaction in calcareous sand of UAE. *Geotechnical & Geological Engineering, 24*, 757–774.
4. Ambily, A. P., & Gandhi, S. R. (2007). Behavior of stone columns based on experimental and FEM analysis. *Journal of Geotechnical and Geoenvironmental Engineering, 133*(4), 405–415.
5. Anderson, E., & Järvinen, H. (1999). Variational quality control. *Quarterly Journal of the Royal Meteorological Society, 125*(554), 697–722.
6. Arson, C., & Juge, B. (2012). A hollow sphere model to dimension soilcrete columns. In *GeoCongress 2012: State of the Art and Practice in Geotechnical Engineering* (Vol. 1, pp. 1006–1015).
7. Arulrajah, A., Abdullah, A., Bo, M. W., & Bouazza, A. (2009). Ground improvement techniques for railway embankments. *Proceedings of the Institution of Civil Engineers-Ground Improvement, 162*(1), 3–14.
8. Ashlock, P. D. (1974). The uses of cladistics. *Annual Review of Ecology and Systematics, 5*(1), 81–99.
9. Ausilio, E., & Conte, E. (2007). Soil compaction by vibro-replacement: A case study. *Proceedings of the Institution of Civil Engineers-Ground Improvement, 11*(3), 117–126.
10. Badalian, A. M., Simonov, Y. A., & Shevchenko, V. I. (2006). Method of vacuum correlation functions: Results and prospects. *Physics of Atomic Nuclei, 69*, 1781–1802.
11. Baldock, J. A., Oades, J. M., Waters, A. G., Peng, X., Vassallo, A. M., & Wilson, M. A. (1992). Aspects of the chemical structure of soil organic materials as revealed by solid-state 13C NMR spectroscopy. *Biogeochemistry, 16*, 1–42.
12. Baral, P., Indraratna, B., Rujikiatkamjorn, C., Kelly, R., & Vincent, P. (2021). Consolidation by vertical drains beneath a circular embankment under surcharge and vacuum preloading. *Journal of Geotechnical and Geoenvironmental Engineering, 147*(8), 0502–1004.
13. Baumann, V., & Bauer, G. E. A. (1974). The performance of foundations

on various soils stabilized by the vibro-compaction method. *Canadian Geotechnical Journal, 11*(4), 509–530.

14. Bhopal, M., & Rowley, C. (2014). Ethnicity as a management issue and resource: Examples from Malaysia. In *Multicultural Organizations in Asia* (Vol. 2, No. 1, pp. 91–112).
15. Blair-Early, A., & Zender, M. (2008). User interface design principles for interaction design. *Design Issues, 24*(3), 85–107.
16. Boggs, S. (1967). A numerical method for sandstone classification. *Journal of Sedimentary Research, 37*(2), 548–555.
17. Brown, R. E. (1977). Vibroflotation compaction of cohesionless soils. *Journal of the Geotechnical Engineering Division, 103*(12), 1437–1451.
18. Burke, A. (2000). Ultracapacitors: Why, how, and where is the technology. *Journal of Power Sources, 91*(1), 37–50.
19. Cai, G., Lin, J., Liu, S., & Puppala, A. J. (2017). Characterization of spatial variability of CPTU data in a liquefaction site improved by vibro-compaction method. *KSCE Journal of Civil Engineering, 21*, 209–219.
20. Calvin, T. (1983). Quality control techniques for "zero defects." *IEEE Transactions on Components, Hybrids, and Manufacturing Technology, 6*(3), 323–328.
21. Chai, J. C., Carter, J. P., & Hayashi, S. (2005). Ground deformation induced by vacuum consolidation. *Journal of Geotechnical and Geoenvironmental Engineering, 131*(12), 1552–1561.
22. D'Appolonia, E., Miller Jr, C. E., & Ware, T. M. (1955). Sand compaction by vibroflotation. *Transactions of the American Society of Civil Engineers, 120*(1), 154–168.
23. Datye, K. R., & Nagaraju, S. S. (1975). Installation and testing of rammed stone columns. In *Proceedings of IGS Specialty Session, 5th Asian Regional Conference on Soil Mechanics and Foundation Engineering, Bangalore, India* (Vol. 4, No. 2, pp. 101–104).
24. De Koninck, R. (1979). The integration of the peasantry: Examples from Malaysia and Indonesia. *Pacific Affairs, 52*(2), 265–293.
25. de Oliveira, L. P., Verstraete, J. J., & Kolb, M. (2013). Development of a general modelling methodology for vacuum residue hydro conversion. *Oil & Gas Science and Technology–Revue d'IFP Energies nouvelles, 68*(6), 1027–1038.
26. Ding, Y., Schuring, J. R., & Chan, P. C. (1999). Parameter determination for engineering applications of pneumatic fracturing analysis. *Practice*

Periodical of Hazardous, Toxic, and Radioactive Waste Management, 3(4), 170–177.

27. Do, D. H., & Pham, T. A. (2018). Investigation of performance of soil-cement pile in support of foundation systems for high-rise buildings. *Civil Engineering Journal, 4*(2), 122–266.
28. Dobrescu, C. F., & Brăguță, E. (2018). Optimization of vibro-compaction technological process considering rheological properties. In *Acoustics and Vibration of Mechanical Structures—AVMS-2017: Proceedings of the 14th AVMS Conference, Timisoara, Romania, May 25–26, 2017* (Vol. 4, No. 2, pp. 287–293).
29. Duan, W., Cai, G., Liu, S., Yuan, J., & Puppala, A. J. (2019). Assessment of ground improvement by vibro-compaction method for liquefiable deposits from in-situ testing data. *International Journal of Civil Engineering, 17*, 723–735.
30. Egan, D., & Slocombe, B. C. (2010). Demonstrating environmental benefits of ground improvement. *Proceedings of the Institution of Civil Engineers-Ground Improvement, 163*(1), 63–69.
31. El Kamash, W., & El Naggar, H. (2021). Numerical simulation of the installation of vibro displacement columns in normally consolidated clay using a field case study. *International Journal of Geosynthetics and Ground Engineering, 7*, 1–19.
32. El Ouni, M. R., Bouassida, M., & Das, B. M. (2009). Vibro compaction improvement of Tunisian liquefiable sands. In *Proceedings of the 17th International Conference on Soil Mechanics and Geotechnical Engineering* (Volumes 1, 2, 3 and 4) (Vol. 2, pp. 2366–2369).
33. El-Kelesh, A. M., Mossaad, M. E., & Basha, I. M. (2001). Model of compaction grouting. *Journal of Geotechnical and Geoenvironmental Engineering, 127*(11), 955–964.
34. Emad, W., Salih, A., & Kurda, R. (2021). Forecasting the mechanical properties of soilcrete using various simulation approaches. In *Structures* (Vol. 34, pp. 653–665).
35. Erol, S., & François, B. (2014). Efficiency of various grouting materials for borehole heat exchangers. *Applied Thermal Engineering, 70*(1), 788–799.
36. Fan, L., Xun, Z., & Peng, S. (2023). A comparative case study on drainage consolidation improvement of soft soil under vacuum preloading and surcharge preloading. *Applied Sciences, 13*(9), 57–82.
37. Fang, Y. S., Liao, J. J., & Lin, T. K. (1994). Mechanical properties of

jet grouted soilcrete. *Quarterly Journal of Engineering Geology and Hydrogeology, 27*(3), 257–265.

38. Fonseca Guerra, C., Handgraaf, J. W., Baerends, E. J., & Bickelhaupt, F. M. (2004). Voronoi deformation density (VDD) charges: Assessment of the Mulliken, Bader, Hirshfeld, Weinhold, and VDD methods for charge analysis. *Journal of Computational Chemistry, 25*(2), 189–210.
39. Friedman, R. M. (2008). Clinical uses of interferons. *British Journal of Clinical Pharmacology, 65*(2), 158–162.
40. Fu, K. K., Yang, M. C., & Wood, K. L. (2016). Design principles: Literature review, analysis, and future directions. *Journal of Mechanical Design, 138*(10), 101103.
41. Gazdek, M., Strelec, S., & Rezo, M. (2011). Estimation of vibro replacement by compression seismic waves. *Tehnicki Vjesnik, 18*(2), 243–252.
42. Gemperle, F., Kasabach, C., Stivoric, J., Bauer, M., & Martin, R. (1998). Design for wearability. In *Digest of Papers. Second International Symposium on Wearable Computers* (Vol. 2, No. 1, pp. 116–122).
43. George, T. I., & Dasaka, S. M. (2023). Mechanics of arch action in soil arching. *Acta Geotechnica, 18*(4), 1991–2009.
44. Gottardi, G., Lionello, A., Marchi, M., & Rossi, P. P. (2015). Monitoring-driven design of a multiphase intervention for the preservation of the Frari bell tower in Venice. *Rivista Italiana di Geotecnica, 49*(1), 45–64.
45. Griesbeck, A. G., Neudörfl, J., Hörauf, A., Specht, S., & Raabe, A. (2009). Antimalarial peroxide dyads from natural artemisinin and hydroxyalkylated 1, 2, 4-trioxanes. *Journal of Medicinal Chemistry, 52*(10), 3420–3423.
46. Hamed Khodadadi, T., Kavazanjian, E., van Paassen, L., & DeJong, J. (2017). Bio-grout materials: A review. *Grouting 2017*, 1–12.
47. Helson, O., Beaucour, A. L., Eslami, J., Noumowe, A., & Gotteland, P. (2017). Physical and mechanical properties of soilcrete mixtures: Soil clay content and formulation parameters. *Construction and Building Materials, 131*, 775–783.
48. Hooper, S., & Rieber, L. P. (1995). Teaching with technology. In *Teaching: Theory into Practice* (pp. 154–170).
49. Hughes, J. M. O., & Withers, N. J. (1974). Reinforcing of soft cohesive soils with stone columns: 18F, 9R. In *Ground Engineering* (Vol. 7, No. 3, pp. 42–49).
50. Hughes, W. J., & Southwood, D. J. (1976). An illustration of modification

of geomagnetic pulsation structure by the ionosphere. *Journal of Geophysical Research, 81*(19), 3241–3247.

51. Indraratna, B., Rujikiatkamjorn, C., Ameratunga, J., & Boyle, P. (2011). Performance and prediction of vacuum combined surcharge consolidation at Port of Brisbane. *Journal of Geotechnical and Geoenvironmental Engineering, 137*(11), 1009–1018.
52. Ivanov, M. V., & Ksenofontov, B. S. (2014). Intensification of flotation treatment by exposure to vibration. *Water Science and Technology, 69*(7), 1434–1439.
53. Jeong, G. S., Park, B. S., Hong, Y. K., & Yoo, N. J. (2004). An experimental study on the characteristics of vibro-compaction of crushed stones. *Journal of Industrial Technology, 24*(B), 89–94.
54. Keller, T., Lamandé, M., Peth, S., Berli, M., Delenne, J. Y., Baumgarten, W., & Or, D. (2013). An interdisciplinary approach towards improved understanding of soil deformation during compaction. *Soil and Tillage Research, 128*, 61–80.
55. Kennedy, J. C. (2002). Leadership in Malaysia: Traditional values, international outlook. *Academy of Management Perspectives, 16*(3), 15–26.
56. Khoshsirat, V., Bayesteh, H., & Sharifi, M. (2019). Effect of high salinity in grout on the performance of cement-stabilized marine clay. *Construction and Building Materials, 217*, 93–107.
57. Kianfar, K., Indraratna, B., & Rujikiatkamjorn, C. (2013). Radial consolidation model incorporating the effects of vacuum preloading and non-Darcian flow. *Géotechnique, 63*(12), 1060–1073.
58. Kılıc, A., Yasar, E., & Celik, A. G. (2002). Effect of grout properties on the pull-out load capacity of fully grouted rock bolt. *Tunnelling and Underground Space Technology, 17*(4), 355–362.
59. Kimball, M. A. (2013). Visual design principles: An empirical study of design lore. *Journal of Technical Writing and Communication, 43*(1), 3–41.
60. Koep, S., Gais, H. J., & Raabe, G. (2003). Asymmetric synthesis of unsaturated, fused bicyclic proline analogues through amino alkylation of cyclic bis (allylsulfoximine) titanium complexes and migratory cyclization of δ-amino alkenyl aminosulfoxonium salts. *Journal of the American Chemical Society, 125*(43), 13243–13251.
61. Kolovos, K. G., Asteris, P. G., Cotsovos, D. M., Badogiannis, E., & Tsivilis, S. (2013). Mechanical properties of soilcrete mixtures modified

with metakaolin. *Construction and Building Materials, 47*, 1026–1036.

62. Kosaka, T., Hayashi, H., Kawaida, M., & Teerachaikulpanich, N. (2016). Performance of vacuum consolidation for reducing a long-term settlement. *Japanese Geotechnical Society Special Publication, 2*(59), 2015–2020.
63. Kurian, J. V. (2005). A new polymer platform for the future—Sorona® from corn derived 1, 3-propanediol. *Journal of Polymers and the Environment, 13*, 159–167.
64. Li, S., Zhang, J., Li, Z., Gao, Y., Qi, Y., Li, H., & Zhang, Q. (2019). Investigation and practical application of a new cementitious anti-washout grouting material. *Construction and Building Materials, 224*, 66–77.
65. Liong, G. T. (2022). Investigation of Dynamic Compaction and Vibro-compaction to Mitigate Liquefaction: A Case Study. *Indonesian Geotechnical Journal, 1*(2), 38–49.
66. Long, P. V., Nguyen, L. V., Bergado, D. T., & Balasubramaniam, A. S. (2015). Performance of PVD improved soft ground using vacuum consolidation methods with and without airtight membrane. *Geotextiles and Geomembranes, 43*(6), 473–483.
67. Lowekamp, B. C., Chen, D. T., Ibáñez, L., & Blezek, D. (2013). The design of SimpleITK. *Frontiers in neuroinformatics, 7*, 45–199.
68. McClements, D. J., Newman, E., & McClements, I. F. (2019). Plant based milks: A review of the science underpinning their design, fabrication, and performance. *Comprehensive Reviews in Food Science and Food Safety, 18*(6), 2047–2067.
69. Mesri, G., & Khan, A. Q. (2012). Ground improvement using vacuum loading together with vertical drains. *Journal of Geotechnical and Geoenvironmental Engineering, 138*(6), 680–689.
70. Mishra, B. (2016). A study on ground improvement techniques and its applications. *International Journal of Innovative Research in Science, Engineering and Technology, 5*(1), 72–86.
71. Modoni, G., & Bzówka, J. (2012). Analysis of foundations reinforced with jet grouting. *Journal of Geotechnical and Geoenvironmental Engineering, 138*(12), 1442–1454.
72. Modoni, G., Croce, P., & Mongiovì, L. (2006). Theoretical modelling of jet grouting. *Géotechnique, 56*(5), 335–347.
73. Muro, T., Kawahara, S., & Mitsubayashi, T. (2001). Comparison between centrifugal and vertical vibro-compaction of high-lifted decomposed weathered granite sandy soil using a tracked vehicle. *Journal of Terramechanics, 38*(1), 15–45.

74. Muro, T., Miyoshi, M., & Mitsubayashi, T. (1998). Effect of a tracked vehicle mounting an oscillator on the vibro-compaction of a high lifted decomposed granite sandy soil. *Soils and Foundations, 38*(4), 129–144.

75. Nagy, P., Adam, D., KOPF, F., & Freitag, P. (2018). Experimental and theoretical investigation of deep vibro compaction. *Ce/Papers, 2*(2–3), 725–730.

76. Ni, P., Xu, K., Mei, G., & Zhao, Y. (2019). Effect of vacuum removal on consolidation settlement under a combined vacuum and surcharge preloading. *Geotextiles and Geomembranes, 47*(1), 12–22.

77. Nikbakhtan, B., & Osanloo, M. (2009). Effect of grout pressure and grout flow on soil physical and mechanical properties in jet grouting operations. *International Journal of Rock Mechanics and Mining Sciences, 46*(3), 498–505.

78. Nikbakhtan, B., Ahangari, K., & Rahmani, N. (2010). Estimation of jet grouting parameters in Shahriar dam, Iran. *Mining Science and Technology (China), 20*(3), 472–477.

79. Ochmański, M., Modoni, G., & Bzówka, J. (2015). Prediction of the diameter of jet grouting columns with artificial neural networks. *Soils and Foundations, 55*(2), 425–436.

80. Okumus, P., Oliva, M. G., & Hoehn, T. (2015). Impact of cast-in-place concrete column construction vibrations on sensitive occupancies. *Practice Periodical on Structural Design and Construction, 20*(1), 0401–4023.

81. Penman, S. H. (1980). An empirical investigation of the voluntary disclosure of corporate earnings forecasts. *Journal of Accounting Research,3*(2), 132–160.

82. Peterson, R. A., & Anand, N. (2004). The production of culture perspective. *Annu. Rev. Sociol., 30*, 311–334.

83. Pettersson, R. (2010). Information design–principles and guidelines. *Journal of Visual Literacy, 29*(2), 167–182.

84. Priebe, H. J. (1995). The design of vibro replacement. *Ground engineering, 28*(10), 31–166.

85. Raju, T. N., Langenberg, P., Bhutani, V., & Quinn, G. E. (1997). Vitamin E prophylaxis to reduce retinopathy of prematurity: A reappraisal of published trials. *The Journal of Pediatrics, 131*(6), 844–850.

86. Renton-Rose, D. G., Bunce, G. C., & Finlay, D. W. (2000). Vibro-replacement for industrial plant on reclaimed land, Bahrain. *Geotechnique, 50*(6), 727–737.

87. Riley, M. W., & Riley Jr, J. W. (2000). Age integration: Conceptual and

historical background. *The Gerontologist, 40*(3), 266–270.

88. Rodbard, D., Rayford, P. L., Cooper, J. A., & Ross, G. T. (1968). Statistical quality control of radioimmunoassays. *The Journal of Clinical Endocrinology & Metabolism, 28*(10), 1412–1418.
89. Sanogo, B., Kocaefe, D., Kocaefe, Y., Bhattacharyay, D., & Côté, J. (2021). Determination of optimum vibro-compaction time using sound analysis. *Journal of Manufacturing Processes, 68*, 118–125.
90. Sarnecki, J., Vinodrai, C., Javed, A., O'Kelly, P., & Dick, K. (1993). Microcell design principles. *IEEE Communications Magazine, 31*(4), 76–82.
91. Seah, T. H., Kim, T. B., & Nguyen, T. D. (2016). Ground improvement via vacuum consolidation method in Vietnam. *Geotechnical Engineering, 47*(4), 80–88.
92. Seedorf, U., Raabe, M., Ellinghaus, P., Kannenberg, F., Fobker, M., Engel, T., & Assmann, G. (1998). Defective peroxisomal catabolism of branched fatty acyl coenzyme A in mice lacking the sterol carrier protein-2/sterol carrier protein-x gene function. *Genes & development, 12*(8), 1189–1201.
93. Sexton, B. G., Sivakumar, V., & McCabe, B. A. (2017). Creep improvement factors for vibro-replacement design. *Proceedings of the Institution of Civil Engineers-Ground Improvement, 170*(1), 35–56.
94. Shaqour, F. M., & Hasan, S. E. (2008). Groundwater control for construction purposes: A case study from Kuwait. *Environmental Geology, 53*, 1603–1612.
95. Shen, S. L., Wang, Z. F., Horpibulsuk, S., & Kim, Y. H. (2013). Jet grouting with a newly developed technology: The Twin-Jet method. *Engineering Geology, 152*(1), 87–95.
96. Shirlaw, J. N. (1995). Observed and calculated pore pressures and deformations induced by an earth balance shield: Discussion. *Canadian Geotechnical Journal, 32*(1), 181–189.
97. Smiley, E. T. (2001). Terravent™: Soil fracture patterns and impact on bulk density. *Journal of Arboriculture, 27*(6), 326–330.
98. Somiya, T., Hirano, S., Kamada, H., & Ogahara, I. (1964). Determination of gases in metals by an improved vacuum fusion method. *Talanta, 11*(3), 581–592.
99. Soukup, D. A., Buck, B. J., & Harris, W. (2008). Preparing soils for mineralogical analyses. *Methods of Soil Analysis Part 5—Mineralogical Methods, 5*, 13–31.
100. Spagnoli, G., Romero, E., Fraccica, A., Arroyo, M., & Gómez, R. (2022).

The effect of curing conditions on the hydromechanical properties of a metakaolin-based soilcrete. *Géotechnique, 72*(5), 455–469.

101. Spallek, J., Zeeb, H., & Razum, O. (2010). Prevention among immigrants: The example of Germany. *BMC Public Health, 10*, 1–6.
102. Stelte, M., & Paßlick, T. (2009). Protection of structures with the SOILFRAC® method in the course of two large tunnel construction projects in Germany. In *Proceedings of the 17th International Conference on Soil Mechanics and Geotechnical Engineering (Volumes 1, 2, 3 and 4)* (pp. 2240–2243). IOS Press.
103. Stiles, W. B. (1993). Quality control in qualitative research. *Clinical Psychology Review, 13*(6), 593–618.
104. Terzis, P. A., & Christodoulakis, T. (2009). The general solution of Bianchi type VII h vacuum cosmology. *General Relativity and Gravitation, 41*, 469–495.
105. Triantafyllidis, T., & Kimmig, I. (2019). A simplified model for vibro compaction of granular soils. *Soil Dynamics and Earthquake Engineering, 122*, 261–273.
106. Wang, S. L., Mu, M. G., Chen, D. W., & Ren, G. (2012). Field design of jet grouting parameters on soilcrete columns. *Applied Mechanics and Materials, 170*, 3068–3071.
107. Webb, D. L., & Hall, R. I. (1969). Effects of vibroflotation on clayey sands. *Journal of the Soil Mechanics and Foundations Division, 95*(6), 1365–1378.
108. Westgard, J. O. (2003). Internal quality control: Planning and implementation strategies. *Annals of Clinical Biochemistry, 40*(6), 593–611.
109. Woo, S. M., Moh, Z. C., Van Weele, A. F., Chotivittayathanin, R., & Trangkarahart, T. (1989). Preconsolidation of soft Bangkok clay by vacuum loading combined with non-displacement sand drains. In *Proc., 12th ICSMFE.Vol. 2* (pp. 1431–1434).
110. Yarkoni, Y., Getahun, A., & Cambier, J. C. (2010). Molecular underpinning of B-cell anergy. *Immunological Reviews, 237*(1), 249–263.
111. Yuan, C. H. E. N.. G., & Song-yu, L. I. U. (2013). Application of resonant vibro-compaction in treatment of liquefiable sites. *Chinese Journal of Geotechnical Engineering, 35*(zk2), 83–87.
112. Zhi-Guang, Q. I. N. G., Xiao-Ming, Y. U. A. N., Ben, N. I. U., & Zhen-Zhong, C. A. O. (2020). Vibroflotation method for foundation treatment of replaced sandy soil in deepwater marine environment. *Chinese Journal of Geotechnical Engineering, 42*(10), 1940–1946.

CHAPTER 6

FOUNDATION ON SOFT AND FILLED-UP-SOIL

Contents

6.1. INTRODUCTION

The first step in building a shallow foundation is laying mud mat, which can be very challenging in areas with soft ground and high water table. Water seeping at the foundation level can remove cement from freshly mixed concrete in a variety of places, making it difficult to control without complex dewatering arrangements. There may not be any guidelines on how to handle such a tough situation when they are most needed (Roy, 2024).

It can be challenging to prepare the founding layer and build a foundation in former hard material dumping sites. A few treatments (pre-construction) on these kinds of sites might fix the problems. The design and construction techniques for "hardstands," which are frequently used to support heavy machinery or as a storage area for bulky items like steel coils and ingots, have been described. Compaction grouting serves to reduce compaction or settling of the rock or soil beneath the foundation and to increase bearing capacity. Grouting is a specialty topic that is designed with a particular site and goals in mind (Gupta et al., 2017).

6.2. CONSTRUCTION OF FOUNDATION IN SOFT SOIL

Large subterranean or partially subterranean/projecting tanks used for the storage of chemicals, oil, or water are frequently constructed on soft soil with a high groundwater table. Why such a tank is to be constructed on soft soils may be a question. Since petrochemical and water supply plants need to store both crude and refined liquids, they need a series of tanks. It is standard procedure to arrange several tanks in various capacities and sizes in a convenient location known as a "tank farm." These tanks typically have moderate bearing pressure requirements and they are situated outside of the plant's core areas (Raithel et al., 2005). For instance, raw water is gathered from a well or river and treated before being sent to the distribution network to supply plants or the township. For cycles of storage and treatment, a number of distribution and storage tanks are required. Water treatment facilities are ideally situated close to river basins because water is obtained from perennial rivers. In basin areas, the subsoil is typically soft and has a high-water table. When building a base slab for a tank, seepage from the excavation's bottom is a common issue that must be managed in order to build rafts and mud mats (Zhao & Zhao, 2021).

6.2.1. Excavations in a Soft Ground

In soft ground, foundation excavation must be done carefully and strategically. The area and depth of excavation must be done in phases, depending on the groundwater level, depth and the area. It is best to carry out backfilling, civil construction, and excavation simultaneously. The earth that has been excavated needs to be stored at least 1.5 times the distance of its original depth. Additionally, the pit shouldn't be allowed to fill with water.

To prevent water from recirculating back into the excavation, it can be extracted using a sump pump and appropriately drained away from the excavation (Yong & Oh, 2016). Deep excavation can be done in two to three steps, with each step providing a berm that is roughly 1.0 m wide and a side slope (V:H) of 1:1.5. It is important to keep in mind that soil loses strength if it is exposed to the elements or allowed to stay in shallow water. It is safer to halt bed excavation roughly 300–500 mm above the design founding level, and to complete final excavation right before mud mat installation. So, all the plans for concrete and other materials should be ready before the final level is excavated (Wang et al., 2010).

6.2.2. Mud Mat on Soft Ground

The bottom of a foundation excavation in soft clay/silt and loose sand strata in the high ground-water areas frequently becomes soft and porous, which makes mud mat installation challenging. Typically, actions to counteract this problem include:

i. Spreading dry granular material (such as slag, brick bats, chips, and leftover building materials)
ii. Increase the mud mat thickness.

In case these methods don't work, the steps below might be useful (Kamali & Hashim, 2010).

6.2.2.1. Boulder Soling Covered in Mud Mat

A layer of slag or stone boulder soling that is about 200 mm thick and is spread out evenly and properly by hand over the surface that has been excavated. The PCC mud mat is then laid as usual (Mahajan et al., 2008).

6.2.2.2. Mud Mat in Soft Ground Suitable for Large Rafts

Lay out a large mud mat in panels. The space will be partitioned into several panels of an appropriate size. First, alternative panels must be cast. Panels that have been cast earlier must be set initially before balancing panels are cast. Mud mat panels must be cast in a specific pattern, such as from one side, end, or corner to the other, from the outside to the inside, or in reverse order. A few sump pits need to be dug at predetermined locations in case there is a greater ingress of groundwater into the excavation (Dunne & Martin, 2017). The water collected in the pits must be pumped out at defined intervals.

This will reduce the hydrostatic pressure-induced inclination to swell. Just before foundation concrete is laid, PCC combined with water-resisting additives will be used to seal the pits. Before shutting, tiny pipe fragments may be put into the pit if needed. Once raft casting is finished, the pipes may be sealed. The construction of a partially or fully underground reservoir, a basement raft, and a clarifier can all benefit from such treatments (Feng et al., 2017).

6.2.2.3. Precast PCC Mud Mat

The aforementioned techniques might not be adequate in places with higher groundwater levels; or loose or soft ground. One might wonder why is a soft, low-bearing capacity area that is susceptible to high settlement chosen for a foundation? When building a pile cap, for example, soil bearing capacity may not be the primary concern. PCC tiles may serve as a mud mat in these situations. The following section discusses how to lay precast mud mats as depicted in Figure 6.1.

i. PCC slabs/tiles, measuring 600 × 400 × 37 to 40 mm in thickness, are correctly cast and cured.

ii. Following excavation, the PCC tiles are then positioned evenly, using a light mortar to fill in any gaps.

iii. Precast tiles should be laid in a second layer in the same pattern as the first, but covering the fir layer joints like brick bonding. Sand packing should be used to level the first layer's undulations so that the top of the second layer is level.

iv. Curved or angular tiles should be installed close to the pile cut-off level.

v. Before installing precast mud mat, pile chipping should be finished (Saeed, 2008).

6.3. FOUNDATION FOR HETEROGENEOUS GROUNDS

Owing to a shortage of space for expansion or new plants construction, plant authorities frequently have to use abandoned dumping sites that have been lying idle for years. These locations have inherent issues that must be resolved before allowing new building (Honjo, 1982).

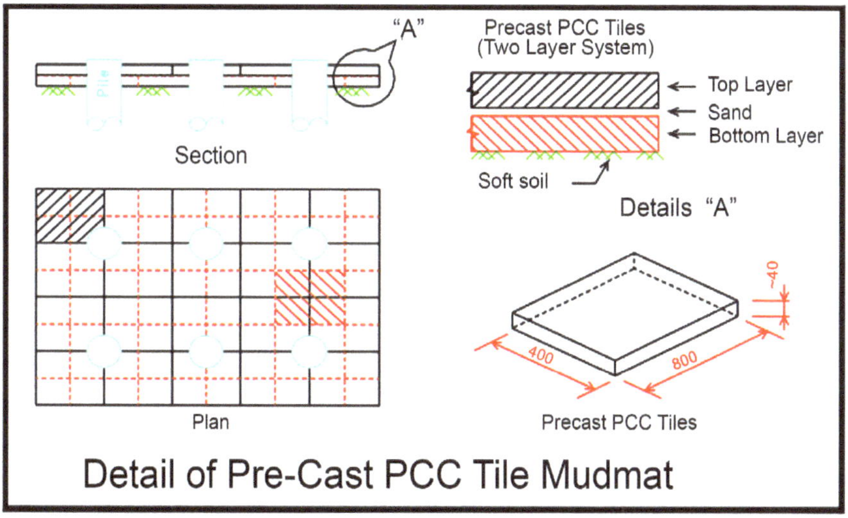

Figure 6.1. *Details of a mud mat containing precast PCC tiles.*

Each dumping site needs to be handled differently because they are all unique. For years, the construction of an open foundation at a highly heterogeneous hard material disposal site has been fraught with uncertainties that make it impossible to accurately assess or develop generalized procedures. For information, the following general rules for open foundation design and construction in such places are outlined (Kolar & Veljković, 1975).

1. Finish leveling the site by making the necessary cuts and fills to the desired formation level.
2. Use multiple passes of a vibratory roller and heavy static traveling in opposite directions to compact the entire area.
3. Label the strip, mat or isolated foundation's footprint areas.
4. Mechanical excavation up to the treatment's thickness plus the design founding level shall be completed.
5. The extent of the excavation needs to be carefully planned in order to enable appropriate bed treatment and compaction. Trench excavation must be done the entire length of any long foundation, such as a track, or a row of isolated foundations. The whole area needs to be excavated for the shop floor or stackyard. The area will be mechanically leveled and compacted following excavation (Varone et al., 2021).
6. To prepare the foundation bed, fill it with carefully chosen, graded granular materials according to the thickness of each layer and mechanically compact each layer to the required level.

7. It is recommended to perform footing load tests on a prepared bed for the foundation of important equipment.
8. Skilled field engineers should oversee the work (Duan & Liu, 2010).

6.4. HARDSTAND

Hardstand is the term used to describe the localized "hardening" of the ground to support heavy loads while preventing settlement. When building heavy-duty cranes or hoist to erect heavy equipment parts, for example, a hardstand is needed. It is also necessary for the long-term storage for heavy finished goods in steel plants, such as coils of steel, joists, rolled steel beams, and iron ingots. Hardstand design and construction call for knowledge and experience (Butler & Kean, 1993). The fundamental idea of a hardstand is to increase the sub-base's in-situ modulus, which enhances bearing capacity and reduces settlement. In this method loose weakbed materials are excavated and replaced with materials of known composition andproperties to ensure adequate bearing capacity and settlement within permissiblelimit. By strengthening the sub-base's low strain modulus, this is accomplished. For the storage of raw materials, long strip footing or raft, rail track and roads used by heavy-duty vehicles such as high-capacity dumpers in soft soil, hardstands can be useful. Key characteristics of a hardstand could be clarified by using case studies from particular initiatives (Vasquez et al., 2004).

6.4.1. Hardstands for Heavy Crane Foundations

At a steel plant, a rotating heavy-duty crane is needed to erect the heavy process equipment for the coal dust injection plant. The crane will help move large equipment components and put them together in stages as they are left hanging for several days. The overall state of the ground was unsuitable for establishing a crane foundation because it only had a 25 mm settlement restriction and a bearing capacity of about 20 t/m^2. The site's expected settlement and estimated bearing capacity were 80 mm and 10–12 t/m^2, respectively. To the satisfaction of the crane supplier, the problem of the crane foundation was resolved by building a designed hardstand. Following the necessary research and design, a hardstand scheme with the specified design parameters was agreed upon (Marquez et al., 2014).

a. Enhancement of the soil: a 1500-mm-thick hardstand filled with four layers of hard granular materials compacted to at least 95% of the maximum dry density in order to obtain $\Phi \sim= 40°$ and $N > 50$ (Hornaday et al., 1993).
b. Settlement after hardstand ($E = 60$ MPa): = 30 mm.

Grain size (mm)	Percent by volume (%)
19.0–9.5	10
9.5–4.75	20
4.75–2.00	30
2.00–1.18	30
Fines 600–75 μm	10

Generally, the top 2.0 to 3.0 meters of soil need to be removed by excavating, and graded granular materials need to be placed, and compacted in a certain way. Stone dust, building-grade stone chips, and clean sand are the non-cohesive materials that need to be used. For reference, the proportions and gradations of granular materials chosen following several trials are listed below in order to guarantee "design compaction" and "minimum void" (Ramaswamy & Pertusier, 1986).

The graded granular materials need to be dry-mixed, then stacked in layers about 250–300 mm thick, and sufficiently watered. The layers should then be compacted by using a vibratory roller several times to reach 95% relative density. The compacted hard bed should have an angle of internal friction (φ) greater than 35° and a low strain modulus of 4500 t/m^2 (450 kg/cm^2) to guarantee a bearing capacity of 20–30 T/m^2 and less settlement, most of which will be immediate. Strict quality control and supervision should be exercised when handling granular materials, mixing, placing, and rolling. It is also advised to perform field and lab tests on the relative density attained at the location. Hardstand surface treatment is possible if necessary for crane movement (Varghese et al., 1997).

6.4.2. Hardstand for a Heavy Storage Floor

In soft ground, floors built for the storage of heavy materials (such as coal, limestone, iron ore, etc.) frequently experience differential and high total settlements, depressions, upheavals, cracks, etc., which have a major impact on the movement of material handling equipment (such as stacker-reclaimers) and waste from "dead storage." Because partial shutdown is required, operating floor damage repair is challenging. It is customary to design expensive, highly reinforced concrete floors with significant thickness (Bischoff & Dickson, 1975). The economic design and building of a floor like this on soft ground is challenging due to factors like inconsistent subbase, inconsistent component materials, etc. Sub-base problems have a low in-situ modulus, but they can be resolved with care. Cross-ribs can be useful if a higher imposed load—say, 12–15 t/m^2 in loose or soft soil—is present. The plan is explained using heavy materials storage facility as an example (Zeng et al., 2010).

6.4.3. Sub-Soil Data and Design Requirements

- Five layers of subsoil have been identified at the site. Table 6.1 displays soil profiles in addition to engineering parameters.
- The storage arrangement: Two-layer stacking of 25-ton steel coil rolls results in a uniformly distributed load (UDL) of approximately 11 t/m^2.

Table 6.1. The Stratigraphy of the Subsurface at the Heavy Material Storage Location

Strata	Description	Thickness	*N*-value	Modulus (MPa)
I	Compacted fill	4.0 m	8	8
II	Loose flyash	5.75 m	1–2	2
III	Stiff sandy clay	3.0 m	20	35
IV	Disintegrated rock	3.0 m	> 100	30
V	Weathered rock	Cont.	Core 20%	> 40

The following techniques, either alone or in combination, can be used to prepare the area for a storage floor:

i. RCC flooring will be built atop a prepared hardstand rather than on the ground.
ii. The hardstand's design was altered by adding compacted granular "grade beams" underneath it.
iii. Create a hardstand that satisfies the sub-base requirements for the floor.
iv. Create an RCC slab with an enhanced sub-base modulus in mind (Ngah & Nwankwoala, 2013).

Steps

i. Excavation totaling up to two meters below the design floor level.
ii. Installation of grid-patterned, compacted granular strips beneath the hardstand.
iii. Divide the floor area into rectangular panels, about 3–4 m by 10–12 m.

iv. Compaction at 1.2–1.5 m c/c using a drop hammer with 2 m drop, 2.3 T W, and 500 to 800 mm diameter (Vähäaho et al., 2011).

v. Hammer blows created craters that were filled with boulders of stone and hard granular material or slag.

vi. Craters were made by hammer blows to be filled with hard granular material, slag, or stone boulders (Livneh et al., 1998).

vii. Continue compacting with a hammer until the penetration is minimal.

viii. The entire floor will be covered in these rolled, dressed, and compacted strips and grids.

ix. The purpose of building the hardstand over the compacted grid system is to increase the sub-base's effective modulus.

x. RCC flooring will be developed and built above a hardstand.

xi. The estimated long-term floor settlement was approximately 30 mm (Denge et al., 2017).

Figure 6.2 illustrates the designed hardstand situated beneath granular grade beams. Granular grade beam dimensions and spacing must be determined using the sub-base geotechnical test results, in accordance with the design of beams on an elastic foundation laid out in a grid pattern. Hard materials for granular beams, however, should have well-graded sizes between 100–75 and 12 mm. The grade beams must be built in 500 mm thickness steps, with each layer being thoroughly compacted. The hardstand is built as described above and is designed in three or four layers with a thickness of 300–250 mm. It is underlined that good material, craftsmanship, and supervision are critical to the success of a hardstand (Uthayopas et al., 2011).

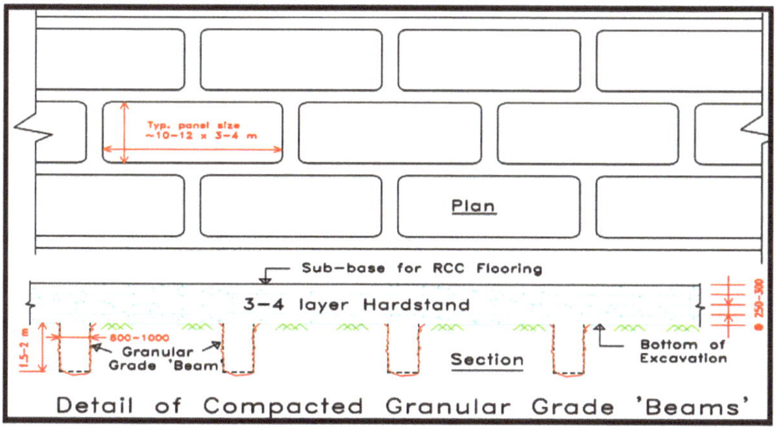

Figure 6.2. *Details of grade beams made of compacted granular material below the hardstand.*

On top of the hardstand, the RCC floor is going to be built. Although building a hardstand might take a little longer, the benefits of economy and storage floor performance will outweigh the costs associated with building a thick and heavily reinforced RCC raft (Filippi et al., 2018).

6.5. COMPACTION GROUT

The entire grouting process is a specialized engineering field that depends on several factors, including geotechnical testing, grout material and equipment selection, injection planning, and, most importantly, proof testing and monitoring (Boulanger & Hayden, 1995).

Grouting is frequently used for a variety of applications, such as filling joints and cracks, controlling seepage, improving bearing capacity, stabilizing slopes, and stabilizing soft soil or fill. Commonly, grouting is accomplished by pressurizing the grouting agent—cement slurry with the chemical additives—into fill, rock, soft soil etc. Grouts fill in gaps, joints, and cracks to strengthen the mass as a whole (El-Kelesh et al., 2001).

The grouting agent and grouted medium determine the grouting pressure. To guarantee a consistent grout flow for complete coverage, the injection process is initially carried out at different pressure stages, ranging from low pressure through medium to the design pressure or refusal. Grouting is carried out in phases, with decisions made during planning regarding depth, pressure, and spacing, pressure, and depth depending on the intended use (Nichols & Goodings, 2000).

Testing grout before implementing the grout scheme on the job site is crucial to ensuring its effectiveness. This section's goal is to give new engineers a general understanding of compaction grouting rather than to go into great detail about grouting. Again, planning, scheme preparation, grout material and equipment selection, injection technique, and overall monitoring all require experience when grouting. When grouting is successfully finished, the results are frequently astonishing (Brown, 1973).

6.5.1. Methodology

To show the extent of the work, the grouting scheme needs to be planned, designed, and prepared with specifications and drawings. Below is a typical example for reference.

Cement and bentonite are typically combined in a 1:2 (by weight) grout mix. A suitable consistency slurry should be created by adding a sufficient amount of water, stirring it well, and injecting it under low pressure at first, then gradually increasing it to a maximum of 2.0 kg/cm^2 or, spillover or refusal, whichever happens earlier (Tellis, 1997).

1. The ground will be marked with grout points at the alignment and spacing indicated in the design sketch.
2. To insert the grout pipe, a rod is first driven or pushed into the ground to the appropriate depth.
3. In the initial stage, grout is injected alternately at different depths (levels). Following steps are taken to ensure that the prior grout has set before grouting the remaining holes and depths.
4. New grout made of cement and bentonite will be mixed in drums. To create a workable slurry, the necessary amount of water is going to be added, and it will be injected right away. The grout in the mixing drums needs to be constantly mixed, and any set grout needs to be thrown out (Lather, 2013).

For the purpose of billing and verification, a record of the grouting work, including materials, mix consistency and proportions, pressure, depth, time, and intake, must be kept. Remember that the performance and condition of the site may require adjustments to the grouting procedure. Before beginning the actual grouting work, it is advisable to confirm the grouting scheme, mix, and equipment on a "test grout" program. It is preferable to finish grouting work while preserving continuity (Howell, 2012).

- Example 1: Tension Grout for Settlement Control Structure to Support Rails

Four RCC columns supporting the 3.2 m square footing, which was founded 2.5 m below FGL, supported the rail supporting structure for the coke oven plant. Column footings' GA is displayed in Figure 6.3. The columns were found to have excessive settlements, ranging in size from 41 — 126 mm, during rail fixing (Morrison, 2012).

The differential and total settlements of about 70 to 80 mm were very alarming. An examination of the site showed that deep excavation—up to 12 meters—had been done for the nearby Junction House's construction. Supporting columns for the rails were set on foundations 2.5 m below FGL on freshly tilled soil that had been compacted using a hand-held plate vibrator (Wu & Shao, 2018).

It is not anticipated that this type of compaction and backfilling will be able to withstand the design-bearing pressure applied to rail support beams. Near the base of each column, SPT were conducted to evaluate the in-situ consistency and the variation with depth (Zhou et al., 2018).

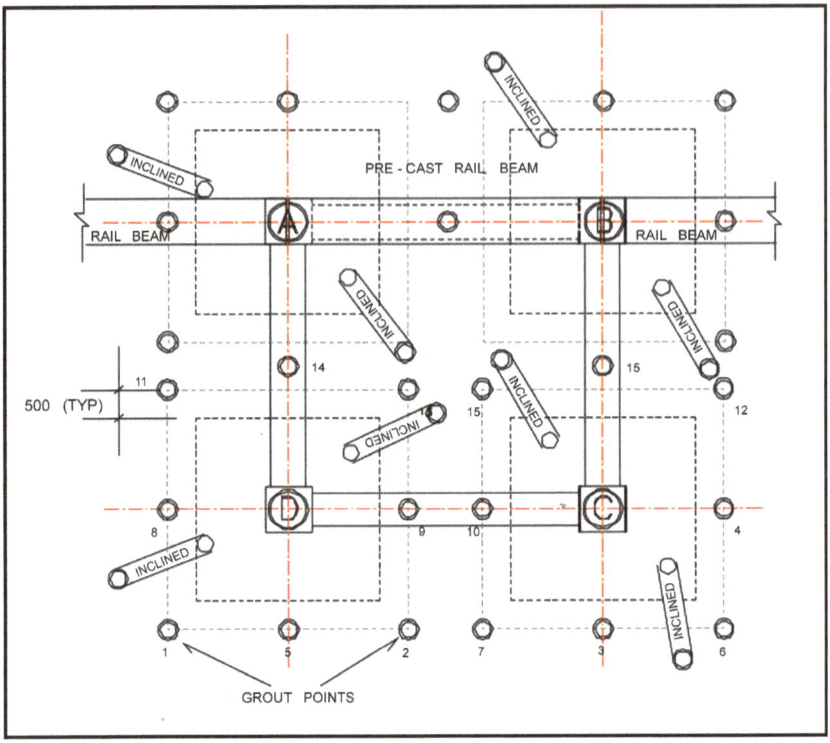

Figure 6.3. *GA with vertical and inclined grout points and isolated footings.*

6.5.2. Treatment

To improve the overall consistency of backfilled soil, pressure grout with admixtures and cement slurry was used as a ground treatment. This resulted in a higher overall modulus, increased bearing capacity, and decreased settlement (Papageorgion et al., 2000).

Nominally 100 mm diameter grout holes are to be bored down to rock, approximately 12 meters below the surface. grout is to be injected at two different depths: Stages 1 and 2 are located 1.0–2.0 m above the rock and 2.5(Gardner & Hill, 2001). –3.50 m below FGL, respectively.

Typically, admixtures are injected into a cement slurry (1 cement: 3 water) at low to medium pressure, with the slurry gradually increasing to approximately 2.5 kg/cm^2. Compaction immediately beneath the footing was achieved by inclined boring to the middle of the raft bottom and grout injection at medium to high pressure until refusal. Every stage and grout hole's maximum pressure as well as the intake of grout were meticulously documented (Hawkes et al., 2018).

For all four columns, a total of 26 grout points—22 vertical and 4 inclined—have been planned. Grout point location and number were altered in response to site conditions. The four-column frame was stabilized and additional settlement was stopped by the treatment. For further erection, the rail supports were leveled and realigned (Gysler et al., 1982).

SUMMARY

"Foundation on Soft and Filled-Up Soil" delves into the complexities of constructing on unstable ground. This chapter explores the challenges of building on soft or filled-up soil, emphasizing the importance of proper foundation design and engineering. It discusses various techniques such as soil stabilization, deep foundations, and ground improvement methods to mitigate risks associated with settlement and structural instability. Additionally, it highlights case studies and practical solutions to address these issues, providing valuable insights for engineers, architects, and construction professionals navigating projects on challenging terrain.

REFERENCES

1. Bischoff, J. L., & Dickson, F. W. (1975). Seawater-basalt interaction at 200 C and 500 bars: Implications for origin of sea-floor heavy-metal deposits and regulation of seawater chemistry. *Earth and Planetary Science Letters, 25*(3), 385–397.
2. Boulanger, R. W., & Hayden, R. F. (1995). Aspects of compaction grouting of liquefiable soil. *Journal of Geotechnical Engineering, 121*(12), 844–855.
3. Brown, D. R., & Warner, J. (1973). Compaction grouting. *Journal of the Soil Mechanics and Foundations Division, 99*(8), 589–601.
4. Butler, D. K., & Kean, T. B. (1993). Investigations of magnetic field disturbances at an air force base compass calibration hardstand. *Geophysics, 58*(3), 434–440.
5. Denge, S. M., Oyawa, W., & Kabubo, C. K. (2017). The effect of quarry dust concrete and underlying sub-soil medium on the structural performance of dwelling house ground floors. *IOSR Journal of Mechanical and Civil Engineering, 14*(02), 52–56.
6. Duan, H., & Liu, S. (2010). Unmanned air/ground vehicles heterogeneous cooperative techniques: Current status and prospects. *Science China Technological Sciences, 53*, 1349–1355.
7. Dunne, H. P., & Martin, C. M. (2017). Capacity of rectangular mudmat foundations on clay under combined loading. *Géotechnique, 67*(2), 168–180.
8. El-Kelesh, A. M., Mossaad, M. E., & Basha, I. M. (2001). Model of compaction grouting. *Journal of Geotechnical and Geoenvironmental Engineering, 127*(11), 955–964.
9. Feng, X., Gourvenec, S., & Shen, Z. (2017). Shape effects on undrained capacity of mudmat foundations under multi-directional loading. *Ocean Engineering, 135*, 221–235.
10. Filippi, P., Cattle, S. R., Bishop, T. F., Odeh, I. O., & Pringle, M. J. (2018). Digital soil monitoring of top-and sub-soil pH with bivariate linear mixed models. *Geoderma, 322*, 149–162.
11. Gardner, T. B., & Hill, D. R. (2001). Treatment of giardiasis. *Clinical Microbiology Reviews, 14*(1), 114–128.
12. Gupta, T., Sridhar, V., & Annam, M. K. (2017). Optimal foundation solution for storage terminal in Mangalore. In *Indian Geotechnical Conference, 2*(1), 14–17.
13. Gysler, M., March, C. M., Mishell Jr, D. R., & Bailey, E. J. (1982). A decade's experience with an individualized clomiphene treatment regimen

including its effect on the postcoital test. *Fertility and Sterility, 37*(2), 161–167.

14. Hawkes, J. E., Yan, B. Y., Chan, T. C., & Krueger, J. G. (2018). Discovery of the IL-23/IL-17 signaling pathway and the treatment of psoriasis. *The Journal of Immunology, 201*(6), 1605–1613.
15. Honjo, Y. (1982). A probabilistic approach to evaluate shear strength of heterogeneous stabilized ground by deep mixing method. *Soils and Foundations, 22*(1), 23–38.
16. Hornaday, W. C., Haas, C. T., O'Connor, J. T., & Wen, J. (1993). Computer-aided planning for heavy lifts. *Journal of Construction Engineering and Management, 119*(3), 498–515.
17. Howell, K. E. (2012). An introduction to the philosophy of methodology. *An Introduction to the Philosophy of Methodology, 5*(3), 1–248.
18. Kamali, B., & Hashim, R. (2010). Bamboo foundation mat for rubble mound breakwaters on mud deposits. *International Journal of the Physical Sciences, 5*(9), 1406–1410.
19. Kolar-Anić, L., & Veljković, S. (1975). Statistical foundations of heterogeneous kinetics. *The Journal of Chemical Physics, 63*(2), 669–673.
20. Lather, P. (2013). Methodology-21: What do we do in the afterward? *International Journal of Qualitative Studies in Education, 26*(6), 634–645.
21. Livneh, M., Livneh, N. A., & Hayati, G. (1998). Site investigation of sub-soil with gypsum lenses for runway construction in an arid zone in Southern Israel. *Engineering Geology, 51*(2), 131–145.
22. Mahajan, P., Biswas, S., & Adhikari, A. (2008). Behavior of reinforced earth structures founded on soft silt deposit in seismically active hilly terrains. In *New Horizons in Earth Reinforcement.4(2)*, 797–802.
23. Marquez, A. A., Venturino, P., & Otegui, J. L. (2014). Common root causes in recent failures of cranes. *Engineering Failure Analysis, 39*, 55–64.
24. Morrison, M. (2012). Understanding methodology. *Research Methods in Educational Leadership and Management, 3*, 14–28.
25. Ngah, S. A., & Nwankwoala, H. O. (2013). Evaluation of sub-soil geotechnical properties for shallow foundation design in Onne, Rivers state, Nigeria. *The International Journal of Engineering and Science (IJES), 2*(11), 08–16.
26. Nichols, S. C., & Goodings, D. J. (2000). Effects of grout composition, depth and injection rate on compaction grouting. In *Advances in Grouting and Ground Modification.5(4)*, 16–31.
27. Papageorgiou, P., Katsambas, A., & Chu, A. (2000). Phototherapy with

blue (415 nm) and red (660 nm) light in the treatment of acne vulgaris. *British Journal of Dermatology, 142*(5), 973–978.

28. Raithel, M., Kirchner, A., Schade, C., & Leusink, E. (2005). Foundation of constructions on very soft soils with geotextile encased columns-state of the art. *Innovations in Grouting and Soil Improvement, V0O 1*, 1–11.
29. Ramaswamy, S. D., & Pertusier, E. M. (1986). Construction of barrettes for high-rise foundations. *Journal of Construction Engineering and Management, 112*(4), 455–462.
30. Roy, M. (2024). Foundation on soft and filled-up soil. In *Geotechnical and Foundation Engineering Practice in Industrial Projects.7(4)*, 93–103.
31. Saeed, A. (2008). United States air force research on airfield pavement repairs using precast Portland cement concrete (PCC) Slabs (Briefing Slides). Vol 3, (p. 0019).
32. Tellis, W. (1997). Application of a case study methodology. *The Qualitative Report, 3*(3), 1–19.
33. Uthayopas, P., Wongvanichtawee, C., & Bhundularp, P. (2011). Geographical information system of soil investigation data for construction purposes. *Advanced Materials Research, 261*, 1529–1533.
34. Vähäaho, I. T., Van Alboom, G., Korpi, J. M., Satola, I. S., & Vergauwen, I. (2011). Geotechnical and geological data management in urban underground areas. In *Proc. World Tunnel Congress and 37th General Assembly.8(5)*, 1–12.
35. Varghese, K., Dharwadkar, P., Wolfhope, J., & O'Connor, J. T. (1997). A heavy lift planning system for crane lifts. *Computer-Aided Civil and Infrastructure Engineering, 12*(1), 31–42.
36. Varone, C., Lenti, L., Martino, S., & Semblat, J. F. (2021). Spatial variability of the urban ground motion in a highly heterogeneous site-city configurations. *Bulletin of Earthquake Engineering, 19*, 27–45.
37. Vasquez, A. P., Regens, J. L., & Gunter, J. T. (2004). Environmental persistence of 2, 3, 7, 8-tetrachlorodibenzo-p-dioxin in soil around hardstand 7 at Eglin Air Force Base, Florida. *Journal of Soils and Sediments, 4*, 151–156.
38. Wang, J. H., Xu, Z. H., & Wang, W. D. (2010). Wall and ground movements due to deep excavations in Shanghai soft soils. *Journal of Geotechnical and Geoenvironmental Engineering, 136*(7), 985–994.
39. Wu, K., & Shao, Z. (2018). Effects of pipe roof support and grouting pre-reinforcement on the track settlement. *Advances in Civil Engineering, 2*(1), 1–9.

40. Yong, C. C., & Oh, E. (2016). Modelling ground response for deep excavation in soft ground. *GEOMATE Journal, 11*(26), 2633–2642.

41. Zeng, R., Wang, X., Xiao, W., Zhang, Y., Zhang, Q., & Di, H. (2010). Thermal performance of phase change material energy storage floor for active solar water-heating system. *Frontiers of Energy and Power Engineering in China, 4*, 185–191.

42. Zhao, C., & Zhao, D. (2021). Application of construction waste in the reinforcement of soft soil foundation in coastal cities. *Environmental Technology & Innovation, 21*, 101–195.

43. Zhou, S., Xiao, J., Di, H., & Zhu, Y. (2018). Differential settlement remediation for new shield metro tunnel in soft soils using corrective grouting method: Case study. *Canadian Geotechnical Journal, 55*(12), 1877–1887.

CHAPTER 7

PILE FOUNDATIONS

Contents

7.1. INTRODUCTION

As illustrated in Figure 7.1, pile foundations, being deep foundations, are frequently employed for bridges, high-rise buildings, and structures subjected to complex loading conditions, like offshore wind turbine foundations and retaining walls. Piles can either be bored or drilled into the ground. Notably, in some regions, like North America, bored piles are typically referred to as drilled shafts. The following are the purposes of piles (Mandolini et al., 2005):

i. Transferring weights from the superstructure via thin, compressible layers.
ii. Water onto rocks or stiffer, more compacted soils that are less compressible.
iii. Transporting uplift loads when utilized to sustain tall buildings vulnerable to wind or wave overturning forces.
iv. Supporting machinery foundations, bridge abutments and piers, and retaining walls by carrying combinations of vertical and horizontal loads (Randolph, 2003).

The loading types, ground conditions, and installation techniques all have a significant impact on pile capacity. This chapter will provide an introduction to pile installation, in situ pile load testing, and the pile load capacity estimation and settlement. Crucially, the approaches for estimating pile load capacity that are presented in the ensuing sections are grounded in model and field tests and theories; ideally, in situ pile load tests will be conducted. The estimation is verified by the pile load test result (Comodromos et al., 2009).

7.2. CLASSIFICATION OF PILES

In general, piles can be categorized according to a variety of factors, including installation technique, material, and function. The ensuing sections will provide a detailed introduction to each classification (Adhikari et al., 2020).

7.2.1. The Functions of Piles

In addition to lateral and vertical (tension and compression) loads, piles offer resistance to superstructure-imposed loads. Therefore, piles can be categorized as follows according to their purposes (Zhong et al., 2011):

7.2.1.1. Compression Piles

Almost all of pile foundations fall into this category. As seen in Figure 7.2, piles withstand the vertical downward vertical loadings from the superstructures. The piles' resistance comes from their side frictional force along with their end-bearing capacity. When loading is applied, settlements happen; therefore, an essential component of the compression pile's design is the allowable settlement (Dash & Pise, 2003).

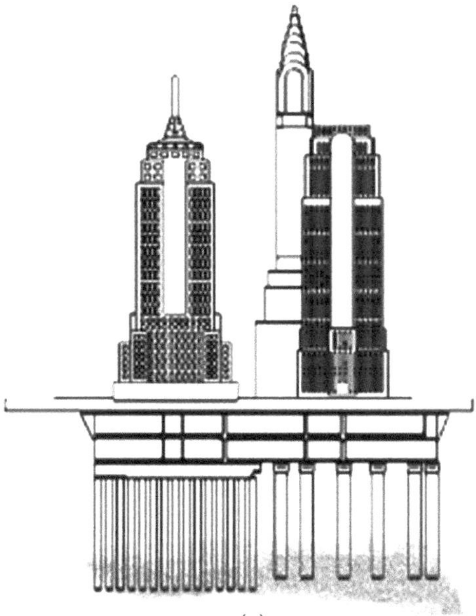

(a)

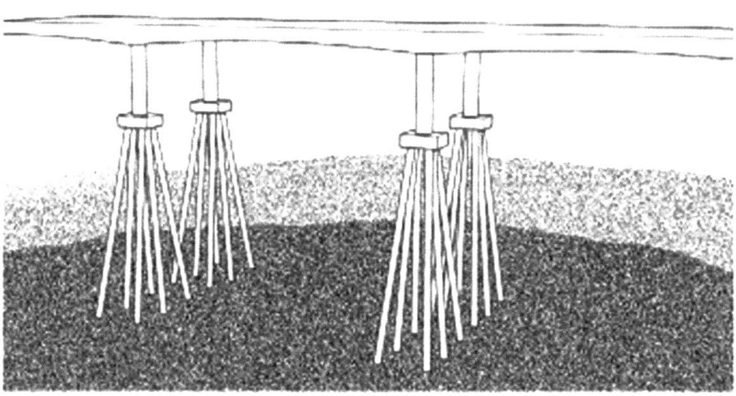

(b)

(c)

Figure 7.1. *Pile functions include (a) high-rise building foundations; (b) bridge foundations; and (c) retaining walls.*

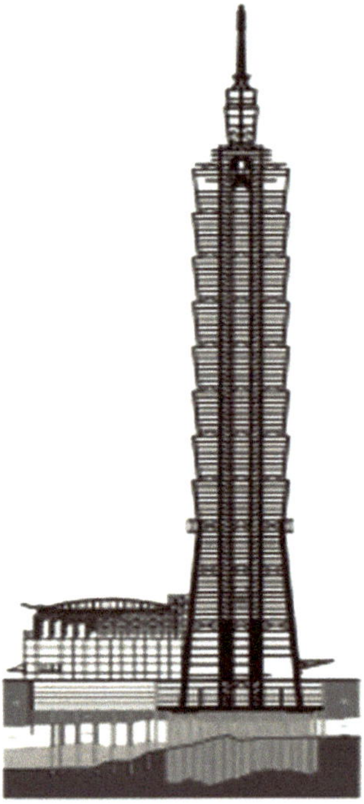

(a)

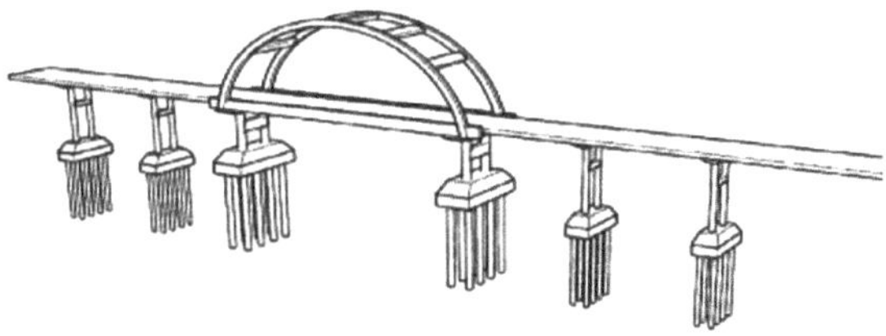

(b)

Figure 7.2. *Piles in compression in (a) bridge foundations; and (b) high-rise building foundations.*

7.2.1.2. Tension Pile

This pile is also known as an uplift pile, and it helps keep transmission lines foundations and towers from being lifted up vertically, as seen in Figure 7.3. The primary source of resistance is the surrounding soils' tensile strength, which is typically lower than their compressive strength. As a result, tension piles typically have a lower design load capacity than compression piles (Levacher & Sieffert, 1984).

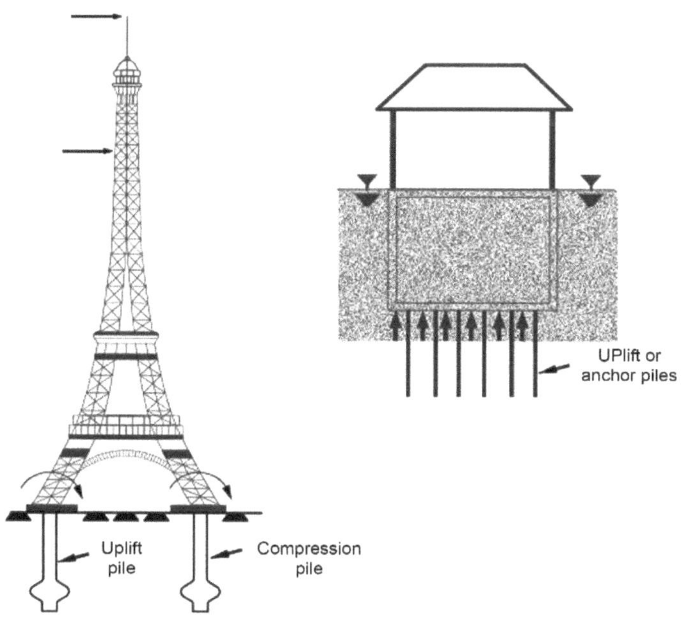

Figure 7.3. *A tension pile in working.*

7.2.1.3. Laterally Loaded Pile

There are two subcategories for laterally loaded piles: pile-to-soil and soil-to-pile. Piles can be used as a wind turbine's foundation and to transfer the wind turbine's lateral loads to the soil, as Figure 7.4a illustrates. In contrast, the retaining structures in Figure 7.4b are made of piles that withstand soil lateral forces. When designing lateral piles, their ultimate horizontal force is based on the deflections and bending moment of the piles as well as the lateral resistance of the soils (Broms, 1965).

7.2.2. Materials

Apart from their intended use, piles can be categorized based on the materials they are made of, including steel, concrete, and timber. The primary factors influencing pile material selection are the intended use, longevity, and availability of local suppliers (Lu et al., 2016). These are some common materials for piles:

a. Timber piles are manageable and lightweight. In areas with abundant timber resources, timber is an excellent natural material and can be fairly competitively priced. Despite being easy to gather and stack, timbers have limited loading capacity and dimension because of their inherent flaws and texture. When driving piles, it is easy to damage a timber pile head. Wood works best when it is completely below the level of groundwater (Fratzl & Weinkamer, 2007).

b. Steel piles are strong, manageable, and able to support heavy compressive loads when pushed into a solid stratum. Steel piles come in a variety of cross sections and shapes, including H-sections, box sections, plain tubes, fluted and tapered tubes, and box piles constructed from sheet piles. But it's more expensive than concrete piles (Park et al., 2010).

c. Concrete piles are widely utilized in practical applications because of their affordability and effectiveness. Concrete pile installation is incredibly adaptable. There are two options: precast and cast-in-place (Allwright, 1981).

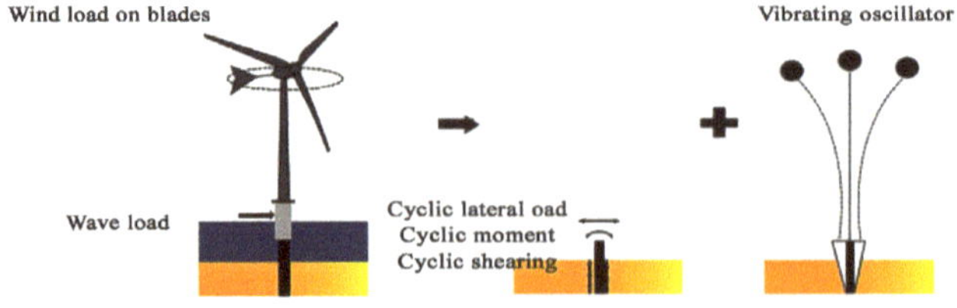

(a)

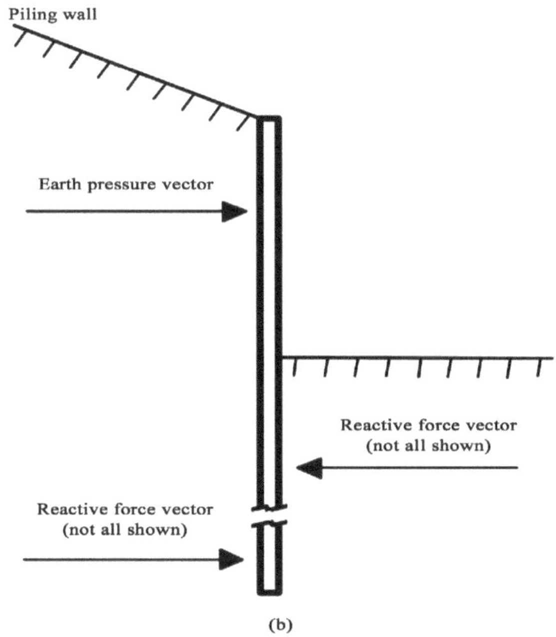

Figure 7.4. *Laterally loaded piles shown as (a) pile-to-soil; and (b) soil-to-pile.*

The installation has a significant impact on the pile load capacity. When a driving device is used for installation, the surrounding soils typically experience greater displacement—that is, displacement piles—which squeezes or compacts the soils. A replacement pile is one that has an opening before the pile is inserted; the soils are removed first, followed by the installation of the pile. Depending on the type of soil and how the piles are installed, the surrounding soils will be affected differently. The effects are explained in the section that follows (Choi et al., 2000). The following provides an explanation of the pile installation classification, which is depicted in Figure 7.5.

1. Displacement pile
   i. **Large Displacement:** The earth is displaced by all forms of driven piles, including the hollow section pile with closed end and solid section piles, which are driven or jackknifed into the ground (Callister & Rethwisch, 2000).
   ii. **Small Displacement:** Similarly jacked or driven into the ground, but with a comparatively small cross section; examples include pipe, box steel piles, H- or I-sections, hollow sections with an open end, and so on (Wang, 2001).
2. Replacement pile (non-displacement pile)

Bore first to remove any soils, and then fill the borehole with concrete (either with or without casing).

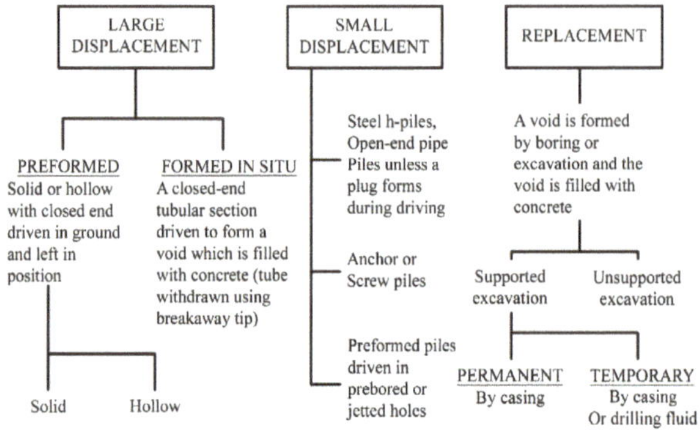

Figure 7.5. *Classification of piles based on their installation method.*

7.3. THE INSTALLATION OF PILES AND THEIR IMPACTS ON THE SOILS

The load capacity of a pile is primarily determined by the installation and state of the ground. The state of the nearby soils is impacted by the installation of piles, which leads to varying load capacities. This section will provide an introduction to installation techniques and how they affect cohesive and non-cohesive soils (Zarrabi & Eslami, 2016).

7.3.1. Driven Pile

Driven installation is frequently utilized for precast concrete and steel piles. The sharp-tipped precast concrete pile and related driven equipment are depicted in Figure 7.6.

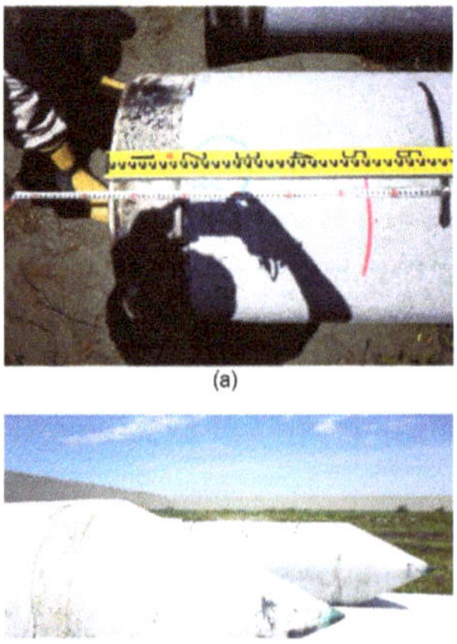

Figure 7.6. *Driven piles and their equipment: the pile head (a); the pile tip (b); and driving equipment for piles (c).*

For a driven pile, the primary operations are transporting, loading, unloading, placing, lifting, arranging, driving, and, if required, welding. Given the aforementioned handle process, handling stress in driven piles needs to be taken into account (Li et al., 2017). Vibrating tools or a drop hammer can be used to drive piles. The benefits of prefabricated driven piles are as follows:

a. Superior pile material, inspected prior to driving.
b. Not susceptible to necking or squeezing, particularly on soft terrain.
c. Groundwater has no effect on the construction process (Randolph & Wroth, 1979).

- Possible Issues Consist of the Following:

a. There is a lot of noise and vibration from driving.
b. They may break while being driven and must be joined for the required length.
c. Could sustain unseen harm.
d. Soil displacement during driving may cause nearby piles and structures to lift.
e. Precast piles with joints might not be appropriate for lateral and tension loads (Duan et al., 2018).

The benefits and drawbacks for cast-in-place and driven displacement piles are as follows.

- Advantages
  - Simple length adjustment.
  - Using a driving tube with a closed end to keep groundwater out.
  - The base can be made larger (Jardine et al., 2005).
- Disadvantages
  - When withdrawable tube kinds are used in soft squeezing soils, concrete is susceptible to defects.
  - There is a lot of noise and vibration from driving.
  - Soil displacement caused by driving may cause nearby piles and structures to rise (Chow et al., 1997).

7.3.2. How Clays Are Affected by Driving Piles

Driving piles into clayey soils will change the soil structure, excess pore water pressure, and the clay's undrained shear strength. This will have an effect on the piles' performance, especially in the short term. According to de (Mello, 1969), pile driving's effects in clays could be divided into four main categories:

a. Soil around the pile may be remolded or partially structurally altered.
b. Modification of the soil's stress condition close to the pile.
c. The extra pore water pressure surrounding the pile dissipating.
d. The soil's long-term recovery of strength (Wersäll & Massarsch, 2013)

The phenomenon will then be thoroughly explained.

7.3.2.1. Affect on the Pile Capacity and Soil Strength

The "remolding" (disturbance caused by pile driving) of clays will initially result in a slight loss of undrained shear strength when piles are driven into it. Following driving, the following factors will result in an increase in pile capacity and soil strength:

i. Partial restoration of structural bonds and thus, undrained shear strength.
ii. Excess pore water pressure dissipation (Arabameri & Eslami, 2021)

The load capacity of piles is restored with the passing of time following pile driving, as seen in Figure 7.7. It can take up to a week or one hundred hours in certain situations to restore the load capacity to 90% of its maximum value. Because of this, the pile load test usually gets carried out a week or two following pile driving (Ng et al., 2010).

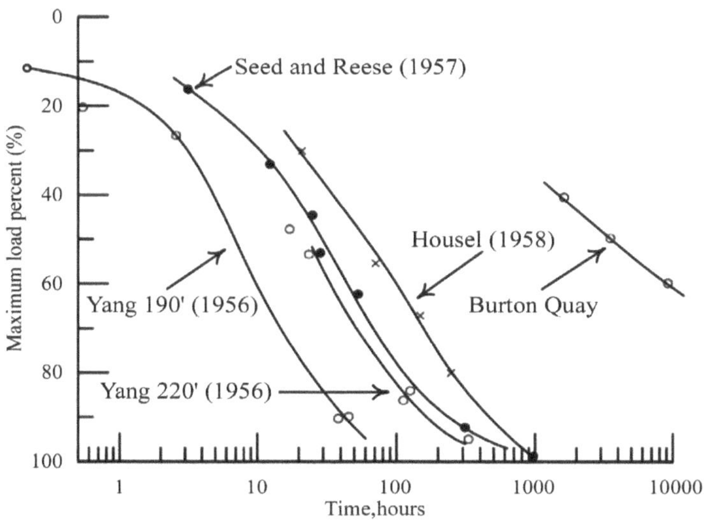

Figure 7.7. *Increase in the percentage load capacity over time (hours).*

7.3.2.2. Driving Induced Pore Pressures

Shear forces were applied to the soils across the pile surface when a pile was driven in clay. When the pile drove into the clay, it squeezed the soils around it, which created pore water pressure.

The maximum excess pore water pressure can be 1–2 times the pile surface effective stresses and 3–5 times the effective stresses at the base of the pile, as indicated by the excess pore water pressure measurement made during pile driving and displayed in Figure 7.8.

The extra pore water pressure slowly goes down as one moves farther away from the pile's center, r. As illustrated in Figure 7.8, the distance 'r' is normalized with respect to the radius of the pile a. Field observations show that within a specific distance, R, related to the type of clay, the excess pore-water pressure stays at its maximum value (Δum). A basic model for the dispersion of excess pore-water pressure (Δu) is postulated in Eqn. (1) and illustrated in Figure 7.9, drawing inspiration from the observation highlighted in Figure 7.8 (Roy et al., 1981).

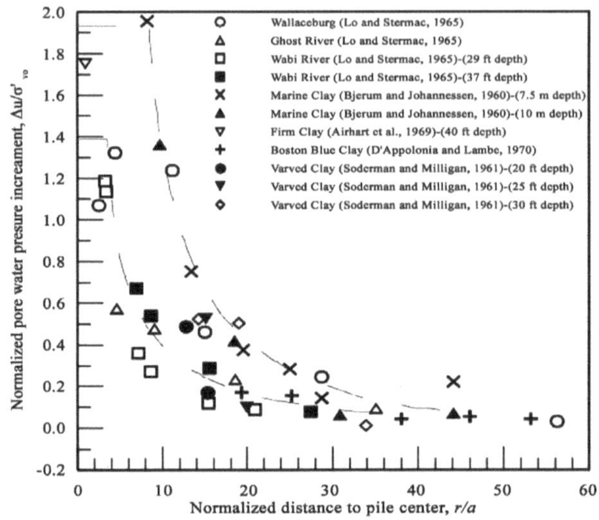

Figure 7.8. *The normalized distribution of the excess pore water pressure at normalized distances to the pile center.*

Within the range of R:

$$\Delta u = \Delta um \quad (1)$$

After the range R:

$$\Delta u = \Delta u_m / (r/R) \quad (2)$$

where: $R \approx 3a \sim 4a$ for the insensitive clay; $R \approx 8a$ for the sensitive clay; and a = the pile radius (Brown & Ge, 2018).

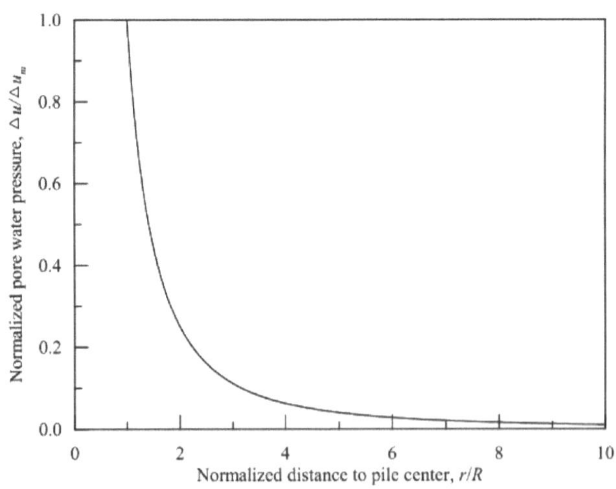

Figure 7.9. *When driving farther than the distance R, pore pressures developed.*

Specifically, for pile groups, the pile capacity may be reduced by superimposition of pore pressure distribution surrounding individual piles (Eigenbrod & Issigonis, 1996).

7.3.3. Impact of Driving Piles in Sand

7.3.3.1. Single piles

Pile driving is a technique where energy is transferred through the soil by dropping or vibrating hammers. The compaction effect in loose sands increases a pile's load capacity.

Additionally, the relative density rises. The results of the field tests indicate that the sands' CPT and SPT-N cone resistances rise following pile driving (Poulos, 1971). The following can be used to estimate the impact of driving piles close to the pile tip:

$$\phi_2' = \frac{\phi_1' + 40^\circ}{2} \qquad (3)$$

where; Figure 7.10 depicts $\emptyset'_1$ and $\emptyset'2$. According to Eqn. (2), pile driving may cause $\emptyset'_2$ to decrease when $\emptyset'_1 \geq 40°$. If the soils' SPT-N following pile driving are identified, φ could be calculated using Eqn. (3) (Novak & El Sharnouby, 1983).

$$\phi' = \sqrt{20N} + 15^\circ \qquad (4)$$

7.3.3.2. Pile Groups

The process of driving groups of piles into loose sand results in a significant compaction of the sand surrounding and separating the piles. If the pile spacing is sufficiently small, the densification effect may also overlap. The efficiency of the group (ξ) will be greater than 1.0 when the pile spacing is not more than six times the diameter of the pile (6d).

This suggests that the group's ultimate capacity is greater than the total of the capacities of the individual piles. The pile-driving sequence and the observation points' locations determine the group effect (El-Marsafawi et al., 1992). By using Kishida's single pile approach, these parameters can be calculated (Eqn. (2) and Figure 7.10).

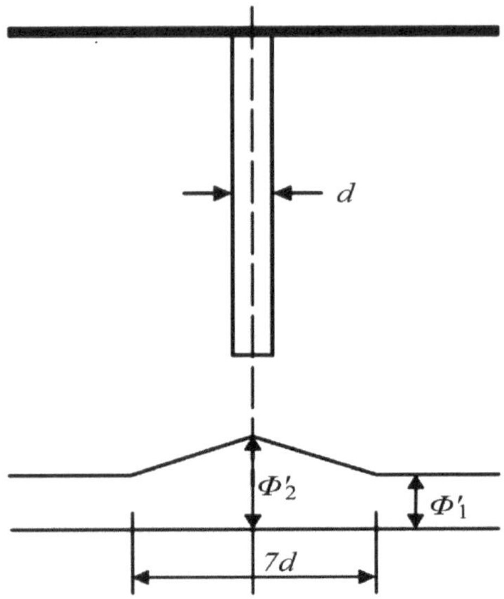

Figure 7.10. *Impact of pile driving on the friction angle within sand.*

7.3.4. Bored and Cast-in-Place Pile

The ground must be opened by cast-in-place and bored piles before the pile body is inserted to the designated depth. If the water level is too high or the ground is not in good condition, the wall of borehole may need to be protected (Rybnikov, 1990). The following are the benefits and drawbacks:

- Advantages

a. The length can be adjusted to accommodate changes in the bearing stratum's level.

b. Neither handling stress nor driving stress controls the formation of material piles.

c. No audible vibration or sound (Yao et al., 2024).

d. There are no ground heaves.

- Disadvantages

a. Bulges, necking, or squeezing (as illustrated in Figure 7.11).

b. Localized open bore slumping brought on by seepage of groundwater (Liu et al., 2013).

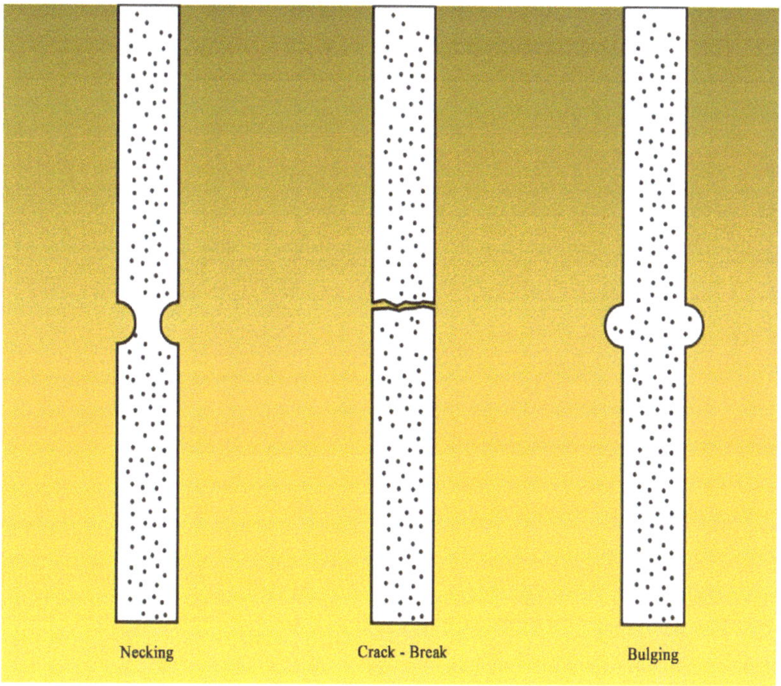

Figure 7.11. *Defects in bored pile installation are shown.*

The full casing piles and the reverse circulation drilling or RCD piles are superior methods of installing bored piles when the stability of the borehole wall presents a challenge during boring. Figure 7.12 depicts the construction of RCD pile. High-pressure air is injected into the annulus between the outer rod and inner tube during the drilling process. The drill tool is powered by air that passes through the drill steel. The air exhausts, carrying the bit cuttings from its surface straight through the drill steel's interior, acting as a circulating medium. The air is directed into a cyclone as it leaves the top of the drill stack, which slows down and separates the cuttings from the air, and gathers them. The residual waste fluids are collected in a separate, airtight containment bin. After the design depth has been reached, a rebar cage is placed into the borehole. The pile forms after concrete is poured from the bottom into the borehole (Mohamad et al., 2016).

As illustrated in Figure 7.11, the full casing (FC) pile is a kind of bored pile that offers adequate protection for the borehole wall and guards against bored pile defects. As seen in Figure 7.13, full casing refers to the protection of the borehole from the surface of ground to the very bottom of the borehole. When the pile's overall length exceeds the length of a single casing, it is welded.

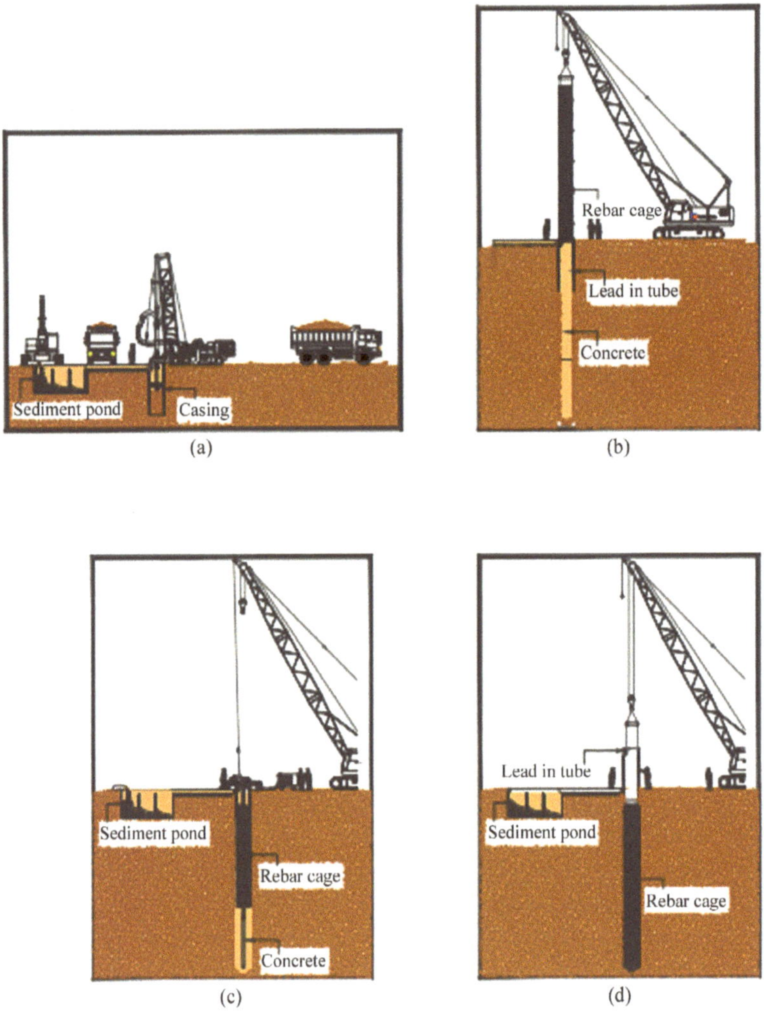

Figure 7.12. *Construction of a RCD pile: RCD ring setup and drilling (a); installing the rebar cage (b); treating the bottom sludge or air lift (c); and removing the casing (d).*

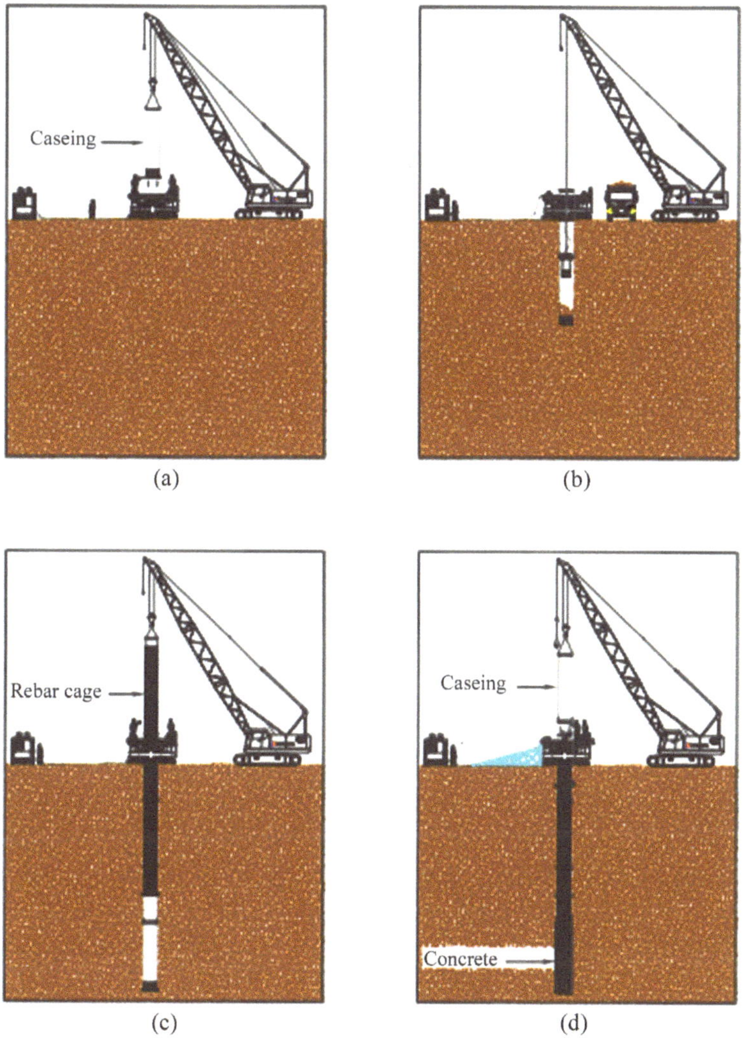

Figure 7.13. *Complete casing pile construction: drilling (a); casing installation (b); rebar cage installation (c); and concrete pouring and casing withdrawal (d).*

7.3.5. Effects of Installing a Bored Pile

7.3.5.1. Effects on the Surrounding Ground

The effects of inserting a bored pile in soil vary depending on the type of soil. The following are the effects for clayey soils (Lande et al., 2020).

1. Softening:
   i. Moisture absorption from the damp concrete.
   ii. Water migrating from the clay body into the less stressed area surrounding the borehole.
   iii. To make it easier for the cutting tool to work, water is poured inside the boring (Nateghi, 2012).
2. **Disturbance:** Particularly in the clay directly below the pile case.

In sandy soils, the absence of stabilized fluid or casing may cause the borehole wall to collapse. If the borehole wall is not protected, piping becomes a problem as well (Xia et al., 2002).

7.3.5.2. Effects on Piles

When installing a bored pile into soils, the effects on piles are:

i. The borehole caves, causing the pile to neck or become misaligned.
ii. Separation of aggregates inside the pile.
iii. The pile reinforcement is buckled (Qian et al., 2003).

7.4. AXIAL LOAD CAPACITY OF PILES

7.4.1. Load Transfer Mechanism

The pile-soil interaction, which is influenced by the installation technique and soil characteristics (such as soil type and/or the presence of a hard stratum), determines a pile's load capacity. The soils absorb loads from the superstructures, and as Figure 7.14 illustrates, end bearing and force of friction provide resistance to loadings. When the pile meets the surrounding soils, frictional force, also known as "shaft resistance," builds up. This force is mainly caused by the soil's shear strength. Whenever the pile pierces through hard strata, the end bearing, also known as base resistance, becomes significant (Wu et al., 2019). A single pile's net ULC (ULC), or Q_u, is typically stated as follows:

$$Q_u = Q_{su} + Q_{bu} - W$$

Where; *W* is the weight of pile; *Qbu* is the ultimate base or point resistance; Qu is the net ULC; and *Qsu* is the ultimate frictional or shaft resistance.

Putting a load Q on top of the pile and slowly raising it will fully engage the maximum frictional resistance when the pile moves 5–10 mm relative to the soil, no matter what size the pile is. The max base resistance won't be reached until the tip of the pile has moved about 10–25% of the width of the pile. Stated differently, at low settlement levels, frictional resistance mobilizes more quickly than base resistance (Nam et al., 2012).

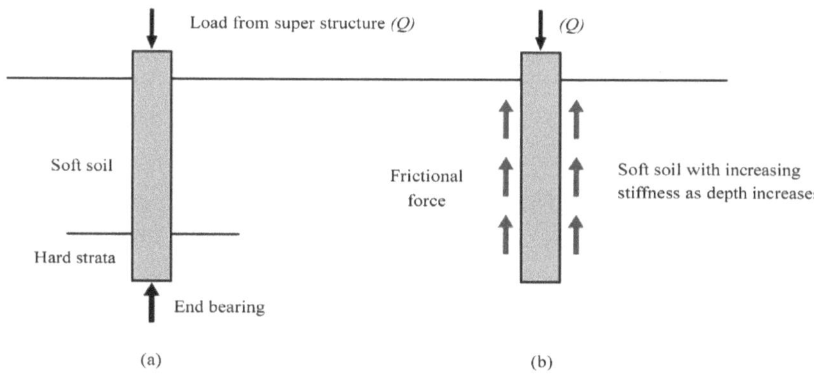

Figure 7.14. *Load transfer mechanisms of (a) end-bearing; and (b) friction piles.*

When the frictional force is much stronger than the end bearing, for example, $Q_{bu} \geq Q_{su}$, and $Q_u \cong Q_{su}$, the pile is called a friction pile. This could happen with a bored pile in soft and thick soils. However, when the end bearing's contribution to load resistance is greater than the frictional force—that means, $Q_{su} \ll Q_{bu}$, and $Q_u \cong Q_{bu}$—the pile is regarded as end-bearing; an example of this would be a pile tip in rocks or gravel.

7.4.2. End-Bearing Resistance

When the soil at the base of the pile reaches ultimate bearing failure, or the failure condition, the ultimate base resistance, or Q_{bu}, is fully activated; the failure surface is depicted in Figure 7.15. In a punching model, the soil mostly fails at the pile foundations, which are deep foundations. First, at the pile tip, a triangular failure zone I forms and is pushed downward without producing another slip surface. A partial development of a radial shear zone II can occur in dense materials. Therefore, bearing capacity theory can be used to evaluate the ultimate base resistance in a manner similar to that of shallow foundations. The various bearing capacity factors

|'\]0

N^*c N^*q $N^*\gamma$, include all required shape and depth factors (Xu & Lehane, 2008).

$$Q_{bu} = A_b\left(cN_c^* + \sigma_{vb}N_a^* + 0.5\gamma dN_\gamma^*\right)$$

Where; σvb is the vertical stress in the soil at the base of the pile; *c* is the Cohesion of the soil; A_b is the Pile base area; γ is the unit weight of soil; N^*c N^*q $N^*\gamma$ is the bearing capacity

The factors for bearing capacity taking the internal friction angle (ϕ) of the soil, the relative soil compressibility, and the geometry of the pile (Kim et al., 2014).

d = pile diameter.

Interestingly, Eqn. (7.5) uses generic notations; for instance, the vertical stress, denoted by σvb, can indicate effective or total stress, depending on the type of ground material encountered. All of the characteristics used to determine the ULC for saturated clay under undrained conditions (short-term) should have values for such condition. For example, all stresses should be 'total.' φ(= 0), and c (= su, undrained shear strength). All parameters should match the drained values for the drained or long-term or drained ULC, and all stresses should be effective stresses. The following will provide an illustration of the differences when calculating the ULC in sand and clay (Malik et al., 2017).

7.4.3. The Uplift Capacity

7.4.3.1. Driven Pile

The nominal uplift capacity of driven piles, $Q_{up,n}$, relies on the side friction capacity only. There is no end bearing, and the contribution from the weight of the pile is typically small. Thus, the selection of appropriate values of the side friction capacity is the most critical parameter in the design of piles subjected to uplift. According to field test results, the side friction capacity in uplift is 70–85% of the corresponding value for downward loading. The possible reasons are as follows (Tomlinson, 1969):

i. The uplift load modifies the soil fabric close to the pile.
ii. The shear surface and uplift load cause the soil to loosen.
iii. The pile diameter is reduced by the Poisson effect under tensile loading, marginally lowering the contact pressure (Duan et al., 2018).

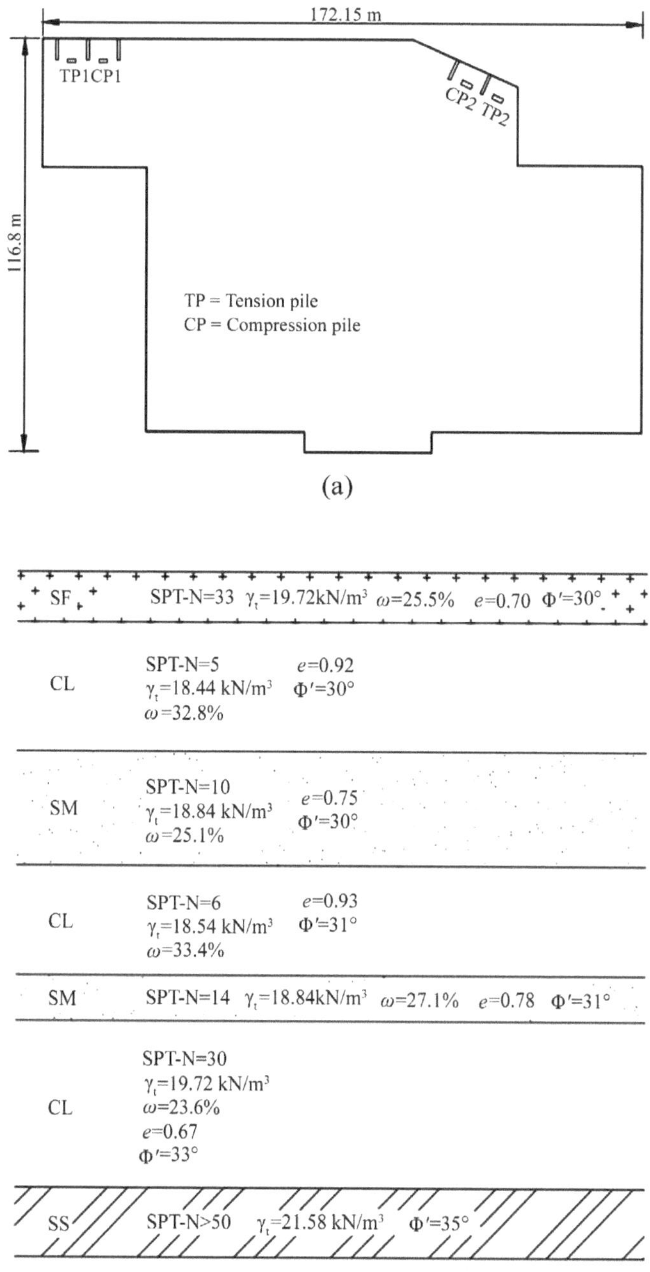

Figure 7.15. *Site plan (a) and soil profile; (b) from the pile load test are shown.*

7.4.3.2. Bored Pile

Bored piles can be constructed with an enlarged (or belled) base. The purpose of an enlarged base is usually to increase the base area, and thus, the downward load is transferred through the end bearing. It also has the advantage of generating a higher upward load capacity by bearing on top of the enlarged portion.

However, because of worker safety concerns and the length of time needed for construction, these bored piles are currently not widely used. Alternatively, instead of building an expanded base, a straight, longer bored pile can be built (Touma & Reese, 1974).

- **Comparing the Load Capacity of Compression and Tension Piles:** There were two rectangular, barrette, piles built at the same location in Taipei City, Taiwan. They were 2.7 m × 1.0 m × 49.2 m (width x thickness x length). Both tension and compression static pile load tests were carried out. The soil profile and the plan for the tested pile are displayed in Figure 7.15. Figure 7.16 displays the results of the pile load test and unequivocally demonstrates that, under a similar settlement, the frictional resistance is higher in the compression pile is higher than in the uplift or tension pile (Ellison et al., 1971).

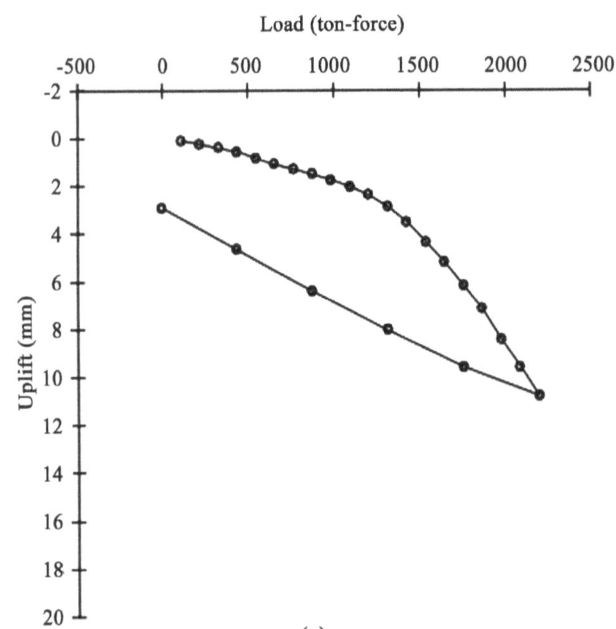

(a)

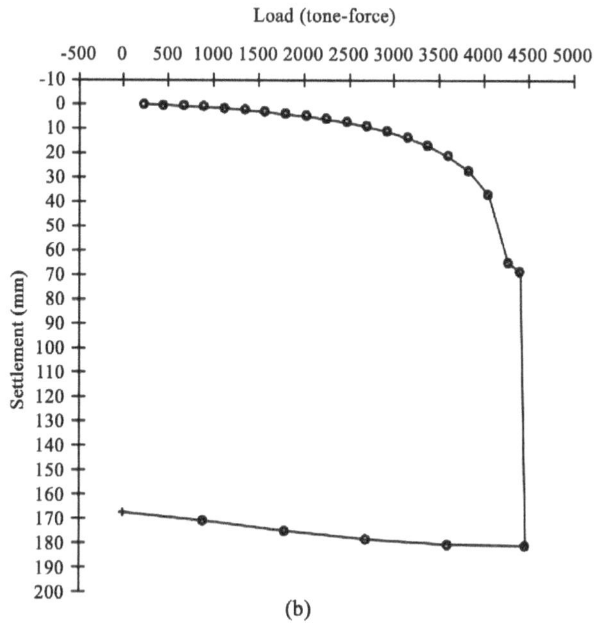

Figure 7.16. *Results of a pile load test Testing for tension (a); and compression (b).*

7.4.4. t-z Curve

The t-z method (Kraft et al., 1981) is a popular static analysis technique for determining vertical load distribution across the length of the pile and pile load settlement response at the head of the pile. It takes into account the interactions between the three main components—end bearing, side friction, and the elastic compression. Various side friction mobilization functions are used in this technique for stratified soil profiles along the pile's length. Notation for the variables z and t is different from its use in this book, and is frequently confusing (Kraft et al., 1981). Here is a list of notation definitions:

- *t*: Resistance of mobilized side friction.
- *q*: Resistance of mobilized net end-bearing.
- *z*: The displacement at the required pile depth.

The pile is divided into finite segments using the *t-z* method. The cross area (A), modulus of elasticity (E), and the segment length, determine the elastic properties of each segment. A non-linear spring is used to model the side friction resistance of soil against each segment. The load-displacement behavior of the spring is determined by the *t-z* curve. A *q-z* curve-defined nonlinear spring is also used to model the end bearing (Bateman et al., 2022).

Functions for the *t-z* and *q-z* curves are typically obtained from numerical analyzes or static load tests, and they are found in published references. The load test result may be extrapolated using a *t-z* analysis after curves have been custom-fitted, provided that in situ load test info is available. An illustration of the static pile load test's *t-z* curve. The axial force the pile absorbs along the pile shaft can be used to calculate the mobilized side friction resistance. The difference between the two sequential depth measurements is the pile segment's mobilized side friction resistance (Walsh et al., 1995).

The settlement, δ, at the top of the pile is designated as an independent variable and an initial value is selected in order to conduct a *t-z* analysis. After that, the numerical model is solved to determine the value of the associated applied load, P, and the force equilibrium. With repeated solutions using different δ values, a load–settlement plot can be generated and used to figure out the ultimate pile load capacity. There is commercial software available that can do t-z analysis. For a more thorough examination (Chin & Poulus, 1991).

7.5. LATERAL LOADED PILES

The surrounding soils will provide lateral resistance to support lateral loads from superstructures imposed on vertical piles. The pile head, or ground surface, should not experience severe movements when significant horizontal loads are applied to vertical piles; that is, the ground's upper part should not experience any failure (Poulos, 1971). Pile caps may be fastened to horizontal concrete beams or pads in order to reduce movement and offer enough lateral support. This section will provide an introduction to single vertical pile responses to lateral forces along with capacity evaluations (Reese, 1977).

7.5.1. Classification of Lateral Piles

As seen in Figure 7.17, piles that experience lateral loads are divided into two categories: long and short piles. When a short pile experiences lateral forces, it acts as a rigid body because it does not have enough embedment length. When a short pile is laterally loaded at its head, its toe has a tendency to rotate. As illustrated in Figure 7.17, a rigid rotation in this pile is the outcome of assuming a rotation point in the pile. The soils surrounding a short pile determine its ultimate lateral load capacity. Prior to the piles experiencing any flexural deflections, some of the soils in front of them yielded, adding to the final lateral resistance. On the other hand, a long pile is one that is embedded deeply in the ground, making it impossible for it to rotate or move horizontally at the pile toe. If the pile is long enough, there will be multiple points with a deflection equal to zero. Thus, the flexural capacity of long piles, which is determined by the structural design of the pile, determines its ultimate lateral capacity (Ashour et al., 2004).

The deflection characteristics controlled by the pile-soil interaction are used to categorize the short and long piles. The pile is loaded laterally and is handled similarly to a beam. In addition to equilibrium and compatibility requirements, a nonlinear differential equation is used to solve the responses of pile and soil interaction.

The displacement and lateral load were typically solved numerically. Eqn. 7.27 can be used to calculate the eigenvalue for β, which governs the deformation pattern of deformation for a laterally loaded pile. Lateral piles are categorized using the dimensionless length factor βL, where L is the pile length (Fan & Long, 2005).

A rigid or short pile is identified by $\beta L \leq 1$, and a long pile is indicated by $\beta L > 3$. $3 \geq \beta L > 1$ denotes a semirigid pile and denotes a nonrigid pile among long piles. Geotechnical engineers are interested in the lateral load capacity, which is discussed in the following section (Rollins et al., 2005).

$$E_p I_p \frac{d^4 y}{dz^4} + E_s y = 0$$

$$\beta = \sqrt[4]{\frac{E_s}{4E_p I_p}}$$

Where; Es is the subgrade modulus of soil reactions; E_p is the modulus of elasticity for the pile; I_p is the moment of inertia of the pile in the bending direction; z is the depth below the surface of ground;

y is the lateral deflection in the pile (Lai et al., 2018).

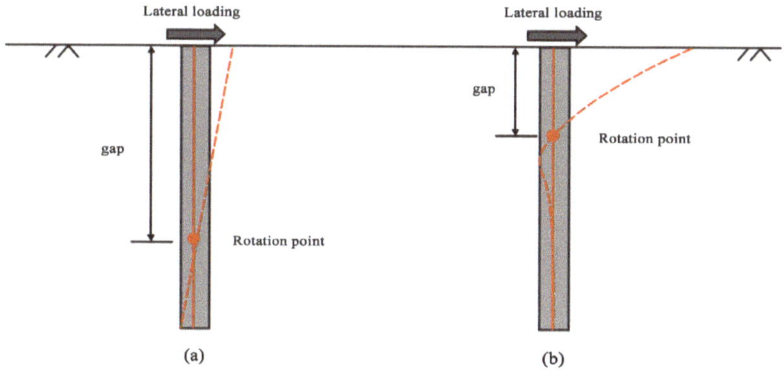

Figure 7.17. *Short (a); and long (b); lateral piles are shown.*

7.5.2. Lateral Load Capacity for Rigid Piles

The methods suggested by (Broms, 1964) are used to evaluate ultimate lateral load capacity of a vertical rigid pile. These methods are predicated on certain assumptions, such as the idea that the pile will behave like a rigid body when pushed by a horizontal force. The field test results and the prediction made using Broms's method agreed fairly, proving the validity of the assumptions. The assessment technique is separated into two categories: granular and cohesive soils with varying pile head conditions, such as free and fixed head. Figure 7.18 illustrates the potential failure mechanisms for the fixed and free head of a short rigid pile under horizontal loading. It is evident that the strength of the soil controls the failure (Georgiadis et al., 2013). As seen in Figure 7.18(a), for the free-head rigid pile, soils on various sides of the pile that are both above and below the rotation point A, all achieve the yielding condition.

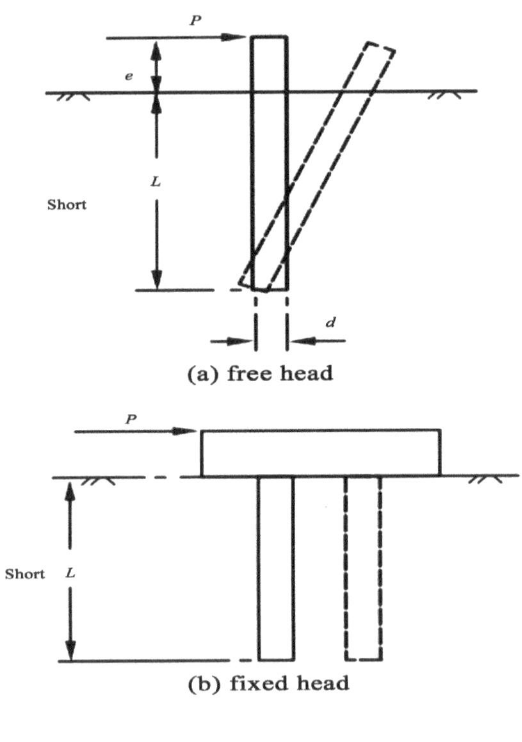

Figure 7.18. *The failure mechanism in short piles is shown.*

- **Rigid Piles in Cohesive Soils: Free Head and Fixed Head:** According to Figure 7.19, cohesive soils pressed by the horizontally loaded pile will ultimately resist 8–12 times their undrained shear strength (su). At a depth equal to three the pile diameter, the lateral resistance achieves its value. Consequently, (Broms, 1964) proposed a soil resistance distribution, which is depicted in Figure 7.20. At a depth 1.5 times the diameter of the pile, an ultimate soil resistance of 9su is visible (Lee et al., 2010).

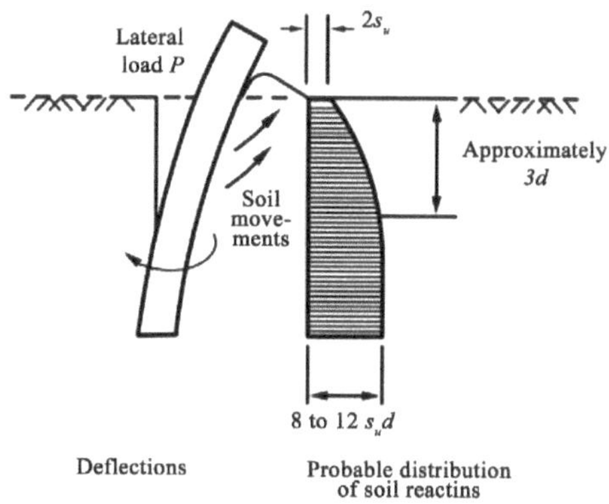

Figure 7.19. *Soil resistance distribution in a laterally loaded pile is shown.*

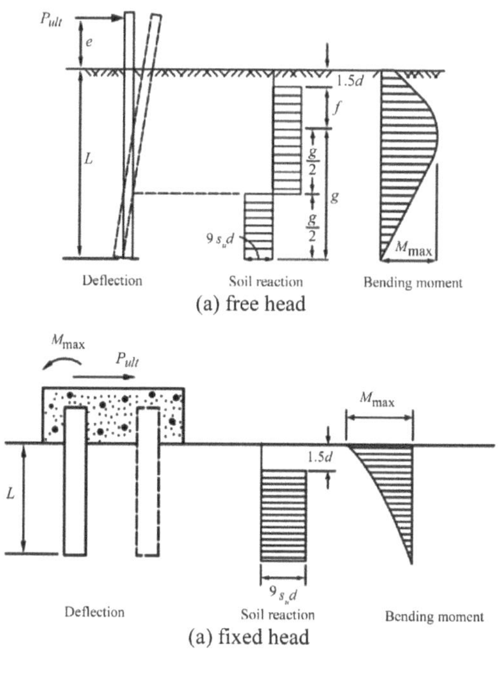

Figure 7.20. *The suggested soil resistance distribution in cohesive soils for a laterally loaded pile.*

7.6. THE PILE LOADING TEST

Certain assumptions, empirical knowledge from the area, and uncertainties were used in the pile load capacity determination methods in Section 7.4 to produce predictions that do not accurately reflect the behavior of installed piles. Therefore, a certain number of pile load tests need to be performed in the majority of large projects. Prediction methods' unreliability is the main cause. In the field, piles can be directly loaded to determine their lateral and vertical load-bearing capacities. It is possible to apply loads greater than the design values—for example, 200% of the design values. This section will provide the pile load test's precise configuration and interpretation (Fellenius, 1980).

7.6.1. Configuration of a Pile Load Test

The components of a pile load test, as illustrated in Figure 7.21, include the monitoring system, a hydraulic jack, a reaction pile and beam, and the tested pile. The purpose of reaction beams and piles is to provide the reaction force needed to load the tested pile. Because of this, the reaction system's capacity should be higher than the tested pile's. During the test, strain gages, load cells, LVDTs, and other instruments were used to measure loadings and displacements (Comodromos et al., 2003).

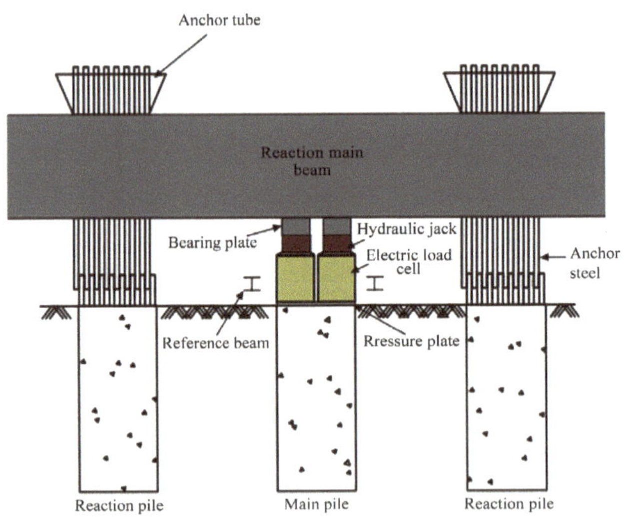

Figure 7.21. *The pile load test configuration is illustrated.*

Stress- and strain-controlled tests are the two categories of static load tests. In the former, measured settlements, considered the dependent variable, and a predefined load, considered the independent variable, are applied. The latter raises the load in order to keep the penetration rate constant, which typically ranges from 0.25–2.5 mm min. Both compression pile testing methods are covered by ASTM D1143, while ASTM D3689 deals with uplift (tensile) pile testing. The quick test, given by ASTM D1143 Procedure A, and the maintained load test, given by ASTM D1143 Procedure B, are the two forms of stress-controlled tests. The test loads have to be applied in accordance with one of the methods outlined below for every test method, as stated in ASTM D1143. If possible, the maximum applied load ought to reach the point of failure where the pile's total static axial compressive loading capacity is reflected. A load-settlement plot is produced by both methods; specifics are discussed here (Dai et al., 2012).

7.6.2. The Maintain Load Test

The increase in load during the maintain load test is about 25% of the design load. Each load is kept for two hours, or until the settlement rate is below 0.25 mm/hour. Until failure occurs or the design load of 200% is reached (that is, for a proof test), the process of loading is continued. In order to observe the creep effect, the max load is maintained for a longer duration during the proof test. After that, the pile is gradually unloaded (Adams & Collin, 1997).

7.6.3. The Quick Test

Compared to the maintain load test, the quick test is quicker. The load increment is about 5% of the maximum load allowed by design, and every load is retained for a set amount of time, no matter how fast the pile displacement occurs at the end of that interval. ASTM allows 4 to 15-minute intervals. Until failure or about 200% of the design load is reached, it will continue. The quick test usually takes 4–6 hours to finish, whereas a maintain load test takes several days. As a result, this test is frequently used to reduce the overall test duration (Ammons & Ammons, 1962).

7.6.4. Interpretation of the Test Results

A load-settlement curve (Figure 7.22) would be obtained following the pile load test in order to ascertain the pile's ULC (Shapiro, 1999). The following introduces and suggests a few ways to determine the ULC:

7.6.4.1. The Davisson Method

It is a commonly used technique, as seen in the international building code (IBC) and the AASHTO, to ascertain a pile's load capacity

$$S = 4 + \frac{d}{120} + \frac{QL}{AE} \text{ (unit: mm)}$$

Here:

S is the pile head settlement; A is the pile cross-sectional area in mm^2; L is the Length of the pile in mm; Q is the Applied load in kN. d = pile diameter of pile in mm; and E = The pile's Youngs Modulus in kN/mm^2.

The 4 mm (about 1.5 in) in equation is the amount of pile head settlement caused by shaft displacement in the ultimate load condition. The $d/120$ (about $0.01d$) is the amount of pile head settlement caused by the base displacement in the ultimate load condition. The offset is the sum of these variables, and the pile's elastic deformation is represented by QL/AE. Therefore, as shown in Figure 7.22, the ULC is located at the intersection of the linear line in Eqn. 7.42 and the settlement curve (Davisson & Beams, 1953).

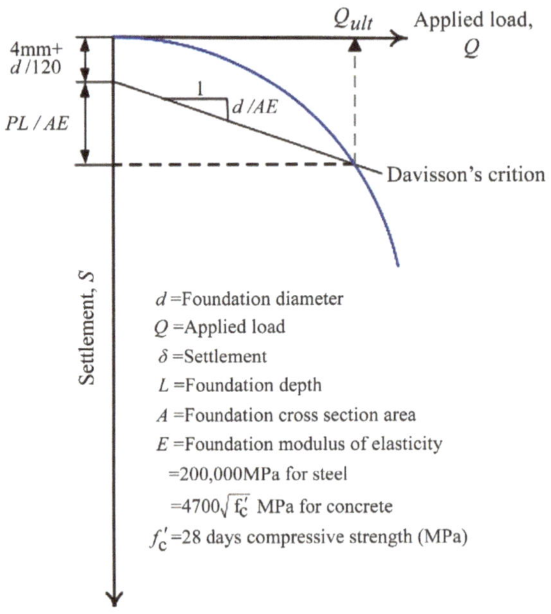

Figure 7.22. *The resulting load-settlement curve from a pile load test.*

7.6.4.2. The Fuller and Hoy Approach

The load at which the load-settlement curve's slope equals 0.140 mm/kN is known as the ultimate load (Davisson & Beams, 1953).

7.6.4.3. The Terzaghi Technique

The load at which 10% of the pile diameter is accounted for by settlement is the ultimate load.

7.6.5. Correlation Between Other In-Situ and the Pile Load Tests Results

The ULC of the pile on the site is the most significant of the detailed behaviors of the pile-soil interaction that engineers can learn from the pile load test. Over 200 data points from pile load tests were gathered by the writers. From the data, the final load capacities calculated using Davisson's offset method were taken (Wang et al., 2021). Other in situ test data, such as those for the CPT and SPT, were also gathered in addition to the pile load test. Thus, the cone resistance q_c and the SPT-N value—two very typical in-situ test results—correlate with the ultimate pile load capacity (Leetsaar et al., 2023). The relationship between ULC, SPT-N, and q_c is depicted in Figure 7.23.

The SPT-N values utilized in Figure 7.33a comprised the mean of the values between 4 and 10 pile diameters below and above the pile tip. Another variable in the figure is the pile's length-to-diameter ratio, or L / d. It is evident that a greater L / d and an average SPT-N result in a higher final load capacity. Nevertheless, the data also show some scatters. The average cone tip resistances, $q_{c,avg}$, are computed along the pile length in Figure 7.33b. It is also evident from the figure that higher $q_{c,avg}$, and larger L / d result in greater ULC (Robertson, 1986).

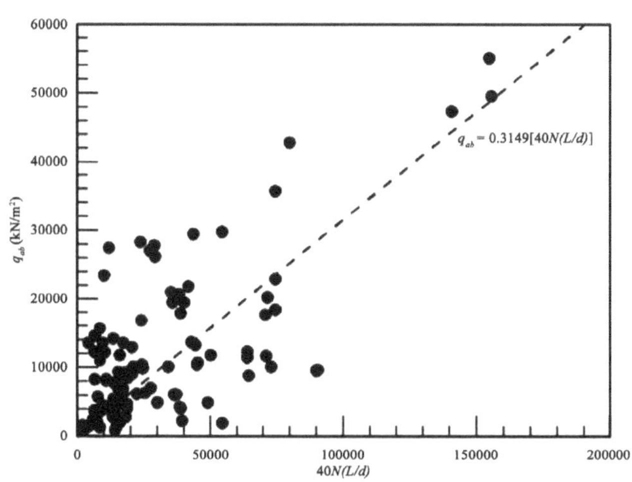

(a)

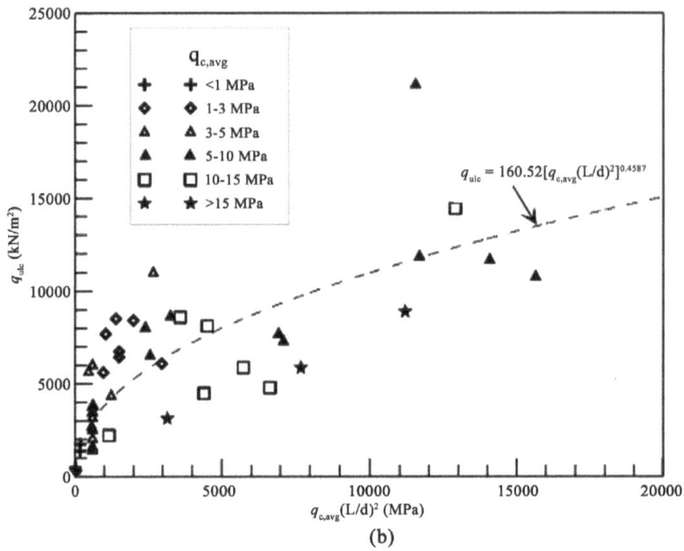

Figure 7.23. *The correlation between the ULC and the in-situ test data, specifically; (a) SPT-N; and (b) qc.*

7.7. NEGATIVE SKIN FRICTION

Negative skin friction adds additional loads to piles driven through soft layers (like soft clay or fill layers) into the more firmer layers, on top of the applied load on the pile head. A downward drag known as negative skin friction is brought on by the surrounding compressible soil's downward movement, or settling, in relation to the pile. The pile is seen passing through a cohesive soil fill that was recently built in Figure 7.24. Under the weight of the fill's overburden, the soil beneath it is totally consolidated. There have been some proposed models, and the negative skin friction determination is still under investigation. A neutral plane model is depicted in Figure 7.35, which makes the assumption that there is a natural plane where the positive and negative friction forces are equal at a given depth (Alonso et al., 1984).

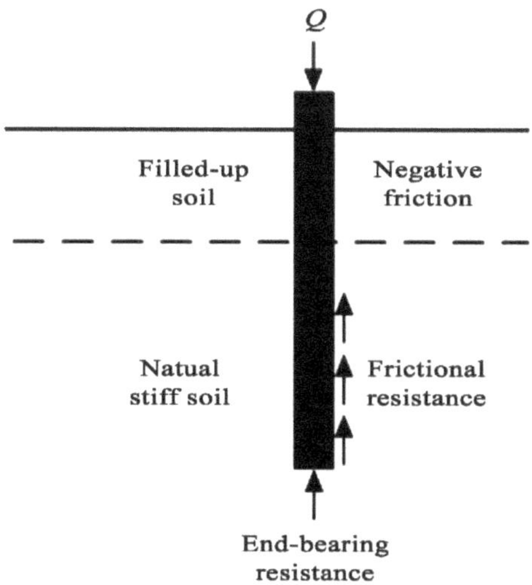

Figure 7.24. *Negative skin friction is illustrated.*

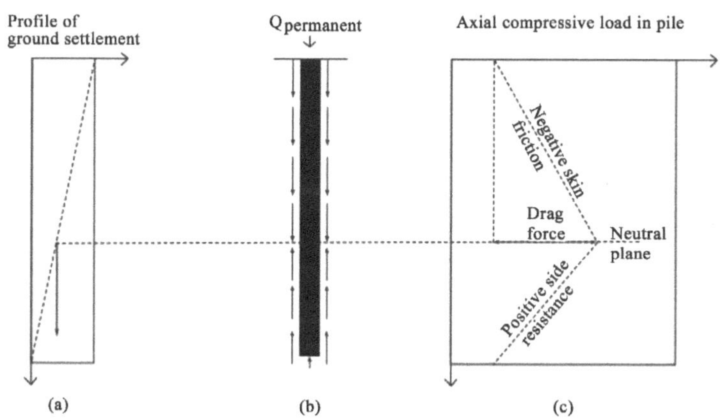

Figure 7.25. *A neutral plane model for negative skin friction is shown: The pile settlement (a); The load transfer (b); and the axial compressive load in pile (c).*

The material of the pile, the soils type, and the rate and amount of relative movement between the pile and the soil all affect the transfer of negative skin friction force to the pile. To completely initiate negative skin friction, there must be a small relative movement of about 10 mm because the soil is actively setting and pulling the pile downward (Comodromos & Bareka, 2005). The additional load on a pile resulting from negative skin friction can be simply estimated as follows. This extra load is related to the soil's adhesion factor and undrained shear strength:

$$Q_{ns} = A_s \times \overline{s_u} \times \alpha$$

Where; Q_{ns} is the Negative skin friction induced load; A_s = The pile circumferential area; s_u is the Undrained shear strength (undrained) across the pile shaft; *a* is the adhesion factor (Chow et al., 1990).

Reducing the negative friction can be achieved by driving the pile inside a casing, and filling it with a viscous material; using a smaller diameter for the pile shaft than the pile tip; and coating the piles with bitumen. Based on a full-scale test published by Bjerrum et al. (1969), bitumen coating has been shown to be an effective way to minimize negative skin friction. On the reference pile i.e., (with no bitumen coating, the negative skin friction recorded a year after pile driving is roughly 3,000 kN, while it is only 150 kN on the pile coated in bitumen. In particular, to guarantee a decrease in negative skin friction, the bitumen coating needs to be adequately shielded throughout the pile driving procedure (Chen et al., 2009).

7.8. GROUP PILE EFFECT

To transfer the structural load into the soil, piles are usually used in groups. Under such circumstances, group piles are covered by a pile cap, as shown in Figure 7.26. The determination of group piles' load-bearing capacity is an exceedingly complex matter that remains unresolved at present. As illustrated in Figure 7.27, it is reasonable to assume that any stresses transmitted from nearby piles to the soil will be overlapping and lower the piles' capacity to support weight. Even in cases where pile load tests conducted on a single pile show satisfactory performance, significant settlement or total general shear failure in the group piles may occur (Bao et al., 2012).

Terzaghi and Peck (1967) provided a classic case study of a group pile failure. The Charity Hospital in New Orleans was built on a foundation made up of about 20,000 7.9-meter-long timber piles. The pile broke through the soft clay and ended in 1.8 meters of dense sand with at least 50 meters of soft soil underneath. Prior to construction, a pile load test had been carried out and the results showed that a 300 kN load would cause a 6.5 mm settlement. Nevertheless, within two years of the building's construction—which is subject to a 150 kN working load—the group piles had settled over 300 mm. As a result, adding the load capacities of individual piles will not yield an estimate of the group pile load capacity (Butterfield & Banerjee, 1971).

The piles in a group are supposed to be spaced such that the group's load-bearing capacity equals the total of the bearing capacities of the individual piles. The minimum pile spacing between centers, or D, is 2.5d in practice and is actually closer to 3 to 3.5d in everyday circumstances. Table 7.1 lists the minimum spacing, taking into account various types of soil and piles (Hussien et al., 2010).

Table 7.1. Minimum Spacing for Group Piles

Pile length (m)	*Friction piles in sand*	*Friction piles in clay*	*End-bearing piles*
Less than 12 m	3*d*	4*d*	3*d*
12 to 24 m	4*d*	5*d*	4*d*
Greater than 24 m	5*d*	6*d*	5*d*

Note that d is the largest side, or the pile diameter. If a racking pile is being used, the pile spacing is measured three meters below the pile cutoff level. Otherwise, the pile spacing is recorded at the cutoff level.

Figure 7.26. *A picture of group piles.*

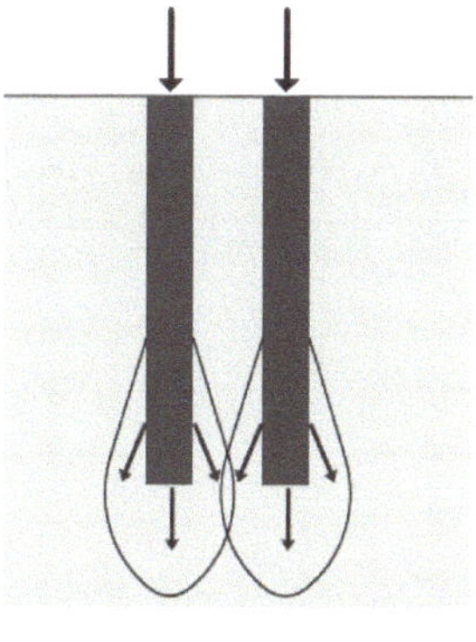

Figure 7.27. *Group pile's impact on load transfer is visualized.*

The "efficiency formula" provides a simple way to calculate the group pile capacity. Equation can be used to determine a group pile's load-bearing capacity efficiency. Because the efficiency factor is seldom empirical, the efficiency formula should be used with extreme caution (Rollins et al., 2006).

$$\eta = \frac{Q_{g(u)}}{\sum Q_u}$$

Where; η is the group efficiency; Q_u is the ultimate load bearing capacity of individual piles without the group effect; $Q_{g(u)}$ is the ultimate load bearing capacity of the group pile.

For a group pile composed of clay, η is typically less than unity, roughly 0.7, as proposed by (Tomlinson, 1969), to indicate the spacing between two pile diameters. In clay, the piles collapse one by one for larger spacing, but the ratio of efficiency equals one at about eight pile diameters. Sand group piles have a tendency to compact the soil surrounding them, which raises the ultimate failure load of the pile groups. Figure 7.28 displays the change in group efficiency for a sand pile with three rows of three pillars. The densification of the sand surrounding the pile is the main reason why the group efficiency for medium and loose sands can be larger than unity. Driven pile groups into dense sand result in an efficiency ratio less than unity (Gazetas et al., 1993).

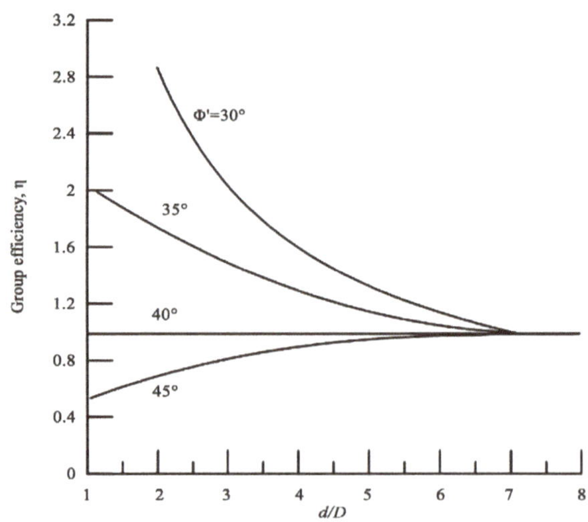

Figure 7.28. *Variation in the sand pile groups' efficiency is shown.*

SUMMARY

The chapter on pile foundations delves into the critical role these structural elements play in supporting various constructions. It elucidates the fundamental principles underlying pile design, emphasizing factors such as soil characteristics, load requirements, and environmental conditions. It discusses different types of piles, including driven piles, drilled shafts, and helical piles, highlighting their respective advantages and applications. Furthermore, the chapter elucidates the installation methods and quality control measures essential for ensuring the integrity and stability of pile foundations. Through clear explanations and illustrative examples, it equips readers with a comprehensive understanding of pile foundation systems, empowering them to make informed decisions in their construction projects.

REFERENCES

1. Adams, M. T., & Collin, J. G. (1997). Large model spread footing load tests on geosynthetic reinforced soil foundations. *Journal of Geotechnical and Geoenvironmental Engineering, 123*(1), 66–72.
2. Adhikari, P., Ng, K. W., Gebreslasie, Y. Z., Wulff, S. S., & Sullivan, T. A. (2020). Geomaterial classification criteria for design and construction of driven steel H-piles. *Canadian Geotechnical Journal, 57*(4), 616–621.
3. Adhikari, P., Ng, K. W., Gebreslasie, Y. Z., Wulff, S. S., & Sullivan, T. A. (2020). Impedance functions of piles in layered media. *Journal of the Engineering Mechanics Division, 104*(3), 643–661.
4. Allwright, R. L. (1981). What do we want teaching materials for? *ELT Journal, 36*(1), 5–18.
5. Alonso, E. E., Josa, A., & Ledesma, A. (1984). Negative skin friction on piles: A simplified analysis and prediction procedure. *Geotechnique, 34*(3), 341–357.
6. Ammons, R. B., & Ammons, C. H. (1962). The quick test (QT): Provisional manual. *Psychological Reports, 11*(1), 111–161.
7. Arabameri, M., & Eslami, A. (2021). Microstructure and strength effect on bearing capacity of helical piles installed in Golestan loess. *International Journal of Civil Engineering, 19*(8), 923–940.
8. Ashour, M., Pilling, P., & Norris, G. (2004). Lateral behavior of pile groups in layered soils. *Journal of Geotechnical and Geoenvironmental Engineering, 130*(6), 580–592.
9. Bao, X., Morikawa, Y., Kondo, Y., Nakamura, K., & Zhang, F. (2012). Shaking table test on reinforcement effect of partial ground improvement for group-pile foundation and its numerical simulation. *Soils and Foundations, 52*(6), 1043–1061.
10. Bateman, A. H., Crispin, J. J., Vardanega, P. J., & Mylonakis, G. E. (2022). Theoretical t-z curves for axially loaded piles. *Journal of Geotechnical and Geoenvironmental Engineering, 148*(7), 0402–2052.
11. Bishop, A. W. (1971). The influence of progressive failure on the choice of the method of stability analysis. *Geotechnique, 21*(2), 168–172.
12. Broms, B. B. (1964). Lateral resistance of piles in cohesionless soils. *Journal of the Soil Mechanics and Foundations Division, 90*(3), 123–156.
13. Broms, B. B. (1965). Design of laterally loaded piles. *Journal of the Soil Mechanics and Foundations Division, 91*(3), 79–99.
14. Brown, M. R., & Ge, S. (2018). Distinguishing fluid flow path from pore

pressure diffusion for induced seismicity. *Bulletin of the Seismological Society of America, 108*(6), 3684–3686.

15. Butterfield, R., & Banerjee, P. K. (1971). The problem of pile group–Pile cap interaction. *Geotechnique, 21*(2), 135–142.
16. Callister, W. D., & Rethwisch, D. G. (2000). Fundamentals of materials science and engineering (Vol. 4, (716–60817)).
17. Chen, R. P., Zhou, W. H., & Chen, Y. M. (2009). Influences of soil consolidation and pile load on the development of negative skin friction of a pile. *Computers and Geotechnics, 36*(8), 1265–1271.
18. Chin, J. T., & Poulus, H. G. (1991). A "TZ" approach for cyclic axial loading analysis of single piles. *Computers and Geotechnics, 12*(4), 289–320.
19. Choi, S. K., Gal, Y. S., Jin, S. H., & Kim, H. K. (2000). Poly (1, 6-heptadiyne)-based materials by metathesis polymerization. *Chemical Reviews, 100*(4), 1645–1682.
20. Chow, F. C., Jardine, R. J., Nauroy, J. F., & Brucy, F. (1997). Time-related increases in the shaft capacities of driven piles in sand. *Geotechnique, 47*(2), 353–361.
21. Chow, Y. K., Chin, J. T., & Lee, S. L. (1990). Negative skin friction on pile groups. *International Journal for Numerical and Analytical Methods in Geomechanics, 14*(2), 75–91.
22. Comodromos, E. M., & Bareka, S. V. (2005). Evaluation of negative skin friction effects in pile foundations using 3D nonlinear analysis. *Computers and Geotechnics, 32*(3), 210–221.
23. Comodromos, E. M., Anagnostopoulos, C. T., & Georgiadis, M. K. (2003). Numerical assessment of axial pile group response based on load test. *Computers and Geotechnics, 30*(6), 505–515.
24. Comodromos, E. M., Papadopoulou, M. C., & Rentzeperis, I. K. (2009). Pile foundation analysis and design using experimental data and 3-D numerical analysis. *Computers and Geotechnics, 36*(5), 819–836.
25. Dai, G., Salgado, R., Gong, W., & Zhang, Y. (2012). Load tests on full-scale bored pile groups. *Canadian Geotechnical Journal, 49*(11), 1293–1308.
26. Dash, B. K., & Pise, P. J. (2003). Effect of compressive load on uplift capacity of model piles. *Journal of Geotechnical and Geoenvironmental Engineering, 129*(11), 987–992.
27. Davisson, C. J., & Beams, J. W. (1953). A new variation of the rotation-by-magnetization method of measuring gyromagnetic ratios. *Reviews of Modern Physics, 25*(1), 2–46.

28. Davisson, C. J., & Beams, J. W. (1953). Prediction of ultimate capacities of bored pile based on routine load tests and static analysis (case study). *Advances in Industrial Engineering and Management, 8*(1), 7–9.
29. De Mello, W. C., Motta, G. E., & Chapeau, M. (1969). A study on the healing-over of myocardial cells of toads. *Circulation Research, 24*(3), 475–487.
30. Duan, N., Cheng, Y. P., & Liu, J. W. (2018). DEM analysis of pile installation effect: Comparing a bored and a driven pile. *Granular Matter, 20*, 1–16.
31. Eigenbrod, K. D., & Issigonis, T. (1996). Pore-water pressures in soft to firm clay during driving of piles into underlying dense sand. *Canadian Geotechnical Journal, 33*(2), 209–218.
32. Ellison, R. D., D'Appolonia, E., & Thiers, G. R. (1971). Load-deformation mechanism for bored piles. *Journal of the Soil Mechanics and Foundations Division, 97*(4), 661–678.
33. El-Marsafawi, H., Han, Y. C., & Novak, M. (1992). Dynamic experiments on two pile groups. *Journal of Geotechnical Engineering, 118*(4), 576–592.
34. Fan, C. C., & Long, J. H. (2005). Assessment of existing methods for predicting soil response of laterally loaded piles in sand. *Computers and Geotechnics, 32*(4), 274–289.
35. Fellenius, B. H. (1980). The analysis of results from routine pile load tests. *Ground Engineering, 13*(6), 19–31.
36. Fratzl, P., & Weinkamer, R. (2007). Nature's hierarchical materials. *Progress in Materials Science, 52*(8), 1263–1334.
37. Gazetas, G., Fan, K., & Kaynia, A. (1993). Dynamic response of pile groups with different configurations. *Soil Dynamics and Earthquake Engineering, 12*(4), 239–257.
38. Georgiadis, K., Georgiadis, M., & Anagnostopoulos, C. (2013). Lateral bearing capacity of rigid piles near clay slopes. *Soils and Foundations, 53*(1), 144–154.
39. Hussien, M. N., Tobita, T., Iai, S., & Rollins, K. M. (2010). Soil–pile separation effect on the performance of a pile group under static and dynamic lateral loads. *Canadian Geotechnical Journal, 47*(11), 1234–1246.
40. Jardine, R., Chow, F., Overy, R., & Standing, J. (2005). ICP design methods for driven piles in sands and clays. *Journal of Geotechnical Engineering, 112*, 33–78.
41. Kim, S. H., Heo, W. H., Woo, K. S., Jung, C. Y., & Park, S. J. (2014). End-bearing resistance of Y-type perfobond rib according to rib width–height

ratio. *Journal of Constructional Steel Research, 103*, 101–116.

42. Kraft Jr, L. M., Ray, R. P., & Kagawa, T. (1981). Theoretical t-z curves. *Journal of the Geotechnical Engineering Division, 107*(11), 1543–1561.
43. Lai, H. J., Zheng, J. J., Zhang, R. J., & Cui, M. J. (2018). Classification and characteristics of soil arching structures in pile-supported embankments. *Computers and Geotechnics, 98*, 153–171.
44. Lande, E. J., Karlsrud, K., Langford, J., & Nordal, S. (2020). Effects of drilling for tieback anchors on surrounding ground: Results from field tests. *Journal of Geotechnical and Geoenvironmental Engineering, 146*(8), 0–502.
45. Lee, J., Kim, M., & Kyung, D. (2010). Estimation of lateral load capacity of rigid short piles in sands using CPT results. *Journal of Geotechnical and Geoenvironmental Engineering, 136*(1), 48–56.
46. Leetsaar, L., Korkiala-Tanttu, L., & Kurnitski, J. (2023). CPT, CPTu and DCPT methods for predicting the ultimate bearing capacity of cast in situ displacement piles in silty soils. *Geotechnical and Geological Engineering, 41*(2), 631–652.
47. Levacher, D. R., & Sieffert, J. G. (1984). Tests on model tension piles. *Journal of Geotechnical Engineering, 110*(12), 1735–1748.
48. Li, L., Li, J., Sun, D. A., & Gong, W. (2017). Analysis of time-dependent bearing capacity of a driven pile in clayey soils by total stress method. *International Journal of Geomechanics, 17*(7), 0401–6156.
49. Liu, N. W., Zhang, Z. M., Zhang, Q. Q., & Fang, K. (2013). Destructive field tests on mobilization of end resistance of cast-in-situ bored piles. *Journal of Central South University, 20*(4), 1071–1078.
50. Lu, Y., Aimetti, A. A., Langer, R., & Gu, Z. (2016). Bio responsive materials. *Nature Reviews Materials, 2*(1), 1–17.
51. Malik, A. A., Kuwano, J., Tachibana, S., & Maejima, T. (2017). End bearing capacity comparison of screw pile with straight pipe pile under similar ground conditions. *Acta Geotechnica, 12*, 415–428.
52. Mandolini, A., Russo, G., & Viggiani, C. (2005). Pile foundations: Experimental investigations, analysis and design. In *Proceedings of the International Conference on Soil Mechanics and Geotechnical Engineering, 16*(1), 1–77.
53. Mohamad, H., Tee, B. P., Ang, K. A., & Chong, M. F. (2016). Characterizing anomalies in distributed strain measurements of cast-in-situ bored piles. *Jurnal Teknologi, 78*(8–5), 75–82.
54. Nam, D. H., Cha, S. I., Lim, B. K., Park, H. M., Han, D. S., & Hong, S. H.

(2012). Synergistic strengthening by load transfer mechanism and grain refinement of CNT/Al–Cu composites. *Carbon, 50*(7), 2417–2423.

55. Nateghi, R. (2012). Evaluation of blast induced ground vibration for minimizing negative effects on surrounding structures. *Soil Dynamics and Earthquake Engineering, 43*, 133–138.
56. Ng, W. K., Selamat, M. R., & Choong, K. K. (2010). Soil/pile set-up effects on driven pile in Malaysian soil. *EJGE, 15*, 1–11.
57. Novak, M., & El Sharnouby, B. (1983). Stiffness constants of single piles. *Journal of Geotechnical Engineering, 109*(7), 961–974.
58. Park, B. J., Fang, F. F., & Choi, H. J. (2010). Magnetorheology: Materials and application. *Soft Matter, 6*(21), 5246–5253.
59. Poulos, H. G. (1971). Behavior of laterally loaded piles: I-single piles. *Journal of the Soil Mechanics and Foundations Division, 97*(5), 711–731.
60. Qian, D., Liu, W. K., & Ruoff, R. S. (2003). Load transfer mechanism in carbon nanotube ropes. *Composites Science and Technology, 63*(11), 1561–1569.
61. Randolph, M. F. (2003). Science and empiricism in pile foundation design. *Géotechnique, 53*(10), 847–875.
62. Randolph, M. F., & Wroth, C. P. (1979). An analytical solution for the consolidation around a driven pile. *International Journal for Numerical and Analytical Methods in Geomechanics, 3*(3), 217–229.
63. Reese, L. C. (1977). Laterally loaded piles: Program documentation. *Journal of the Geotechnical Engineering Division, 103*(4), 287–305.
64. Robertson, P. K. (1986). In situ testing and its application to foundation engineering. *Canadian Geotechnical Journal, 23*(4), 573–594.
65. Rollins, K. M., Lane, J. D., & Gerber, T. M. (2005). Measured and computed lateral response of a pile group in sand. *Journal of Geotechnical and Geoenvironmental Engineering, 131*(1), 103–114.
66. Rollins, K. M., Olsen, K. G., Jensen, D. H., Garrett, B. H., Olsen, R. J., & Egbert, J. J. (2006). Pile spacing effects on lateral pile group behavior: Analysis. *Journal of Geotechnical and Geoenvironmental Engineering, 132*(10), 1272–1283.
67. Ronold, K. O. (1990). Reliability analysis of tension pile. *Journal of Geotechnical Engineering, 116*(5), 760–773.
68. Roy, M., Blanchet, R., Tavenas, F., & Rochelle, P. L. (1981). Behavior of a sensitive clay during pile driving. *Canadian Geotechnical Journal, 18*(1), 67–85.
69. Rybnikov, A. M. (1990). Experimental investigations of bearing capacity of bored-cast-in-place tapered piles. *Soil Mechanics and Foundation*

Engineering, 27(2), 48–52.

70. Shapiro, D. E. (1999). The interpretation of diagnostic tests. *Statistical Methods in Medical Research, 8*(2), 113–134.
71. Tomlinson, R. A. (1969). Perachora: The remains outside the two sanctuaries. *The Annual of the British School at Athens, 64*, 155–258.
72. Touma, F. T., & Reese, L. C. (1974). Behavior of bored piles in sand. *Journal of the Geotechnical Engineering Division, 100*(7), 749–761.
73. Walsh, K. D., Houston, S. L., & Houston, W. N. (1995). Development of tz curves for cemented fine-grained soil deposits. *Journal of Geotechnical Engineering, 121*(12), 886–895.
74. Wang, Y., Sang, S., Zhang, M., Bai, X., & Su, L. (2021). Investigation on in-situ test of penetration characteristics of open and closed PHC pipe piles. *Soils and Foundations, 61*(4), 960–973.
75. Wang, Z. L. (2001). Characterization of nanophase materials. *Particle & Particle Systems Characterization: Measurement and Description of Particle Properties and Behavior in Powders and Other Disperse Systems, 18*(3), 142–165.
76. Wersäll, C., & Massarsch, K. R. (2013). Soil heave due to pile driving in clay. In R. D. Holtz II (Ed.), *Sound geotechnical research to practice: Honoring Robert D. Holtz* (pp. 480–498). ASTM International.
77. Wu, K., Chen, F., Chen, C., Zheng, H., & Xu, J. (2019). Load-transfer mechanism and bond-stress components in steel and steel fiber–Reinforced concrete structure. *Journal of Structural Engineering, 145*(12), 0401–9160.
78. Xia, H. E., Cao, Y., Zhang, N., & Qu, J. (2002). Vibration effects of light-rail train-viaduct system on surrounding environment. *International Journal of Structural Stability and Dynamics, 2*(02), 227–240.
79. Xu, X., & Lehane, B. M. (2008). Pile and penetrometer end bearing resistance in two-layered soil profiles. *Géotechnique, 58*(3), 187–197.
80. Yao, H., Ma, L., Duan, Z., & Guo, W. (2024). Reliability analysis of the cast-in-place bored pile with different defects. *Frontiers in Built Environment, 10*, 133–7986.
81. Zarrabi, M., & Eslami, A. (2016). Behavior of piles under different installation effects by physical modeling. *International Journal of Geomechanics, 16*(5), 0401–6014.
82. Zhong-Miao, Z., Qian-Qing, Z., & Feng, Y. (2011). A destructive field study on the behavior of piles under tension and compression. *Journal of Zhejiang University-Science A, 12*(4), 291–300.

CHAPTER 8

GEOSYNTHETICS AND THEIR APPLICATIONS IN GEOTECHNICAL ENGINEERING

Contents

8.1. INTRODUCTION

In the last few decades, the most important event in the geotechnical engineering history has been the dramatic rise of different types of geosynthetic products into the field. It has essentially transformed this area of civil engineering practice to a degree never seen before. Consequently, it was deemed appropriate to dedicate a chapter in this book to this topic, attempting to illustrate its various applications through case studies whenever feasible. As previously mentioned in earlier texts, geosynthetics are "inclusions" and perform a variety of functions, including reinforcement, drainage, separation, and filtration (Kurian, 2005). Among them, geotextiles in particular play a number of roles.

Globally, there are numerous companies producing geosynthetic goods. In order to give these companies' claims credibility, a manufacturer-wise overview is thus attempted, primarily based on the technical literature they have published. This is evident from the instances of their usage in various regions of the world. The objective is to portray the topic from the perspective of these influential figures in a field that they dominate (Bayraktar, 2020).

8.2. GEOSYNTHETICS – AN OVERVIEW

The concept has been historically applied as early as 3000 B.C. in China (the Great Wall of China) and Iraq (Ziggurat). In South Carolina, United States of America, heavy cotton fabric was first used to stabilize roads in 1926 as a separator. Approximately 10 million square meters of geotextiles were used for coastal protection as part of the Dutch Delta Works project in 1956, marking a significant milestone in the history of the material (Touze et al., 2016).

The first woven geotextiles appeared in the early 1960s (see two examples in Figure 8.1). One noteworthy instance of its application as a separator and filter material—rather than granular filters—was carried out by the United States Army Corps of Engineers in Memphis in 1962. In the 1960s, new geotextile products for soil reinforcement applications, such as geonets or geotextile nets, and geogrids, etc., were developed, mostly in the United Kingdom (Giroud, 2005).

France's first needle-punched geotextile (nonwoven) was made by this company. After then, a variety of heat-bonded nonwoven geotextiles were produced by Imperial Chemical Industries in the United Kingdom. One of the pioneers of geotextile technology, Chemie Linz in Austria produced needle-punched nonwoven geotextiles using the brand name Polyfelt (Fannin & Sigurdsson, 1996).

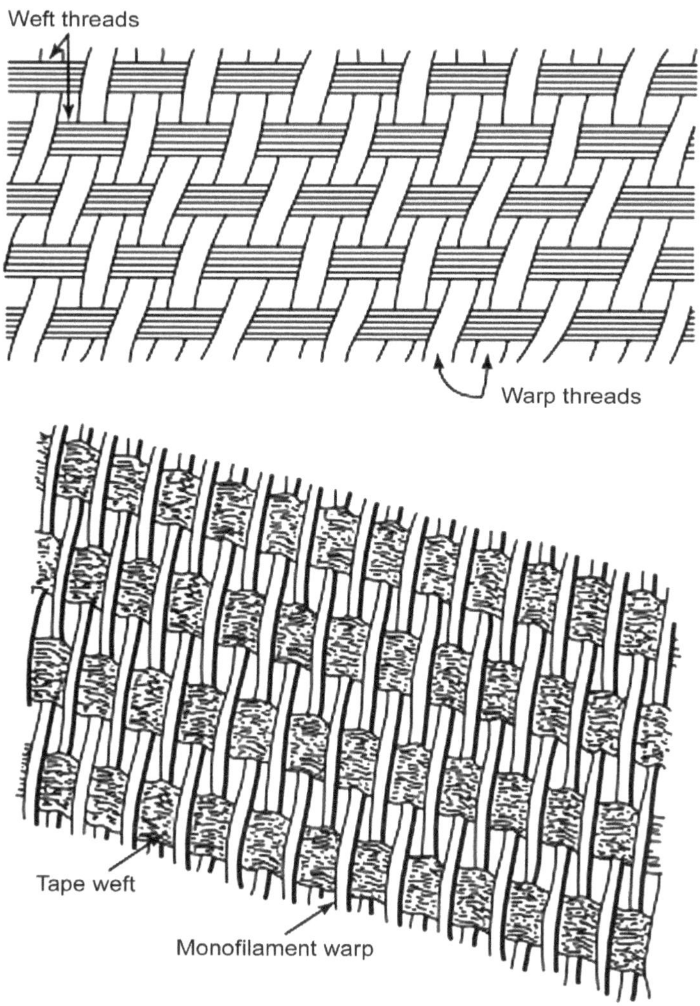

Figure 8.1. *Woven geotextiles – two types.*

The International Geotextiles Society or IGS was founded as a result of the growth of geotextiles being reported in a number of international, regional, and national conferences. In Drexel, USA, the Geosynthetics Research Institute (GRI) also formed (Palmeira et al., 2021).

Geotextiles serve the following six basic purposes: 1) separation; 2) filtration; 3) drainage; 4) reinforcement; 5) waterproofing; and 6) as a protective layer. Figure 8.2 provides a schematic representation of these functions. The functions of geotextiles against different applications are displayed in Table 8.1.

Four synthetic polymers are used to make geotextiles and related products: (1) Polyethylene (PE) in 1931, (2) Polyamide (PA) in 1935, (3) Polyester (PES) in 1941, and (4) Polypropylene (PP) in 1954. These polymers become available in that order. A woven geotextile can be made with monofilament, multifilament, or combinations of the two types of yarn. Slit film tapes, on the other hand, are now the most widely used type of yarn when creating woven geotextiles.

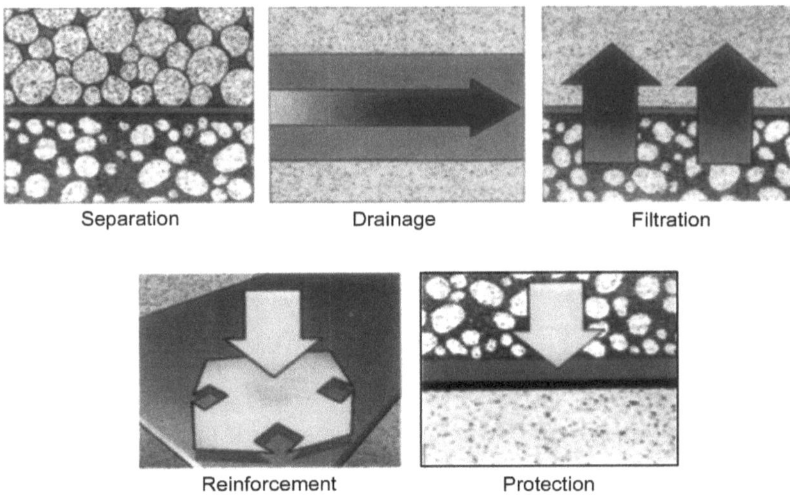

Figure 8.2. *Geotextile functions – schematic.*

One method of creating nonwoven geotextiles is 1) needle-punching (mechanical bonding; refer to Kurian 2005), Chemical bonding, which is the least popular technique, or thermal bonding (through heat; see Figure 8.3).

8.2.1. Tests on Geotextiles

At first, the textile industry provided some of the test procedures, if not the specifications. The Standard Designation along with the nation in question are indicated for a few of these tests below. They are primarily DIN (Germany), BS (UK), and ASTM (USA) (Blair et al., 1980).

Table 8.1. Relative Importance of the Basic Geotextile Functions in Various Applications

Application	Function			
	Separation	Fluid transmission (drainage)	Reinforcement	Filtration
Unpaved roads	◯	o	○	○
Coastal and river protection	◯	o	○	◯
Aread of granular fill	◯	o	○	∘
Retaining wall drains	o	◯	∘	o
Below geomembranes	∘	◯	○	∘
Nearly horizontal drains	o	◯	○	∘
Embankment basal reinforcement	◯	o	◯	∘
Reinforced soil walls	∘	∘	◯	∘
Embankment piles	○	∘	◯	∘
Rockfall nets	∘	o	◯	∘
Encapsulated hydraulic fill	○	o	◯	◯
Erosion control	o	o	o	◯
Flexible formwork	∘	o	○	◯
Trench drains	○	o	∘	◯

◯ *dominant function* ∘ *minor function*

8.2.1.1. Physical properties

Mass per unit area: DIN 53855/3, ASTM D 5261, and ASTM D 3776 Nominal thickness: DIN 53855, ASTM D 5199 and ASTM D 1777 Amount of apparent pore size: BS 6906/2, ASTM D 4751 (Hale, 1976).

8.2.1.2. Mechanical properties

Wide width tensile test: ASTM D 4594, BS 6906/1, Strip elongation and breaking force – ASTM D 5035 Puncture strength: DIN 54307, BS 6906/3, and ASTM D 4833 ASTM D 4533 Trapezoidal Tear – ASTM D 4533 and Direct shear coefficient of friction: ASTM D 5321(Metsger et al., 1999).

8.2.1.3. Hydraulic Properties

Gradient ratio: ASTM D 5101; Water permeability: BS 6906/3, ASTM D 4491; and Constant head hydraulic transmissivity: BS 6906/7, ASTM D 4716 (Van Genuchten & Nielsen, 1985).

8.2.2. Geotextile as Filter

The filtering function of geotextile is one of its main uses. This feature of geotextile function has become widely accepted as a replacement for traditional graded filters in riverbank and coastal erosion control projects.

Permeability and soil retention are the two primary factors to consider when choosing a geotextile for a filter. The geotextile pore size has a maximum limit determined by the criterion of soil retention and a minimum limit determined by the permeability criterion. According to Loke and Wee (1992), the latter needs to be both large enough to maintain adequate permeability and small enough to hold the soil (Luettich et al., 1992).

After more than five years of use, the needle punched nonwoven (NPN) geotextiles from two Malaysian coastal project sites were excavated by Loke et al. (1994) in an attempt to assess the long-term performance of the geotextiles as a filter on marine clays. The mechanical and filtration properties of the dug-up geotextiles were determined by testing samples to come up with good long-term design guidelines for geotextile filters (Koerner & Koerner, 2015).

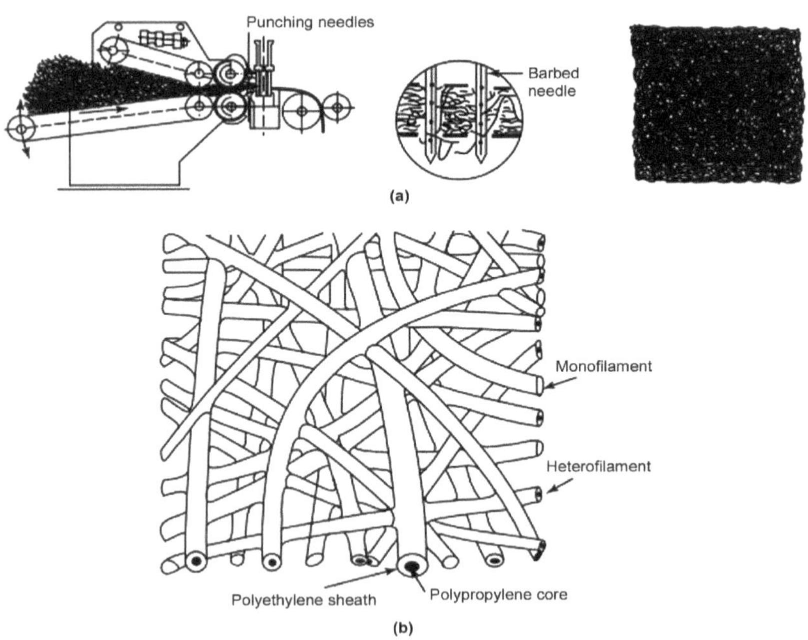

Figure 8.3. *Non-woven geotextiles made by (a) Needle-punching; and (b) Thermal bonding.*

8.2.3. Reinforced Earth as Retaining Structure

There are two categories of stability criteria that come up when designing a reinforced earth wall retaining structure: internal and external stability. Similar to traditional gravity retaining structures, overturning, sliding, and base pressure are the designing criteria for external stability. However, internal stability refers to the tension failure of the reinforcing strip due to insufficient tensile strength and reinforcement pull-out caused by insufficient interfacial friction between the backfill soil and reinforcement (Masini et al., 2015).

According to (Loke, 1996), using geotextile to reinforce reinforced earth works with the idea of Vidal's reinforced earth. However, sand reinforced with extensible inclusions doesn't work with the idea. Given the high cost and limited supply of granular backfill material, using nonwoven geotextiles to cover poorly draining backfills presents an appealing alternative due to their ability to improve in-plane drainage. The former could account for 40–60% of the total price of earth retaining structures that are reinforced. In this context, it's interesting to remember that the French Highway Administration used cohesive soil with poor drainage as the foundation material for the first geotextile reinforced wall, which was built in Rouen in 1971.

However, in the case of clay, the uncertainty surrounding contact friction—especially if the soil is wet—combined with the worry about corrosion, render metallic reinforcements unfavorable options in such soils. In reinforced earth, poorly draining backfills are typically avoided. Chew and Loke (1996) try to accurately evaluate the design requirements for employing these backfills in places where high-quality free draining backfills are not easily accessible (Joshi et al., 1990).

8.3. POLYFELT GEOTEXTILES AND THEIR USES

8.3.1. Subgrade Stabilization

Not only do some geotextiles, like needle-punched and continuous filament polyfelt, let the subgrade drain freely in the vertical direction, but they allow it to drain freely in the horizontal direction. This is different from heat-bonded and woven and heat-bonded geotextiles, that can't drain laterally.

To keep the geotextile from getting clogged, the size of its openings needs to match the subgrade grain size that needs to be filtered (Zimar et al., 2022).

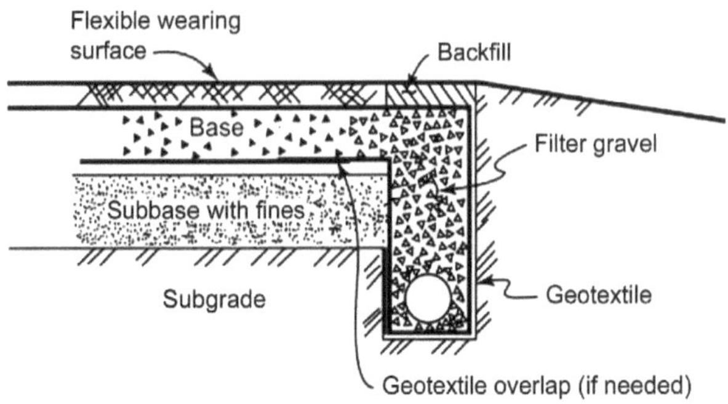

Figure 8.4. *Roadside drain with geotextile.*

All geotextiles degrade when they are left out in the sun for a long time. In order to achieve the highest level of ultraviolet light resistance, polymers formulated specifically with Hindered Amine Light Stabilisers (HALS) are used to create Polyfelt geotextiles (Amakye & Abbey, 2021).

Table 8.2. Comparative pH Resistance of Common Geotextile Polymers

Properties \ Polymers	Polypropylene*	Polyethylene*	Polyester
Resistant against: Acids, diluted	Excellent	Excellent	Good
Acids, concentrated	Requires assessment	Requires assessment	Not achieved
Alkali, diluted	Excellent	Good	Requires assessment
Alkali, concentrated	Requires assessment	Requires assessment	Not advised
Micro-organisms	Very good	Very good	Very good
Hydrolysis (humidity and high temperature	Excellent	Excellent	Not advised
Sunlight	Good	Very good	Good

*U.V. stabilised

Legend: Excellent; Very good; Good; Requires assessment; Not achieved; Not advised

8.3.2. Drainage and Filtration

The biggest threat to a drainage system that can quickly impair its operation is the drainage path clogging. By serving as both a filter and a separator, geotextiles can actively prevent clogging and ensure that the drainage system runs smoothly and continuously. The subsurface drain depicted in Figure 8.4 has a geotextile layer that covers the gravel layer keeping the natural soil intact while letting water enter the drainage system. To achieve this, the soil gradation to be filtered needs to mesh well with the geotextile's pore size distribution. In general, higher performance is seen in Nonwoven, needle-punched, and mechanically bonded continuous filament geotextiles additionally, they offer greater installation flexibility by adapting to various soil profiles (Chang & Zeng, 2016).

8.3.3. Embankments and Slopes Reinforcement

Putting down several layers of geotextiles on a slope during either re-construction or the initial construction makes the slope more stable by making the soil stronger. With this kind of reinforcement, steeper slopes of about 1V:1H can be built. Stability is increased when seepage forces are reduced as a result of better drainage. Regarding longevity, Table 8.2 compares the pH resistance of the polymers used to make geotextiles (Farooq et al., 2020).

Table 8.3. Soil/Soil, Geotextile/Soil Interface Friction Angle – Degrees

ASTM D3080 (100 mm Box)	Clayey silt	Silty clay	Clay	Sandy clay	Angular sand	Rounded sand
Soil to soil	34	38	30	22	40	36
Geotextile to soil	32	32	30	22	30	26

To guarantee that the load is effectively transferred from soil to reinforcement, the interface friction between the geotextile and the soil must be taken into consideration during the design process. The soil-geotextile interface and soil-soil friction are given in Table 8.3.

Figure 8.5. *Bamboo mat platform.*

Figure 8.6. *Geotextile on bamboo mat.*

8.3.4. Soft Soil Stabilization Through Bamboo Containing Geotextile

In many Asian countries, there is a common feature of peat and extremely soft clay (CBR < 1). A bamboo mat is placed upon the soft soil to provide a sturdy working platform, an innovation that Polyfelt spearheaded. The geotextile is placed over this mat, and then the fill is layered on top of it. One of the many effective projects using this technique is the reclamation of 20-meter-deep mine tailing ponds in Malaysia, where the soil shear strength ranges from 1 to 5 kN/m^2.

Loke et al. (1997) report their experience stabilizing very soft soil deposits, such as mine slime tailings, marine clays, and peat that are widely distributed in Malaysia, employing geotextile as well as bamboo fascine mattress—the first as a separator and the second as reinforcement (Khatib, 2005).

Peat makes up 8% of Malaysia's total land area and is primarily found in low-lying and poorly drained coastal basins. A significant portion of it is accounted for by the State of Sarawak. Its low bulk density is of the order of 10 kN/m^3, its shear strength is in the range of 5 to 10 kN/m^2, and its water content is high and varies between 200 and 2000%. That nation's coastal plains are covered in soft marine clays, which have shear strengths between 8 and 15 kN/m^2. The unconsolidated silty clay soil, or slime, are the byproduct of tin mining operations (Talib et al., 021). They are more common in Kuala Lumpur and Ipoh because of the previous heavy mining activity in these areas. Their shear strength is extremely low, rarely surpassing 5 kN/m^2. It is impossible to build embankments on these types of soils for projects like building roads without also causing the subgrades to fail. The subgrade was stabilized by horizontally placing bamboo poles over it and tying them in a cross-hatch pattern to create a sturdy frame that served as a working platform for the geotextile separator to be laid over. The bamboo mattres provides a tensile and bending reinforcement effect when the fill is placed on top of the geotextile, and the geotextile serves as a separator and filter and separator between the subgrade and the fill material. It allows for the efficient distribution of the traffic load across the fill by preventing the soft subgrade from intermixing and contaminating the fill. Simultaneously, it prevents fill loss from entering the soft subgrade and preserves the intended fill thickness (Ready & Nurtjahjaningtyas, 2020).

In 1991, Sungei Besi in Kuala Lumpur's 3000 m^2 slime pond was stabilized through the use of geotextile and a bamboo mattress, marking the first such application. The slime had an 5 kN/m^2 undisturbed vane shear strength and a thickness of roughly 20 m. The continuous filament needle-punched nonwoven geotextile, which had a low, 18 kN/m, tensile strength (width), was utilized. The usage of comparatively low tensile strength nonwoven geotextile was made possible by the presence of bamboo. This composite construction turned out to be the most economical option; high strength geotextile would have been required if bamboo hadn't been used (Datye & Gore, 1994).

8.3.5. Railway Engineering

Ballast integrity loss is a common cause of track failures. By acting as a separator between the ballast and the subgrade, geotextile keeps the ballast's integrity intact by preventing fines from pumping upward and contaminating it while also preventing its loss into the soft subgrade (Iwnicki, 2009). A portion of the Malaysian Railways' track rehabilitation project is depicted in Figure 8.7.

Figure 8.7. *Track rehabilitation using geotextile in Malaysian railways.*

8.3.6. Geomembrane Protection

Applications for geomembrane liners include basement waterproofing, treatment ponds, waste containment systems, and water reservoirs. When used in conjunction with geomembrane liners, polyfelt geotextiles have been found to significantly improve their performance by functioning as a cushioning layer that protects against both dynamic and static puncture stresses and efficiently dissipates gas in soils that have a high gaseous content (Tognon et al., 2000).

Microbiological attack does not degrade Polyfelt geotextiles in soils. Long-term exposure to acidic and alkaline solutions with a pH range of 2–13 does not affect them as well. When subjected to tropical climate conditions, they do not undergo hydrolysis. Strong and functional seams are ensured via thermal bonding of overlaps (Zanzinger, 1999).

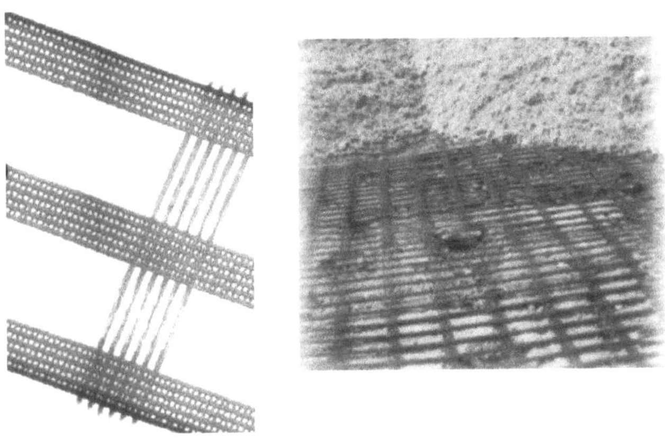

Figure 8.8. *Polymer Rock Strata geogrid.*

8.3.7. Polyfelt 'Rock Strata' Geogrids

The "Rock Strata" geogrid (Figure 8.8) is an open polymer grid structure that Polyfelt has developed especially to work with granular soils in a various applications of civil engineering. It is a polyester with a high strength-to-weight ratio and high tenacity that is designed for extended service life. As reinforcement, it absorbs higher tensile forces at low elongations, especially in granular soils (Collin et al., 2021).

A stable gravity mass is produced behind the retaining wall by strengthening the backfill soil. Polyfelt provides the most sophisticated design software for reinforced earth slopes, called STRATASLOPE, along with an extensive software set labeled STRATAWALL.

8.3.8. Design of Geotextile-Stabilized Earth Structures Over Soft Soils

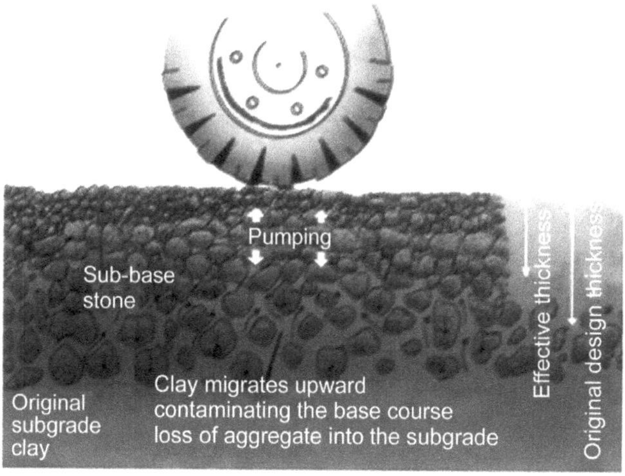

Figure 8.9. *Loss of aggregate into subgrade.*

When building a sub-base over a soft saturated subgrade, some of the sub-base material is typically lost into the subgrade (Figure 8.9). This means that extra material is needed to compensate for the full design thickness and compensate for the lost material.

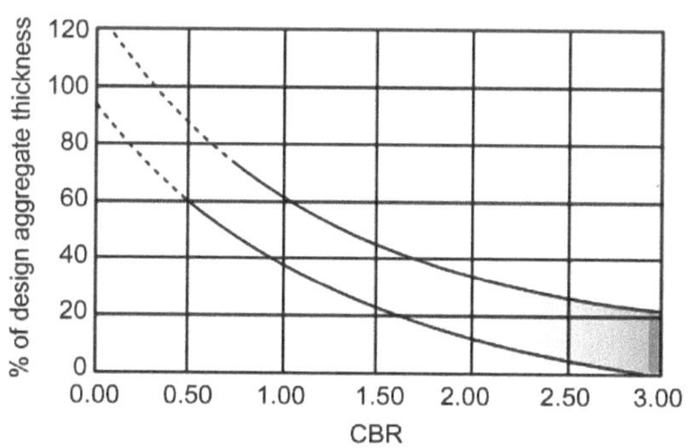

Figure 8.10. *Range of aggregate thickness loss vs. subgrade strength.*

Figure 8.10 makes an estimation of this loss. This loss is avoided by placing an appropriate geotextile layer at the sub base layer-soft subgrade interface. As a result, the sub-base layer retains its integrity to withstand traffic loads as intended by the design. That's what Figure 8.11 shows. In resistance of wheel loads, the geotextile's reinforcing effect is minimal, if present at all. On the other hand, the geotextile helps stabilize slopes against potential slip failure in the case of bridge or high embankments approaches (Rawal & Saraswat, 2011).

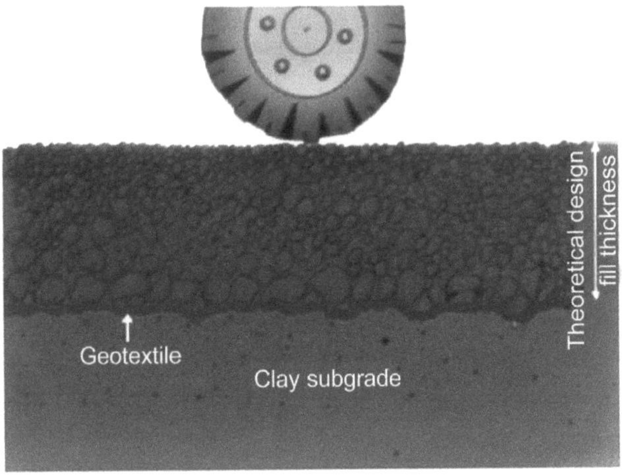

Figure 8.11. *Geotextile as separator preventing intermixing of aggregate and subgrade.*

8.3.9. Drainage and Filtration

Besides acting as a drainage layer, the geotextile placed above stops the buildup of pore water pressure by efficiently releasing it. This keeps the saturated soil below from flooding and allowing water to drain both in and across the layer. Only dimensionally thick nonwoven needle-punched geotextiles are capable of this; heat-bonded nonwovens or thin woven geotextiles, on the other hand, eventually turn into an impermeable barrier of fines that obstructs drainage and precipitation (Palmeira & Gardoni, 2002).

8.3.10. Construction

Experience in the field has demonstrated that a geotextile's capacity to withstand the stresses of construction is the most important factor in determining how well it performs in a geotextile design assessment. Before laying the geotextile, it is advised to remove any objects that protrude more than 100 mm, even in the case of strong and robust geotextiles (Plotkin, 1976).

Depending on the circumstances at the site, the sheets may be connected by heat welding, sewing, or overlapping. When dealing with exceptionally soft soils (CBR < 1), the bamboo mat acts as a working platform for the installation of geotextiles and other operations. This system was invented by Polyfelt, who successfully applied it to many projects, including the reclamation of Malaysian tailing ponds as mentioned before.

Loke (1996) ends with Guideline Specifications for Geotextiles in the construction of roads. This document includes a list of properties that need to be determined as well as the relevant ISO Standards. The user is guided by a helpful chart regarding issues and their geotextile solutions.

8.3.11. Riverbank and Coastal Erosion

Soil erosion is a persistent concern on riverbanks, beaches, earth berms, and other locations that are frequently subjected to turbulent water currents, wave action and fluctuating water levels. The conventional approach suggests placing concrete block rivets or stone armor above layered granular aggregate filters above the subsoil to provide protection (Van Tho, 2020).

Researchers have found that geotextile filters can be used instead of layered granular filters to save money by cutting down on the thickness and number of the granular filter layers. In fact, the introduction of geotextiles into this field has completely changed how erosion control is done today (Mallick & Mallick, 2021).

8.4. HUESKER GEOTEXTILES (GERMANY)

Germany's top producer of geotextiles is Huesker. This section begins with a few noteworthy instances of the usage of geotextiles in projects (Koerner, 2016).

8.4.1. Examples

i. In Unterägeri, Switzerland, a 40 m long and 5 m high retaining wall was built around an apartment complex utilizing a "noise protection planted retaining wall" concept created by Zug-based LS-Larmschutzban AG. In order to hold the loamy backfill material in place, pressure-impregnated Swiss beach wood pieces are placed on the slope using a geogrid. Four to five plants per square meter were planted on the front to create a visually pleasing green cover (Detert et al., 2013).

ii. A steep slope in Pineville, the state of North Carolina USA, is depicted in Figure 8.12. It is faced with "Keystone" modular blocks and has a set-back of 20 mm for every meter of wall height. The slope is reinforced with geogrids. A 300-mm-wide drainage course was built behind the wall, and a system of perforated collector pipes was set up at the base.

Figure 8.12. *Keystone modular block wall with fortrac geogrids.*

iii. In order to span a railway line below a road overbridge, the Danish State Railways selected reinforced soil abutments supported by a glacial clay subsoil. The fill material is made up of a mixture of gravel and sand that has been compressed to 100% Proctor density. It has been strengthened with Fortrac geogrid from Huesker, spaced 500 mm vertically, allowing for an 81° frontal inclination, or almost vertical. The reinforced soil abutments set by only about 2 mm, according to measurements, whereas the subsoil settled by 40–50 mm (Kirschner & Witte, 2020).

iv. To stop reactive cracking at the surface, asphalt overlays applied over pavement concrete on airport runways must be quite thick. Reinforcing geogrids can be used to prevent this, allowing for lower overlay thicknesses (usually 150 to 100 mm). Material with low creep and high modulus is best for strengthening asphalt layers in roads. Generally, polyester have these qualities. Additionally, they have outstanding temperature resistance up to 190°C and minimal shrinkage. Huesker created the "HaTelit" grid specifically for this purpose, which is composed of high-modulus polyester filaments. They have a bituminous coating that makes sure the asphalt sticks to them well and keeps the reinforcing grid in a position to take and distribute tensile stresses. The size of the grid mesh should be 2–2.5 times larger than the largest aggregate size (Kempfert et al., 2002).

v. A two-layered geotextile mattress containing impermeable concrete was used to fix up a sealing asphalt layer on the canal bank near Bamberg, Germany. The concrete utilized in this instance had a w/c ratio of 0.5 and was composed of slow-hardening furnace cement and sand, allowing for exceptional pumpability of the mixture. Huesker employed "Incomat" mattresses, which are made with two fabric layers woven from polyethylene and high-tenacity polyamide yarns. The two layers of fabric were sewn together, and the distance between the stitches determined how thick the mattress was (Garber & Rasmussen, 2010).

vi. Additionally, the synthetic landfill liner at Damsdorf, Germany, was shielded by an Incomat mattress. The first lining system had a base liner made of 2.5 mm thick HDPE, with a HDPE (500 g/m^2) nonwoven geotextile to protect it, and 150 mm thick lean–mix concrete on top of that. On the 1:1.5 (34°) slopes, the mattress solution was very flexible.

vii. The composite known as "NaBento' geosynthetic clay liner" (GCL) consists of two layered stitch-bonded fabric encasing bentonite. The in-situ soil becomes less permeable as a result of bentonite's low permeability. It is a calcium bentonite that is activated by sodium (Na) and has a montmorillonite content of over 70%. Additionally, the "NaBento" GCL can be utilized to cap landfill sites or enclose contaminated areas. Ponds that retain water can make use of them. In one instance, it was positioned on top of a 0.5 m-deep aggregate drainage layer that was covered by a 1 m-thick compacted cohesive layer in the Yorkshire region of the United Kingdom. Intriguing uses of geotextile lining systems for landfills in Leipzig, Germany's Kleinleifringhausen and Augsburg are detailed by Huesker (Ebbert et al., 2023).

viii. The high quality woven fabric known as "Stabilenka" from Husker is utilized as reinforcement. It is constructed with polyester yarns running longitudinally (warp) and polyamide (or optionally, polyester) running transversely (weft). It is made using a patented straight warp process, which allows it to produce higher tensile stresses at low elongation, in contrast to the conventional method. Its high modulus polyester multifilament yarn limits the creep strain to 1 percent after two years of use due to its low creep potential. Huesker Synthetic provides a large selection of knitted fabrics made from raw materials like polyester, aramid, polyethylene, polyamide, and polypropylene as well as synthetic non-wovens and wovens, geogrids, and geocomposites (Alkhorshid et al., 2019).

ix. 'Comptrac' is produced through a knitting process utilizing high-modulus polyester yarn. The yarns are depicted in Figure 8.13 in their tied-up positions, and it differs from the interlayering that results from the standard weaving procedure. Geocomposites can also be created by combining it with non-woven materials. "Comtrac" is produced utilizing a revolutionary new kind of production technique. In this, the longitudinal and transverse yarns are layered on top of one another and secured via knitting yarn (Assinder et al., 2016).

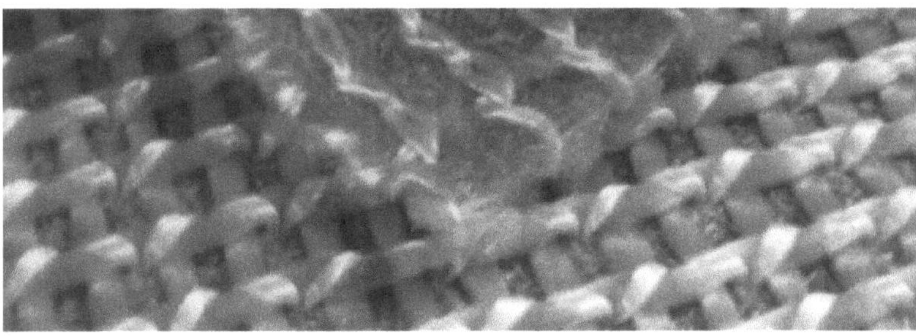

Figure 8.13. *'Comtrac' knitted geotextile.*

x. Incomat is a cheaper alternative to traditional systems for protecting slopes and beds that use geotextiles and rock fill. It is made up of two incredibly durable synthetic fabric plies that have been specially bonded together to serve as the formwork. This is filled with a high strength and impermeable mortar or concrete that is made by mixing sand, cement, and water. Either a fixed platform or a floating platform is utilized to house the mixing plant and concrete pump. Concrete is continuously pumped through the hoses to fill and lay the fabric formwork. There exist various variations of the product, including Incomat tube, slab, crib, flex and so on. Incomat crib, one of them, has holes in it that are filled with topsoil so that seeds or plants can be planted (Almeida et al., 2015).

xi. Developed by Huesker in 1986, "HaTelit" is a reinforcing grid coated with a bituminous material that ensures a good bond with adjacent asphalt layers. HaTelit keeps cracks from refracting into the wearing course from binder and base courses, as seen in Figure 8.14. "HaTelit" is equally helpful in preventing reflection cracking within the asphalt overlays of airport runways, taxiways, and aprons. It's interesting to note that the setting of reinforcement grids in a slab can prevent cracking in a slab that isn't reinforced. Prior to laying the mix, it is advised to have the reinforcing grid be positioned flat and undisturbed using a tensioning device (Blume et al., 2006).

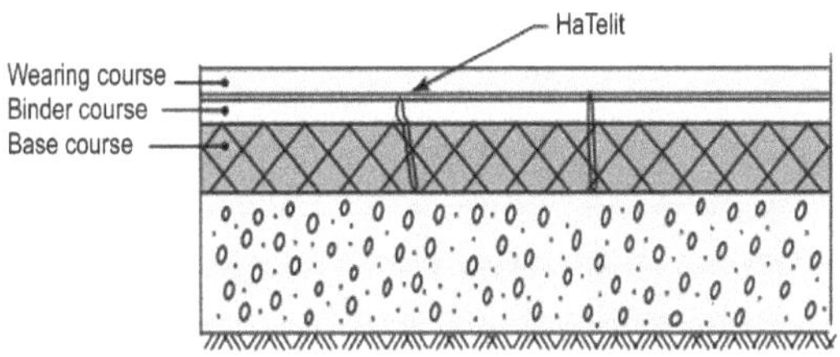

Figure 8.14. *Reinforcing new wearing course (to prevent reflective cracking).*

8.4.2. Geotextiles in Landfill Construction

The landfill must be kept isolated from the surrounding environment at all costs. The "multibarrier" concept, which calls for multiple impermeable barrier layers that can successfully incorporate geotextiles and geomembranes, effectively achieves this in modern landfalls. In fact, geotextiles are now a crucial component of such sealing systems. For instance, they shield geomembranes from harm or separate the subgrade from the mineral sealing layer. They could act as barriers to keep the drainage layer and the garbage apart. They fortify steep embankments and surface sealing. Geotextiles made of high density polyethylene (HDPE) can be used without restriction to provide chemical resistance. Since a land-fill's base sealing channel is the most crucial barrier for keeping groundwater safe, it is particularly significant (Dixon et al., 2016).

It is best to use nonwovens for separation. Geomembanes are more effectively protected by reinforced nonwovens. One way to reinforce something is by using fabric or by placing a grid on the top. The polyester yarns used to make "Fortrac" geogrids are comparable to those found in seat belts, automobile tires, and other products where flawless operation is essential. They have added protection from UV rays and mechanical harm thanks to the PVC coating that has been applied. They have a strong bond to their surrounding soil, which leads to short anchorages. It is a byproduct of Huesker Synthetics and Akzo and Huesker Synthetics' joint venture (Koda et al., 2016).

8.4.3. Hydraulic Engineering

High-strength synthetic fabric "HaTelit" sandbags have demonstrated their worth in emergency situations time and time again.

Colcrete, a type of colloidal mortar, is used to make formwork for hydraulic engineering purposes. For underwater construction, this type of formwork works ideally. In order to safeguard river and sea beds and banks and stabilize the subsurface for the construction of dykes, groins and similar structures, mattresses and sea carpets are utilized (Heller, 2011).

8.5. NETLON (U.K.) – PRODUCTS AND USES

The net (geogrid) flexibility allows closer identification with the ground contours (McNitt & Landschoot, 2003).

8.5.1. Properties

Weight — 5.5 N/m^2

Mesh Size — 60 × 60 mm (nominal)

Tranversal yield strength — 5.33 kN/m F

Longitudinal yield strength — 4.82 kN/m

Lexural modulus — 112 kN/m^2

This product has finely divided carbon black added to it as an additive that protects it from UV damage for a half-life of more than 15 years (Ooi & Tee, 2011).

8.5.2. Stabilization of Railway Track Beds with Geogrids

Geogrids are made of high-strength polymer grids or meshes with integrally fused joints and have very stable dimensions and high junction strength. Unlike geotextiles, the majority of the tensile strength in such high modulus products is developed within 2% to 4% of their elongation. Because they are made of HDPE, geogrids are resistant to biological and chemical soil attack as well as corrosion. These are stable in a temperature range of -60°C to +100°C and shielded from environmental and UV degradation. These qualities allow engineers to satisfactorily design long-term durable and survivable geogrid reinforced soil structures (Das, 2016).

Geogrids make sure that the ballast fits tightly into the grid holes, which stops mud pumping and ballast loss. The former is brought on by mud that gets sucked up when stones separate. Under dynamic loading, these ballast particles that interlock with the geogrid openings keep the soil from moving laterally and cause oscillatory movements, which stops the pumping action (Esmaeili et al., 2018).

The main cause of track maintenance expenses is the deformation of the soil subgrade beneath the track's formation level. Increased traffic speeds and heavier loads have made this issue more severe. By placing geogrid reinforcement beneath the ballast, track settlements are significantly reduced. Additionally, mud pumping and ballast loss into soft subgrades are prevented, significantly lowering the need for track attention and line and level maintenance. Geogrids directly reduce costs by doing away with the need for moorum blankets or deep sand layers. The grids create a layer of separation that helps contain the ballast and keeps it from mixing with the soft subgrade soil, extending the track's overall life (Fischer & Horvát, 2011).

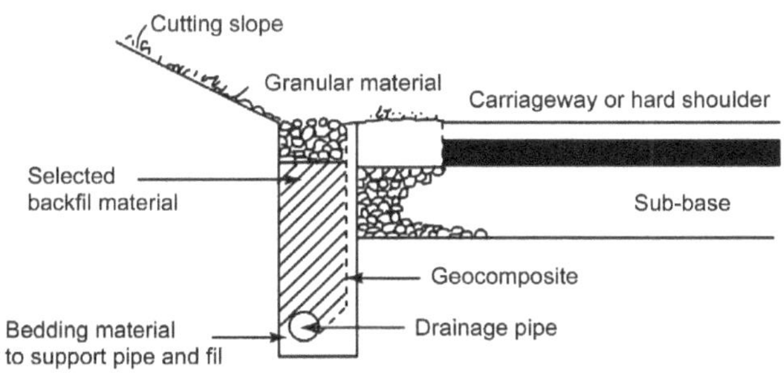

Figure 8.15. *Drainage along roads.*

8.5.3. Drainage

Many civil engineering problems can be solved with effective long-term drainage. Subsurface and surface drainage are both involved in this. Flexible of perforated rigid pipes and sheet-form geocomposites that blend geomembranes, geotextiles, and/or geonets are examples of Netlon drainage products. The geocomposites' application for drainage to a retaining wall and along major roads, respectively, is depicted in Figures 8.15 and 8.16. The vertical geocomposite that begins at the surface in both situations ends in drain pipes that are buried at the bottom (Horton, 1932). Using core drainage laminates, a slope can be drained as seen in Figure 8.17.

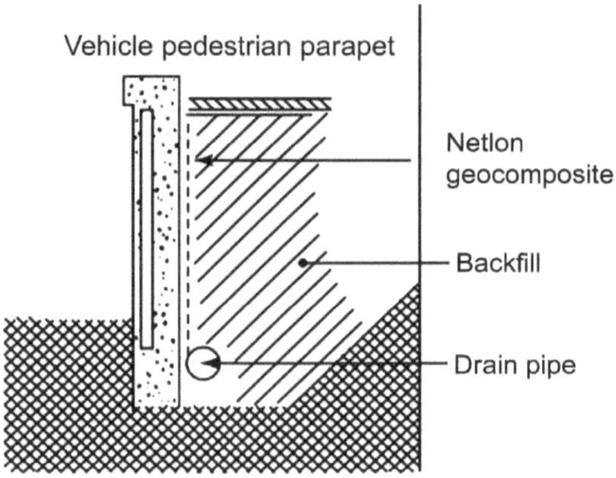

Figure 8.16. *Drainage for retaining walls.*

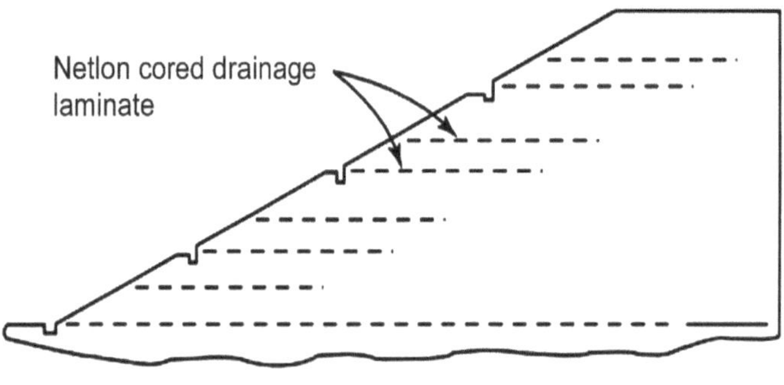

Figure 8.17. *Slope drainage using core drainage laminates.*

8.5.4. Geogrid Applications for Slope Protection

High embankment slopes, cuttings for roads and railways and roads, river bank, reservoir and hill slopes frequently worry civil engineers because of soil erosion from the surface of slope during rainy seasons and the ensuing instability, which frequently results in failure.

First and foremost, installing geogrids on sloping surfaces will stop soil erosion by slowing down water runoff and allowing water to percolate, which encourages the development and growth of native grass. By permitting the roots of the natural grass to entwine in the meshes, the grid serves as a constant reinforcing medium for the grass once it has taken root on the slope surface (Wang et al., 2014).

In fact, natural vegetation has been shown to be an effective option for long-term erosion control; however, in order to maintain the integrity of its root matrix, it needs external reinforcement. For this reason, slope-based geogrids serve as a permanent reinforcement. It quickens the growth of vegetation, which covers earth slopes permanently with a protective layer. Geogrids shield cliff faces and rock slopes from weathering-induced dislodgement, as seen in Figure 8.18. The grids are fastened to the rock faces by spikes that resemble rock bolts (Rai et al., 2012).

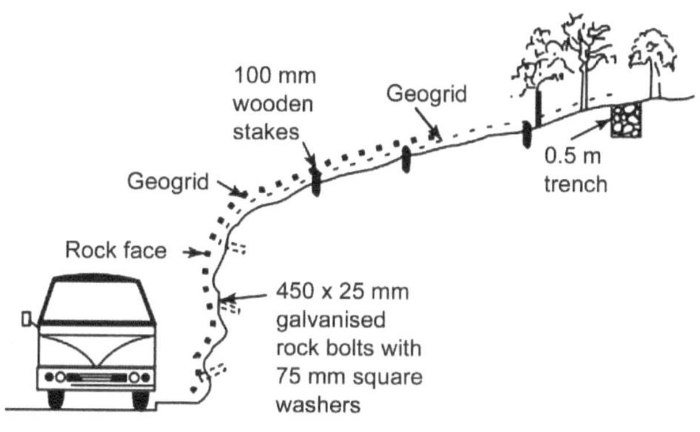

Figure 8.18. *Cliff protection with geogrids.*

8.5.5. Waterfront Protection

Lower-cost flexible protection schemes and techniques that absorb wave energy instead of reflecting it and cause less erosive damage are gradually replacing the traditional method of building impermeable, rigid, mass concrete or reinforced works in the revetments form for the water front protection. During floods, they recover from impact and dynamic loads brought on by wave action and excessive currents (Weaver & Stehno, 2024). One type of flexible armored riveting uses mattresses and gabions inside high-tensile strength and lightweight geogrid cages. Utilizing smaller stones and locally sourced fill materials to achieve superior results with minimal environmental impact, as well as a more expedient installation process, are a few of the distinctive benefits associated with the application of geogrids in waterway and maritime construction. A water front shielded by mattresses and gabions is depicted in Figure 8.19.

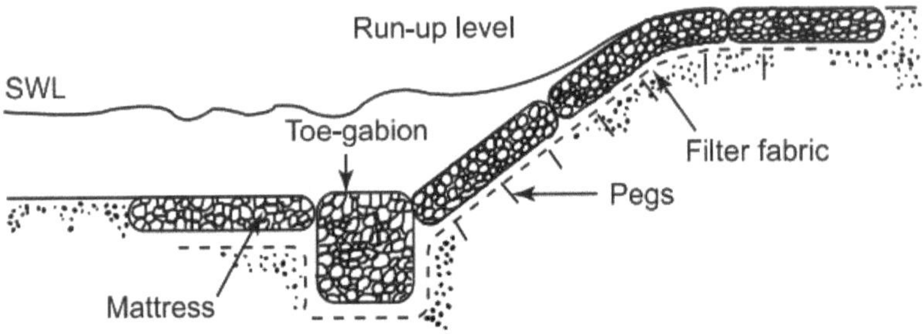

Figure 8.19. Illustration of *Shallow slope.*

It shields shallow slopes against the sea waves, which release energy when they break and cause a subsequent run-up of the slope (Yevheniia, 2021). A breakwater with geocell and geogrid acting as core reinforcement is depicted in Figure 8.20.

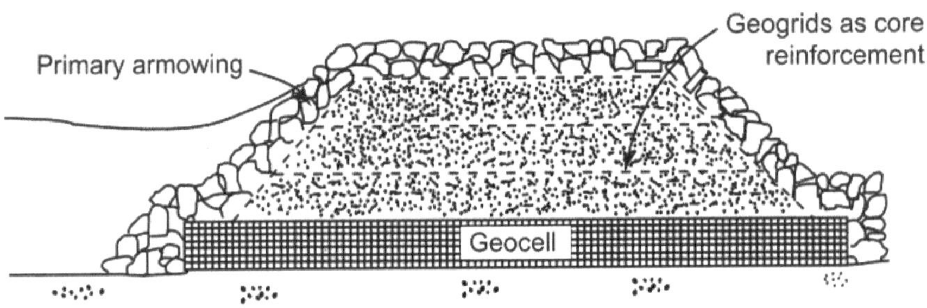

Figure 8.20. *Breakwater with geocell mattress and geogrid reinforcement.*

8.6. FIBERTEX (DENMARK) GEOTEXTILES

"Fibertex" is a well-known producer of nonwoven geotextiles with headquarters in Denmark's Aalborg St. In addition to separation, filtration, and drainage, these geotextiles are composed of needle-punched polypropylene fibers to create robust, pliable non-woven textiles having exceptional elongation properties that enable it to fulfill its protective role. The former results from its capacity to withstand significant local stresses through the required elongation (Koerner, 2016).

The geotextiles are shielded from delamination by needle-punching from external actions like traffic loads. Both in wet and dry conditions, they hold their integrity and strength. They don't rot or get attacked by fungus. They are also free of chemicals and binders, non-toxic, and resistant to alkalis and acids. To accommodate a range of uses, they are made in different surface textures, weights and thicknesses. Their products are certified by ISO.

8.7. NYLEX (MALAYSIA) DRAINS

NYLEX (Malaysia) Sdn. Bhd., a company situated in Shah Alam, Selangor, produces a range of geotextile drains. Their "Cordrain" consists of a cuspated plastic core that is held up against a basement or retaining wall, etc., on one side and on the other side, bonded to a geotextile filter fabric to facilitate easy horizontal drainage flow. It is biologically inert, lightweight, flexible, simple to install, and is resistant to the chemical attacks that are likely to occur in soils (Parsa et al., 2014).

A wick-type vertical drain known as the "Flodrain" is implemented to expedite the consolidation process of saturated compressible soils, including silty clays and clays that are saturated. The domains in which this technology finds application encompass reclaimed coastal regions, storage tank farms, highways, railways, bridge approaches traversing swampy terrain and airfields, among others (Krishnan et al., 2008).

In essence, it is a geocomposite comprised of a geotextile filter fabric encircling a cuspated plastic water channeling core. It is packaged in bands, each measuring approximately 100 x 4 mm. Because of the soil's progressive consolidation and settlement, a wick drain keeps draining even when bent, and a mandrel and anchor plate are used to install it. The mandrel is used to hold the anchor plate that has the wick drain attached to it. The mandrel is then driven down, pushing the wick and the anchor plate with it. The mandrel is removed after the necessary depth is reached, leaving the anchor plate and wick in position. The wick is cut at the necessary height at the surface after being fed, from a roll, in bands. The aforementioned "Stripdrain" is also produced by Nylex and is laid out horizontally in sheets. Smaller projects like gardens, playgrounds, golf courses, etc., can also benefit from their use. Geomembranes, another product line from Nylex, are useful for a range of applications, including sanitary landfills, landscaping, canals, aquaculture ponds, and industrial lining requirements (Long & Covo, 1994).

8.8. TECHFAB INDIA–GEOSYNTHETICS

Since its founding in 2003, Techfab India has been one of the top producers of geosynthetics in India. Their product line consists of high strength geogrids, gabions, woven and nonwoven geotextiles, and geodrains. These products are used in transportation, environment protection, marine structures, pavement rehabilitation, soil reinforcement, and filtration among other areas (Rao, 2016).

8.8.1. Geogrids

Planar polymeric structures called geogrids are made up of an integrally connected network of tensile elements arranged in a regular pattern. These elements can be joined by bonding, extrusion, or interlacing, and their openings are bigger than their constituent parts.

The first geogrid was made in the early 1980s by punching holes in a polyethylene or polypropylene sheet produced through extrusion, then strengthening the sheet. "Stiff" geogrids are geogrids that have been extruded and oriented in this way. In the mid-1980s, a different type of geogrid was introduced: the "flexible" or textile grids. These grids are made of high-tenacity polyester filament yarn that is knitted or woven and coated with PVC or other polymeric compound (Al-Qadi et al., 2008).

The first geogrid made in India, called "Techgrid," is made of knitted polyester with a polymeric coating. A highly complex warp-knitting process forms polyester yarns of the highest quality of high tenacity, high molecular weight and tenacity; and a low carboxyl end group into a grid structure, which is then coated with a formulated polymeric compound to create a durable, dimensionally stable, tough, flexible and strong geogrid (Das, 2016).

8.8.2. TechDrain' – Prefabricated Vertical Drains (PVD)

In the early 1970s, prefabricated vertical drains—also known as PVD, "band drains," or "wick drains"—were introduced with the intention of expediting the consolidation of fine-grained soils.

The two-piece composite grain has a polymer core inside and an unwoven outer fabric layer. The innermost core of "TechDrain" is constructed using continuous extrusion technology using polypropylene material. Despite undergoing lateral displacements, vertical settlements, or high pressures, the extruded corrugated or fin-shaped profiles on both sides of the core serve as numerous conduits to facilitate substantial hydraulic flow in the longitudinal course of the substance (Chai et al., 2013). The "TechDrain" core is securely encased in a geotextile filter jacket made of polyester or spun-bonded polypropylene that has been chosen for its excellent filtration and water permeability qualities. It guarantees an unobstructed drainage channel and is impervious to blockages (Vyas & Shah, 2020).

8.8.2.1. Installing the TechDrain

Similar to installing Flodrain, the TechDrain is housed within a rectangular tubular steel mandrel measuring 50 ×125 mm. A sacrificial anchor plate, marginally larger than the mandrel, holds it to the bottom drain. Now, a static or vibratory rig drives the mandrel along the drain down to the necessary depth. Now, the mandrel is retracted, thereby leaving the drain anchored in the soil at its base. After the mandrel is completely extracted, the drain is cut off. Now the procedure is carried out again at a different site. Such band drain installations can be made to a depth of more than 65 meters. They are typically arranged 1 to 1.5 meters apart in a triangle or square configuration. It is possible to install up to 5,000 meters of drain in a single day (Bergado et al., 2002).

8.9. TECHBOX' GABIONS

The "TechBox" gabion is a kind of metal wire mesh (MWM) product composed of zinc-galvanized, PVC-coated, double-twisted, hexagonally woven mesh (steel wire) that is formed into compartmented baskets that resemble rectangular boxes (Toprak et al., 2016).

The MWM is perfect for use as netting to stop debris and rock fragments from falling off the rock faces into the roads and railroads that pass through the base of the slopes, in addition to its function as gabions. In the event that a wire breaks, the mesh's double twist keeps everything together.

The earliest documented application of the gabion technique dates back thousands of years, when reed basket systems were used to protect the banks of the Nile River. This idea has developed into a sophisticated technology that uses a scientific approach and modern materials for design and construction, from such an archaic state characterized by the use of natural materials and empirical knowledge (Mohamed, 2010).

Energy dissipation capacity, permeability, flexibility and large mass are their qualities that have allowed them to expand their use from gravity retaining structures to scour protection and erosion control works (Ramli et al., 2013).

8.10. STRATA GEOSYSTEMS (INDIA)

Formally founded in 2004, Strata Geosystems (India) Pvt. Ltd. is a collaborative effort with Strata Systems Inc., U.S.A., a member of the 130-year-old Glen Raven group of companies (Qasim et al., 2018).

8.10.1. Products

Strata products include wall panel units with an eye-catching stripe` design and "knitted" geogrids with the facing blocks. These geogrids are used to form a Reinforced Earth system. One unique product is the "Strata Web," a three-dimensional cellular confinement system composed of HDPE strips that have been ultrasonically welded and can be expanded on location to create a structure resembling a honeycomb that will be filled with soil. It is similar to the idea of "skirted foot-ings" in bearing (see Figure 4.59), which results in a notable increase in bearing capacity. It works very well for slope protection, retention of soil, and other applications (Rajesh et al., 2016).

8.10.2. Soil Encapsulation

Impeding the flow of moisture into or out of expansive soils is a way of stabilizing it and keeping its negative effects at bay. Geomembranes are recommended by Datye (1985) as a medium to encase these types of soils. Prior to the development of geomembranes, Kurian (1982) describes the use of bituminous membranes as water barriers to stabilize pavement subgrade layers (Fayer & Hillel, 1986).

8.11. GEOTEXTILES IN NATURAL FIBERS

"Jute" and "coir" belong in the category of natural fibers. Natural fibers can come from plants (vegetable fibers) or animals (wool and silk, for example). The restricted amount of animal fibers has clearly defined applications in the apparel industry and is unthinkable as a geotextiles raw material (Ghosh et al., 2009).

Three classes of vegetable fibers exist: "fruit" or "seed" fibers (like cotton, coir), "leaf" fibers (like sisal), and "bast" fibers (like jute, hemp). Of all the above-mentioned fibers, cotton is the most produced and has the greatest commercial value; it is primarily used in clothing. Jute is primarily used for carpeting and sacking, but coir has a wider range of applications. Geotextiles frequently employ both jute and coir (Leão et al., 2012).

8.11.1. Geotextiles in Jute

Textile technologists have done some research on jute fabrics, which are one type of textile material. They have shed some light on topics like the chemical composition of jute and the microstructural alterations that cause biodegradation in it. Despite the fact that jute geotextiles have been used for a while, not much is known about their behavior in civil engineering. It is clear that cooperation between the jute industry and civil engineers is necessary for both parties to benefit (Datta, 2007). The biodegradability of jute under various soil and environmental conditions across the nation is a pressing concern. The United Nations has funded Jansen Consultants and Manstock Geotechnical Services, located in Geneva and Manchester, respectively, to encourage the use of jute. UNCTAD/GATT and the International Trade Center and UNCTAD/GATT are actively considering promotional trials for the use of jute geotextiles. At Indian Jute Industries Research Association (IJIRA) located in Calcutta, several trials are currently underway in India (Saha et al., 2012).

8.11.1.1. Erosion Control

In a tropical nation like India, surface erosion occurs primarily on hill and embankment slopes, and even plains because of rainwater that is aided by wind. Massive volumes of fertile topsoil are lost as a result of this erosion. In addition, it destroys canal slopes, dams, railway and road embankments, and even puts river banks and other natural slopes in jeopardy. It has been discovered that the use of geotextiles along with the encouragement of re-establishment or growth of the vegetation cover can effectively control land erosions (Rao & Dutta, 2002).

Rao and Balan (1994) give a good account of how coir and jute geotextiles were first used, mostly in and around Calcutta (The State of West Bengal produces the majority of the jute used in India.) Additionally, they present the findings of the pilot research on fiber drains and jute textiles provided by IJIRA that was carried out at the Indian Institute of Technology (IIT), Delhi. According to (Pandey,1989), treating jute with cobalt and manganese hydroxides and oxides can significantly inhibit its biodegradability. This resistance may be rendered permanent by adding the appropriate resin (Dabral et al., 2008).

Since the 1950s, exposed soil surfaces in Europe and the USA have been covered with geotextiles made exclusively of jute fibers, or "geojute," as an open-meshed woven fabric designed to stop soil erosion by encouraging vegetation growth. To date, there is only one jute geotextile, "Soil Saver," that is sold commercially. Despite being a latecomer to this field, coir has demonstrated its value in filtration. Tests carried out over an extended period of time in a highly fertile soil in Germany, maintained at a moderate temperature (27°C) and high humidity (90%) revealed that (Posthumus et al., 2015).

i. Coir held 20% of its strength after a year, whereas cotton and jute completely degrade in six and eight weeks, respectively.
ii. Coir degrades 15 times more slowly than cotton and 7 times slower than jute.

The Ludlow Jute Mills, a jute mill established in Calcutta by the British in the early 1940s, had a dedicated unit for producing jute mesh, popularly known as "soil saver." Originally, this item was exported, mostly to the UK, but several mills in Calcutta are still producing it and exporting it abroad today (Chen & Chen, 2017).

For the purpose of controlling erosion, jute has ventured into the USA. They were employed to control erosion along the cutting slopes in the building of Kingsway Road in Dundee, Scotland.

Jute is the only erosion control material that has been used for as long. The best justification for using jute mesh is due to its effectiveness that has been demonstrated for almost thirty years in the Asia, Europe, Canada and the US. Jute is trusted and used by highway administrators, revegetation contractors, landscape architects, and others because they know what to expect from a high-quality product that performs well in a variety of environmental conditions.

8.11.1.2. Soil Saver

In order to prevent soil erosion during intense rainstorms, Soil Saver is a heavy-duty woven jute mesh (mulch) which stays in place, holding soil and seed secure on the slopes along with other concentrated water flow areas. It breaks down naturally as a vegetable fiber over time, enriching the soil with rich organic matter. While water flows over the soil and seed, it holds them in place. Owing to its mesh structure, the gap between the strands provides ample room for the grass to grow. It gives the option to choose any type of ground or grass cover that suits the local soil and climate the best (Ghosh et al., 2017).

The owners of Ludlow Jute Mills, one of the top producers of jute geotextiles, Aekta Limited, (Kolkata, India), have released comprehensive usage guidelines covering steps from soil preparation through soil saver laying to compaction and inspection (Neupane et al., 2018) etc.

8.11.2. Geotextiles in Coir

Coir is a robust cellulose fiber with a high lignin content that is entirely organic. It is inherently resistant to moisture, mold, and decay and no chemical preservative treatment is necessary. The fiber degrades slowly. For nearly five years, it offers strong support on slopes, giving natural vegetation enough time to grow and take over the stabilizing role. Only humus remains after the coir geotextile finally disintegrates. Because it degrades much more slowly in water, using it underwater is especially advised (Banerjee & Mukherjee, 1996).

Approximately 60% of the erosion control materials are made of natural materials, according to a 1993 market survey. Given that India produces roughly

44% and 66% of the world's coir and jute fibers, respectively, it should hold a leading position in both the manufacturing and global marketing of natural fiber geotextiles. In general, coir yarn is stronger and more durable compared to jute yarn (Subaida et al., 2009).

Natural geotextiles help control erosion until vegetation eventually takes over by absorbing water; by then, the seeds will have matured into plants and could take over the role. This means that where vegetation is thought to be the long-term solution to erosion control and slope protection, natural geotextiles should be chosen, sourced, and used. When it comes to erosion control, this key characteristic of geotextiles gives them an advantage over synthetic geotextiles.

Balan and Rao (1996) list the following as some of their intrinsic extra benefits:

i. They protect against erosion caused by rain and splashes;
ii. They can absorb water up to five times their own weight;
iii. They lower run-off velocity and its erosive effect;
iv. They maintain soil and atmosphere humidity;
v. They lessen temperature extremes;
vi. They biodegrade, adding beneficial mulch to the soil; and
vii. They hold seeds and other vegetation securely even on steep terrain.

Geotextiles made of coir offer the benefits of biodegradability as well as the durability needed in situations where plant establishment may be sluggish. Geotextiles made of coir, which have a distinct advantage in this area, have been demonstrated to persist and serve to control erosion in UK upland environments for a minimum of three years.

Balan & Rao (1996) cite several case studies from Germany, Australia, and the United States, including those from Colorado, Mississippi, and Ohio. The study concludes with field trials conducted at Pullangode Plantation Area, which is close to Nilambur located in Malappuram district of Kerala, India. In these investigations, mattings of coir geotextiles with various densities and aperture sizes were supplied by Aspinwall & Co. (Travancore) Ltd., Alleppey, Kerala (Sumi et al., 2018).

Walter Kjellman attempted to create prefabricated drains for the first time in 1939. The prototype for every contemporary band drain on the market today originates from the cardboard wick drain created by Kjellman. It was constructed from three cardboard strips glued to one another, with the middle strip divided to create longitudinal drainage channels. Only five years had passed since the invention of the conventional sand drains when the cardboard wick drain was put into service.

Prefabricated drains come in two primary varieties. The most common type has a thin geotextile filter sleeve that keeps fine soil particles out but makes it easy for pore water to flow into the central core, which serves as a drainage channel. In contrast, the thin geotextile cover of the alternative variety features unitized continuous drain holes (Lal et al., 2017).

Jute fiber has very little elongation at break and a somewhat high modulus and tenacity. Contrarily, coir fiber has a high elongation at break, a low modulus, and tenacity. Because of their completely opposite behavior, (Banerjee, 1996) came to the conclusion that they would make perfect matches in a blended product. Therefore, yarn made from a combination of suitable coir fibers and much softer jute fibers should be much finer than a typical coir yarn. At IIT Delhi, a blended product known as "Brecodrain"—a prefabricated strip drain with triaxial braiding and eco-friendliness—was created (Vishnudas et al., 2005). Strip drains combine the two functions of pore water drainage (by the core) and soil particle filtration (by the sleeve).

The Government of India set up the Coir Board, which is part of the Ministry of Industry located in Kerala. Its job is to develop, diversify, and come up with novel ways to use coir. The world consumption of coir geotextiles was 1400 million m^2 in 2000.. A mere 5 percent of all geotextiles are made using natural fibers; the remaining 95% are made using synthetic materials. When considering the Indian context, it becomes evident that coir geotextiles have a lot of unrealized potential. The German company DEKOWE was the one who first promoted them in 1985. 70% of the geotextile market is made up of nonwoven materials (Vivek et al., 2019).

Coir mesh matting with varying mesh sizes is the most well-known variety of coir geotextile. The majority of nonwoven mats are constructed from loose fibers that are rubberized or needled to create interlocks (similar to coir mattresses). Currently, Kerala is the primary location for the production of coir geotextiles, although the markets are primarily found in northern and eastern India. This indicates a geographic drawback with regard to the requirement for long-distance transportation. Germany's DEKOWE is selling its geotextile under the "Geocoir" brand. Moreover, RoLanka International of Sri Lanka markets their product under the name "Bio D-Mat." It seems like it's about time for the Indian companies that produce geotextiles made of coir to choose appropriate brand names to represent themselves (Lekha, 2004).

8.12. DESIGNING WITH GEOTEXTILES

For those working with geotextiles and, to a lesser degree, geogrids, "A Geotextiles Design Guide" provides helpful guidelines.

It addresses how geotextiles are used in railways, drainage, unpaved, and paved roads, as well as in soil reinforcement and erosion control. Aside from corrective slope work, soft ground embankments, walls, and slopes are covered under reinforcement (Giroud, 1981).

8.12.1. Unpaved Roads

The pavement material on these roads is not bound. It also consists of paved surface, the laying of which was done after trafficked, unbound granular subbase. Unpaved roads can also be constructed with the intent to provide permanent or temporary access to particular areas (Oloo et al., 1997).

The remolded CBR-Value of a clay is associated with its remolded undrained shear strength, c_u:

$$c_u(\text{kN/m}^2) = 23 \times CBR(\%)$$

Three distinct benefits occur when a geotextile is applied at the interface between the fill above and the subgrade (refer to Figures 8.9 and 8.11).

i. The granular fill's original thickness and integrity are preserved by the geotextile's separative function. The granular fill material will intermix and be punched in the formation soil by the wheel load in the absence of the geotextile, reducing the effective fill thickness.
ii. The geotextile will completely or partially absorb the horizontal shear stress that is directed outward and prevent it from entering the formation (Shtayat et al., 2020).
iii. The geotextile's "membrane action" beneath a rut creates an upward tensile component that helps to partially offset downward pressure, at the formation, brought on by the wheel load transmission.

8.12.2. Paved Roads

Paved roads are made up of two types of pavement: flexible and rigid, and flexible pavement is more prevalent. It is a bituminous bound material "layered system," while the latter is a flexural system composed of RC or plain slab. Both are built upon an unbound granular sub-base that is positioned above the natural subgrade's prepared formation level. Occasionally, a bituminous course taking the shape of a "overlay" is applied to the top of a rigid pavement. Geotextiles present three different avenues for integration. 1) at the interface between the subbase and the formation to stop migration; 2) (Conlon, 2022). inside the pavement, between the surfacing and base course, to increase the pavement's fatigue life; and 3) on the pavement's surface prior to applying the asphaltic overlay, for increasing the overlay's lifespan in addition to making it impermeable by preventing cracking by acting as reinforcement (Norman & Johansson, 2006).

- **Geotextile at 1st Position:** It has been demonstrated that an inverse power function links pavement life to its deflection.
- Geotextile at the 2nd Position: The geotextile would stop reactive cracking in this position (see Figure 8.21), extending the pavement's service life.

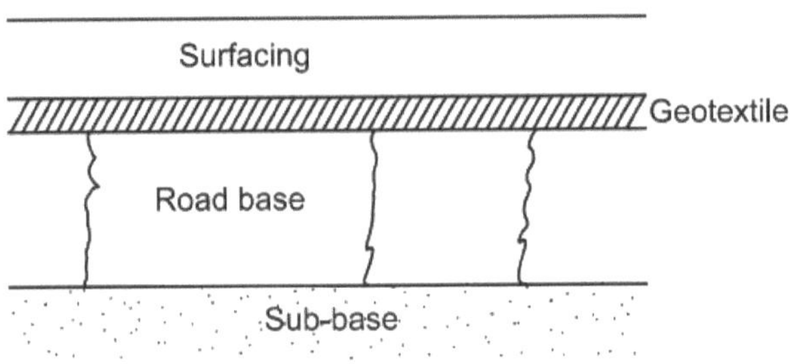

Figure 8.21. *Geotextile location for arresting reflective cracking.*

- Geotextile in the 3rd Position: In this position, the geotextile stops old pavement cracks from migrating to the surface overlay of the pavement. For better bonding, firstly, a tack coat is applied on the current road surface before the geotextile is laid (Denning & Jennissen, 2016).

8.12.3. Railways

Similar functions are performed by geotextiles in railways and highways, with the main one being that of "anti-contaminant layers," which stop ballast from migrating into the soil or the soil from migrating into the ballast. Here, "contamination" means that soil and ballast have come into contact. On the other hand, this issue usually affects railways more severely than highways. Ballast for railroad tracks is usually evenly graded and has particles that are less than 50 mm in diameter (Oum & Yu, 1994). A condition known as "pumping" results from water and fine particles slurry penetrating the ballast. Large permanent strains form under the dynamic loads as a result of the track bed's dynamic modulus being reduced by the slurry filling the void. Occasionally, replacing or cleaning the ballast may require costly track realignment. Putting a sand blanket on top of the sub-grade is the best way to stop the erosion pumping at in interface where ballast meets a silty or clayey subgrade. While this approach is well-known to be effective in managing erosion pumping, it is unable to stop the bearing capacity failure that occurs in cohesive subgrades with the ingress of rainwater (Sañudo et al., 2016). British Rail has chosen to address this issue by adding a geomembrane to the sand blanket's central plane. Nonwoven geotextiles of the needle-punched filament type are preferred over wovens when utilized as separators in the subgrade level.

8.13. GEOTEXTILES FOR HILLSIDE DEVELOPMENT

When developing a hillside, it is always necessary to evaluate how stable the current slopes are, as this could require building new slopes that are more resilient to failure. For stabilizing such slopes, geotextile reinforcement in the form of reinforced earth is the perfect solution (Broda et al., 2020).

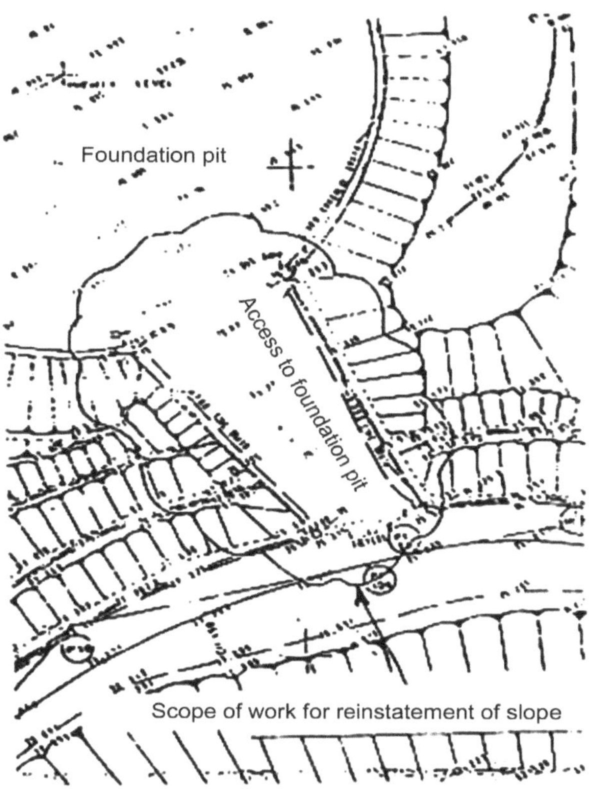

Figure 8.22. *Excavated slope – plan.*

Lee et al. (1994) give the case study of a project that involved building a high rise on top of a hill. The existing slope needed to be cut open to allow access to the foundation pit that was excavated as seen in Figure 8.22. But once the foundation work was finished, the slopes had to be reinstated back to the former steep gradient up to a 17 meter height. Polywoven PW70, a slit film woven geotextile composed of polypropylene, was used for this purpose. The usual wrapping of the geotextile around the facing wasn't done. The backfill was compacted to a standard Proctor density of 95%. Eventually, the slopes were turfed after being trimmed to the design slopes. The pull-out and sliding charts provided by (Jewell et al., 1989) were used in the design process. Using the STABGM program, a global stability analysis using Bishop's slip circles was performed to complement the internal stability analysis (Duncan et al. 1985).

8.14. NEHEMIAH (MALAYSIA) CASE HISTORIES

When embankments are wide, a pair of reinforcements are positioned, each ending at an anchor block, rather than using common reinforcements that run from one wall to the other. A wall designed by Nehemiah AE is depicted in Figure 8.23 while the completed wall is shown in Figure 8.24.

Figure 8.23. *AE wall with embossed hexagonal units under construction.*

Figure 8.24. *AE wall with embossed hexagonal facing blocks.*

The Figure 8.24 illustrates how the center of every hexagonal unit in the facing panels is made to project via a short width (embossed), which significantly improves the wall's aesthetics by breaking up the monotony.

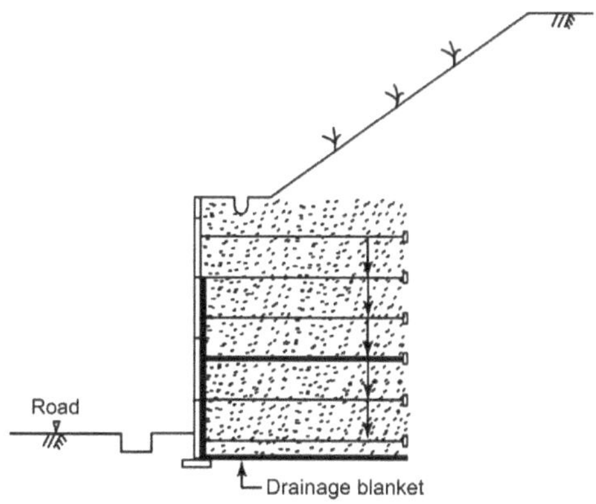

Figure 8.25. *AE wall with drainage blanket under a slope.*

Figure 8.25 displays an anchored earth wall with drainage blankets placed under a slope next to a road. The drainage blanket beneath HWL that is placed behind a wall with water on one side is demonstrated in Figure 8.26.

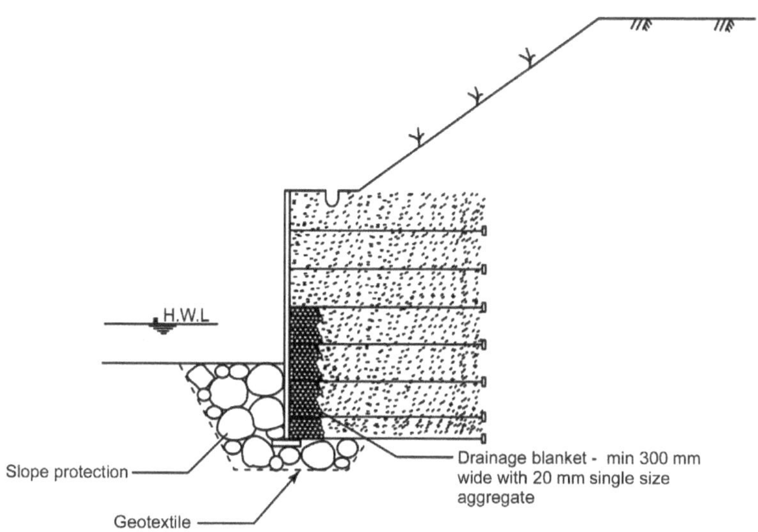

Figure 8.26. *AE wall with stone drainage blanket.*

Nehemiah favors overlapping tendons with each terminates at its own anchor block, even in cases where the width is minimal. Nehemiah specifies that cushion pads and joint fillers be used to separate the facing panels, creating a watertight wall. The minimum tensile strength requirements for the reinforcing tendons are 250 MN/m^2 for mild steel bars and 460 MN/m^2 for high yield strength bars (I N/mm^2 = I MN/m^2). 70 μm should be the minimum thickness for the hot-dipped galvanization (1 μm = 10–6 m = 1/1,000 mm). After being galvanized, the connections are covered with a layer of anti-corrosive tar epoxy that is at least 75 μm thick. Flexible open cell polyurethane foam strips with a 40 mm square section make up the panel joint fillers. After the footing has been cast, at least 24 hours must pass before panels can be erected on the plain concrete footing (YAP, 2024). In terms of backfill, no fill lift should be higher than 375 mm when compacted. The vehicles used to spread and compact the fill should always move parallel to the wall. However, only a vibratory plate compactor should be used to compact material for a width of 1.5 meters from the panel's rear face. Subsurface drainage must use perforated PVC pipes with a diameter of 100 mm (Ali et al., 2020). Turkiye is said to have utilized reinforced earth extensively for industrial structures, railroads, sea walls bridge abutments, retaining walls, industrial buildings, and civilian and military installations (Murugesan et al., 2021).

8.14.1. Green Mur

Nehemiah called the green approach to slope protection "Green Mur." It consists of steel fabric facing panels supported by counterford hooks on the incline and geotextile layers positioned horizontally to act as reinforcement. Geojute, a biodegradable woven jute fabric, covers the facing panel and acts as a foundation for close turfing. In addition to adding aesthetic value with their greenery, the roots shield the slope against soil erosion. A Green Mur slope is depicted in Figure 8.27, with a road below and residences above.

Figure 8.27. *Finished Green Mur Slope.*

SUMMARY

This chapter explores the vital role of geosynthetics in modern geotechnical engineering practices. Geosynthetics, including materials like geotextiles, geomembranes, and geogrids, offer a wide array of applications in soil stabilization, drainage, erosion control, and reinforcement. The chapter delves into the diverse functions of geosynthetics, such as enhancing soil strength, preventing soil erosion, and providing barriers against contaminants. Through case studies and examples, it illustrates how geosynthetics contribute to cost-effective and sustainable solutions in various geotechnical projects, ranging from road construction to environmental remediation. Understanding the properties and applications of geosynthetics is crucial for engineers and practitioners to optimize the performance and longevity of geotechnical structures.

REFERENCES

1. Ali, N. E. H., Rahman, N. A., Zainol, H., Abdullah, K., & Saleh, A. A. (2020). Urban Households Solid Waste Generated and Characteristics: The case of Shah Alam City Hall, Selangor. *Jurnal Arsitektur dan Perencanaan (JUARA), 3*(1), 53–64.
2. Alkhorshid, N. R., Araujo, G. L., Palmeira, E. M., & Zornberg, J. G. (2019). Large-scale load capacity tests on a geosynthetic encased column. *Geotextiles and Geomembranes, 47*(5), 632–641.
3. Almeida, M. S., Hosseinpour, I., Riccio, M., & Alexiew, D. (2015). Behavior of geotextile-encased granular columns supporting test embankment on soft deposit. *Journal of Geotechnical and Geoenvironmental Engineering, 141*(3), 0401–4116.
4. Al-Qadi, I. L., Dessouky, S. H., Kwon, J., & Tutumluer, E. (2008). Geogrid in flexible pavements: Validated mechanism. *Transportation Research Record, 2045*(1), 102–109.
5. Amakye, S. Y., & Abbey, S. J. (2021). Understanding the performance of expansive subgrade materials treated with non-traditional stabilizers: A review. *Cleaner Engineering and Technology, 4*, 100–159.
6. Assinder, P. J., Breytenbach, M., Wiemers, J., & Hörtkorn, F. (2016). Utilizing geotextile tubes to extend the life of a Tailings Storage Facility. In *Proceeding of the First Southern African Geotechnical Conference*, Vol. 1, (pp. 373–379).
7. Banerjee, D., & Mukherjee, B. (1996). A practical approach for routing and wavelength assignment in large wavelength-routed optical networks. *IEEE Journal on Selected Areas in Communications, 14*(5), 903–908.
8. Bayraktar, O. Y. (2020). Use of geosynthetics in road construction. *Kastamonu University Journal of Engineering and Sciences, 6*(2), 107–113.
9. Bergado, D. T., Balasubramaniam, A. S., Fannin, R. J., & Holtz, R. D. (2002). Prefabricated vertical drains (PVDs) in soft Bangkok clay: A case study of the new Bangkok International Airport project. *Canadian Geotechnical Journal, 39*(2), 304–315.
10. Blair, J. C., Bell, J. R., & Hicks, R. G. (1980). Permeability testing of geotextiles. *Transportation Research Record, 827*, 1–100.
11. Blanco-Fernandez, E., Castro-Fresno, D., Díaz, J. D. C., & Lopez-Quijada, L. (2011). Flexible systems anchored to the ground for slope stabilization: Critical review of existing design methods. *Engineering Geology, 122*(3–4), 129–145.

12. Blume, K. H., Alexiew, D., & Glötzl, F. (2006). The new federal highway (Autobahn) A26 in Germany with high geosynthetic reinforced embankments on soft soils. *Geosynthetics, Milpress, Rotterdam, 2*(1), 913–916.
13. Broda, J., Franitza, P., Herrmann, U., Helbig, R., Große, A., Grzybowska-Pietras, J., & Rom, M. (2020). Reclamation of abandoned open mines with innovative meandrically arranged geotextiles. *Geotextiles and Geomembranes, 48*(3), 236–242.
14. Chai, J., Bergado, D. T., & Shen, S. L. (2013). Modelling prefabricated vertical drain improved ground in plane strain analysis. *Proceedings of the Institution of Civil Engineers-Ground Improvement, 166*(2), 65–77.
15. Chang, C., Ji, Z., & Zeng, F. (2016). The effect of a drainage layer on filtration performance of coalescing filters. *Separation and Purification Technology, 170*, 370–376.
16. Chen, D., Wei, W., & Chen, L. (2017). Effects of terracing practices on water erosion control in China: A meta-analysis. *Earth-Science Reviews, 173*, 109–121.
17. Collin, J. G., Stark, T. D., Lucarelli, A., Taylor, T. P., & Berg, R. R. (2021). Stability and stress-deformation analyses of reinforced slope failure at Yeager Airport. *Journal of Geotechnical and Geoenvironmental Engineering, 147*(3), 040–20179.
18. Conlon, K. (2022). Plastic roads: Not all they're paved up to be. *International Journal of Sustainable Development & World Ecology, 29*(1), 80–83.
19. Dabral, P. P., Baithuri, N., & Pandey, A. (2008). Soil erosion assessment in a hilly catchment of North Eastern India using USLE, GIS and remote sensing. *Water Resources Management, 22*, 1783–1798.
20. Das, B. M. (2016). Use of geogrid in the construction of railroads. *Innovative Infrastructure Solutions, 1*, 1–12.
21. Datta, U. (2007). Application of jute geotextiles. *Journal of Natural Fibers, 4*(3), 67–82.
22. Datye, K. R., & Gore, V. N. (1994). Application of natural geotextiles and related products. *Geotextiles and Geomembranes, 13*(6–7), 371–388.
23. Denning, G. M., & Jennissen, C. A. (2016). All-terrain vehicle fatalities on paved roads, unpaved roads, and off-road: Evidence for informed roadway safety warnings and legislation. *Traffic Injury Prevention, 17*(4), 406–412.
24. Detert, O., Alexiew, D., Schanz, T., & König, D. (2013). Analysis of an adaptive foundation system for embankments on soft soils by means of physical and

numerical modeling. In *Proc. 18th Intl. Conference on Soil Mechanics and Geotechnical Engineering, Paris, France*, Vol. 2, (pp. 907–910).

25. Dixon, N., Raja, J., Fowmes, G., & Frost, M. (2016). Sustainability aspects of using geotextiles. In *Geotextiles, Vol. 2*, (pp. 577–596).
26. Ebbert, S., Derksen, J., & Harnisch, J. (2023). Performance of concrete filled geotextile mattresses. In *Geosynthetics: Leading the Way to a Resilient Planet, 3*(2), (pp. 1628–1634).
27. Esmaeili, M., Naderi, B., Neyestanaki, H. K., & Khodaverdian, A. (2018). Investigating the effect of geogrid on stabilization of high railway embankments. *Soils and Foundations, 58*(2), 319–332.
28. Fannin, R. J., & Sigurdsson, O. (1996). Field observations on stabilization of unpaved roads with geosynthetics. *Journal of Geotechnical Engineering, 122*(7), 544–553.
29. Farooq, M. U., Mujtaba, H., Farooq, K., Sivakugan, N., & Das, B. M. (2020). Evaluation of stability and erosion characteristics of soil embankment slope reinforced with different natural additives. *Iranian Journal of Science and Technology, Transactions of Civil Engineering, 44*(Suppl 1), 515–524.
30. Fayer, M. J., & Hillel, D. (1986). Air encapsulation: I. Measurement in a field soil. *Soil Science Society of America Journal, 50*(3), 568–572.
31. Fischer, S., & Horvát, F. (2011). Superstructure stabilization of ballast bedded railway tracks with geogrids. *Hungarian Journal of Industrial Chemistry, 39*(1), 101–106.
32. Garber, S., & Rasmussen, R. O. (2010). Nonwoven geotextile interlayers in concrete pavements. *Transportation Research Record, 2152*(1), 11–15.
33. Ghosh, M., Choudhury, P. K., & Sanyal, T. (2009). Suitability of natural fibers in geotextile applications. *IGC Geotide, 5*(2), 497–501.
34. Ghosh, S. K., Bhattacharyya, R., & Mondal, M. M. (2017). Potential applications of open weave jute geotextile (soil saver) in meeting geotechnical difficulties. *Procedia Engineering, 200*, 200–205.
35. Giroud, J. P. (1981). Designing with geotextiles. *Matériaux et Construction, 14*, 257–272.
36. Giroud, J. P. (2005). Quantification of geosynthetic behavior. *Geosynthetics International, 12*(1), 2–27.
37. Hale, D. K. (1976). The physical properties of composite materials. *Journal of Materials Science, 11*, 2105–2141.
38. Heller, V. (2011). Scale effects in physical hydraulic engineering models. *Journal of Hydraulic Research, 49*(3), 293–306.

39. Horton, R. E. (1932). Drainage-basin characteristics. *Transactions, American Geophysical Union, 13*(1), 350–361.
40. Iwnicki, S. (2009). Future trends in railway engineering. *Proceedings of the Institution of Mechanical Engineers, Part C: Journal of Mechanical Engineering Science, 223*(12), 2743–2750.
41. Joshi, V. H., Lavania, B. V. K., & Verma, A. (1990). Reinforced earth retaining structures for earthquake loading conditions. *Bulletin of Indian Society of Earthquake Technology, June*, 65–80.
42. Kempfert, H. G., Mobius, W., Wallis, P., Raithel, M., Geduhn, M., & McClinton, R. G. (2002). Reclaiming land with geotextile-encased columns. *Geotechnical Fabrics Report, 20*(6), 34–39.
43. Khatib, A. (2005). Bearing Capacity of Soft Soil Using Bamboo-Geotextile Composite. *Jurnal Saintis, 8*(2), 61–69.
44. Kirschner, R., & Witte, R. (2020). Use of Geotextiles for the Protection of Geomembranes. In *Landfilling of Waste, 4*(2), (pp. 475–486).
45. Koda, E., Miszkowska, A., & Stępień, S. (2016). Quality control of non-woven geotextiles used in a drainage system in an old remedial landfill. In *Geo-Chicago 2016, 5*(3), (pp. 254–263).
46. Koerner, R. M. (2016). Early background and history of geotextiles. In *Geotextiles*, (pp. 3–15).
47. Koerner, R. M., & Koerner, G. R. (2015). Lessons learned from geotextile filter failures under challenging field conditions. *Geotextiles and Geomembranes, 43*(3), 272–281.
48. Krishnan, S., Nakanishi, T., & Goh, A. H. (2008). Performance of PVD and surcharging in reclamation works for a power plant in Negeri Sembilan, Malaysia. *Geotechnical Engineering Journal SEAGS & AGSSEA, 39*(2), 63–75.
49. Lal, D., Sankar, N., & Chandrakaran, S. (2017). Effect of reinforcement form on the behavior of coir geotextile reinforced sand beds. *Soils and Foundations, 57*(2), 227–236.
50. Leão, A. L., Cherian, B. M., De Souza, S. F., Kozłowski, R. M., Thomas, S., & Kottaisamy, M. (2012). Natural fibers for geotextiles. In *Handbook of Natural Fibers, Vol. 3*, (pp. 280–311).
51. Lekha, K. R. (2004). Field instrumentation and monitoring of soil erosion in coir geotextile stabilized slopes—A case study. *Geotextiles and Geomembranes, 22*(5), 399–413.
52. Loke, M. H., & Barker, R. D. (1996). Rapid least-squares inversion of apparent resistivity pseudosections by a quasi-Newton method. *Geophysical Prospecting, 44*(1), 131–152.

53. Long, R. P., & Covo, A. (1994). Equivalent diameter of vertical drains with an oblong cross section. *Journal of Geotechnical Engineering, 120*(9), 1625–1630.
54. Luettich, S. M., Giroud, J. P., & Bachus, R. C. (1992). Geotextile filter design guide. In *Geosynthetics in Filtration, Drainage and Erosion Control, Vol. 3*, (pp. 19–34).
55. Mallick, A., & Mallick, B. (2021). Staying despite riverbank erosion: Evidence of coastal Bangladesh. *SN Social Sciences, 1*(6), 1–55.
56. Masini, L., Callisto, L., & Rampello, S. (2015). An interpretation of the seismic behavior of reinforced-earth retaining structures. In *Geotechnical Earthquake Engineering: Géotechnique Symposium in Print 2015, 5*(3), (pp. 25–34).
57. McNitt, A. S., & Landschoot, P. J. (2003). Effects of soil reinforcing materials on the surface hardness, soil bulk density, and water content of a sand root zone. *Crop Science, 43*(3), 957–966.
58. Metsger, D. S., Rieger, M. R., & Foreman, D. W. (1999). Mechanical properties of. *Journal of Materials Science: Materials in Medicine, 10*, 9–17.
59. Mohamed, H. I. (2010). Flow over gabion weirs. *Journal of Irrigation and Drainage Engineering, 136*(8), 573–577.
60. Müller, W. W., & Saathoff, F. (2015). Geosynthetics in geoenvironmental engineering. *Science and Technology of Advanced Materials, 16*(3), 034–605.
61. Murugesan, S., Bhuvaneswaran, R. S., Khanna Nehemiah, H., Keerthana Sankari, S., & Nancy Jane, Y. (2021). Feature selection and classification of clinical datasets using bioinspired algorithms and super learner. *Computational and Mathematical Methods in Medicine, 2021*, 1–18.
62. Neupane, M., Parsons, R. L., Han, J., Parr, D. A., & Jacobe, J. A. (2018). Evaluation of soil saver walls on aquatic organism passage through box culverts. *Journal of Testing and Evaluation, 46*(4), 1313–1320.
63. Nilaweera, N. S., & Nutalaya, P. (1999). Role of tree roots in slope stabilization. *Bulletin of Engineering Geology and the Environment, 57*, 337–342.
64. Norman, M., & Johansson, C. (2006). Studies of some measures to reduce road dust emissions from paved roads in Scandinavia. *Atmospheric Environment, 40*(32), 6154–6164.
65. Oloo, S. Y., Fredlund, D. G., & Gan, J. K. (1997). Bearing capacity of unpaved roads. *Canadian Geotechnical Journal, 34*(3), 398–407.

66. Ooi, T. A., & Tee, C. H. (2011). Advance in geogrid reinforced slopes in Malaysia. *Geotechnical Engineering, 42*(1), 9–12.
67. Oum, T. H., & Yu, C. (1994). Economic efficiency of railways and implications for public policy: A comparative study of the OECD countries' railways. *Journal of Transport Economics and Policy*, 121–138.
68. Palmeira, E. M., & Gardoni, M. G. (2002). Drainage and filtration properties of non-woven geotextiles under confinement using different experimental techniques. *Geotextiles and Geomembranes, 20*(2), 97–115.
69. Palmeira, E. M., Araújo, G. L., & Santos, E. C. (2021). Sustainable solutions with geosynthetics and alternative construction materials—A review. *Sustainability, 13*(22), 12756.
70. Pandey, A., Mathur, A., Mishra, S. K., & Mal, B. C. (2009). Soil erosion modeling of a Himalayan watershed using RS and GIS. *Environmental Earth Sciences, 59*, 399–410.
71. Parsa-Pajouh, A., Fatahi, B., Vincent, P., & Khabbaz, H. (2014). Trial embankment analysis to predict smear zone characteristics induced by prefabricated vertical drain installation. *Geotechnical and Geological Engineering, 32*, 1187–1210.
72. Plotkin, G. D. (1976). A power domain construction. *SIAM Journal on Computing, 5*(3), 452–487.
73. Posthumus, H., Deeks, L. K., Rickson, R. J., & Quinton, J. N. (2015). Costs and benefits of erosion control measures in the UK. *Soil Use and Management, 31*, 16–33.
74. Qasim, M., Ding, L., Khan, M. A., Jadoon, I. A., Haneef, M., Baral, U., & Yue, Y. (2018). Tectonic implications of detrital zircon ages from lesser Himalayan Mesozoic-Cenozoic strata, Pakistan. *Geochemistry, Geophysics, Geosystems, 19*(5), 1636–1659.
75. Rai, R., Khandelwal, M., & Jaiswal, A. (2012). Application of geogrids in waste dump stability: A numerical modeling approach. *Environmental Earth Sciences, 66*, 1459–1465.
76. Rajesh, U., Sajja, S., & Chakravarthi, V. K. (2016). Studies on engineering performance of geogrid reinforced soft subgrade. *Transportation Research Procedia, 17*, 164–173.
77. Ramli, M., Karasu, T., & Dawood, E. T. (2013). The stability of gabion walls for earth retaining structures. *Alexandria Engineering Journal, 52*(4), 705–710.
78. Rao, G. V. (2016). Three decades of geosynthetics in India. *Water and Energy International, 59*(4), 42–49.

79. Rao, G. V., & Dutta, R. K. (2002). Testing and application of coir-based geotextiles. In *Proc. of Seventh International Conference on Geosynthetics Sept France, 5*(3), (pp. 22–27).
80. Rao, K. M. M., Prasad, A. R., Babu, M. R., Rao, K. M., & Gupta, A. V. S. S. K. S. (2007). Tensile properties of elephant grass fiber reinforced polyester composites. *Journal of Materials Science, 42*, 3266–3272.
81. Rawal, A., & Saraswat, H. (2011). Stabilization of soil using hybrid needle punched nonwoven geotextiles. *Geotextiles and Geomembranes, 29*(2), 197–200.
82. Ready, B., & Nurtjahjaningtyas, I. (2020). Analysis of slope stability in soft soil using hardening soil modeling and strengthening of bamboo mattress. *GEOMATE Journal, 19*(73), 226–234.
83. Rickson, R. J. (2003). Simulated vegetation and geotextiles. In *Slope Stabilization and Erosion Control: A Bioengineering Approach, 5*(3), (pp. 100–138).
84. Saha, P., Roy, D., Manna, S., Adhikari, B., Sen, R., & Roy, S. (2012). Durability of trans esterified jute geotextiles. *Geotextiles and Geomembranes, 35*, 69–75.
85. Sañudo, R., Dell'Olio, L., Casado, J. A., Carrascal, I. A., & Diego, S. (2016). Track transitions in railways: A review. *Construction and Building Materials, 112*, 140–157.
86. Sathananthan, I., Indraratna, B., & Rujikiatkamjorn, C. (2008). Evaluation of smear zone extent surrounding mandrel driven vertical drains using the cavity expansion theory. *International Journal of Geomechanics, 8*(6), 355–365.
87. Shtayat, A., Moridpour, S., Best, B., Shroff, A., & Raol, D. (2020). A review of monitoring systems of pavement condition in paved and unpaved roads. *Journal of Traffic and Transportation Engineering (English Edition), 7*(5), 629–638.
88. Subaida, E. A., Chandrakaran, S., & Sankar, N. (2009). Laboratory performance of unpaved roads reinforced with woven coir geotextiles. *Geotextiles and Geomembranes, 27*(3), 204–210.
89. Sumi, S., Unnikrishnan, N., & Mathew, L. (2018). Durability studies of surface-modified coir geotextiles. *Geotextiles and Geomembranes, 46*(6), 699–706.
90. Talib, F. M., Mohamad, H. M., & Mustafa, M. N. (2021). Peat soil improvement with bamboo reinforcement technology: A review. *GEOMATE Journal, 21*(88), 75–85.

91. Tognon, A. R., Rowe, R. K., & Moore, I. D. (2000). Geomembrane strain observed in large-scale testing of protection layers. *Journal of Geotechnical and Geoenvironmental Engineering, 126*(12), 1194–1208.

92. Toprak, B., Sevim, O., & Kalkan, I. (2016). Gabion walls and their use. *International Journal of Advances in Mechanical and Civil Engineering, 3*(4), 56–58.

93. Touze-Foltz, N., Bannour, H., Barral, C., & Stoltz, G. (2016). A review of the performance of geosynthetics for environmental protection. *Geotextiles and Geomembranes, 44*(5), 656–672.

94. Van Genuchten, M. T., & Nielsen, D. R. (1985). On describing and predicting the hydraulic properties. *Annales Geophysicae, 3*(5), 615–628.

95. Van Tho, N. (2020). Coastal erosion, river bank erosion and landslides in the Mekong Delta: Causes, effects and solutions. In *Geotechnics for Sustainable Infrastructure Development, 5*(3), 957–962.

96. Vishnudas, S., Savenije, H. H. G., Van der Zaag, P., Anil, K. R., & Balan, K. (2005). Experimental study using coir geotextiles in watershed management. *Hydrology and Earth System Sciences Discussions, 2*(6), 2327–2348.

97. Vivek, Dutta, R. K., & Parti, R. (2019). Effect of chemical treatment of the coir geotextiles on the interface properties of sand–/clay–coir geotextile interface. *Journal of the Institution of Engineers (India): Series A, 100*, 357–365.

98. Vyas, S. D., & Shah, P. J. (2020). Review of formula for consolidation with prefabricated vertical drains, for a project. *Indian Journal of Geosynthetics and Ground Improvement, 9*(2), 30–37.

99. Wang, Q., Wen, X., Jiang, J., Zhang, C., & Shi, Z. (2014). Experimental study on performance of multidirectional geogrid and its application in engineering of high slope. *Journal of Wuhan University of Technology-Mater. Sci. Ed., 29*(4), 704–711.

100. Weaver, R. J., & Stehno, A. L. (2024). Mangroves as Coastal Protection for Restoring Low-Energy Waterfront Property. *Journal of Marine Science and Engineering, 12*(3), 470.

101. Yap, F. H. (2024). Phrasal alternation and Kerinci demonstrative (i) NEH: Implications for spatial and socio-interactional deixis. *Journal of Pragmatics, 222*, 40–59.

102. Yevheniia, S. (2021). Revitalization of the urban waterfront areas. *World Science, 2*(1), 33–88.

103. Zanzinger, H. (1999). Efficiency of geosynthetic protection layers

for geomembrane liners: Performance in a large-scale model test. *Geosynthetics International, 6*(4), 303–317.

104. Zimar, Z., Robert, D., Zhou, A., Giustozzi, F., Setunge, S., & Kodikara, J. (2022). Application of coal fly ash in pavement subgrade stabilization: A review. *Journal of Environmental Management, 312*, 114–926.

INDEX

D

E

F

G

H

I

J

R

S

W

FSC
www.fsc.org
MIX
Papier | Fördert gute Waldnutzung
FSC® C083411

Kolibri 360 GmbH
Ferdinand-Jühlke-Straße 7
99095 Erfurt, Deutschland
produktsicherheit@kolibri360.de